Handbook of Irrigation Hydrology and Management

Ever-increasing population growth has caused a proportional increased demand for water, and existing water sources are depleting day by day. Moreover, with the impact of climate change, the rates of rainfall in many regions have experienced a higher degree of variability. In many cities, government utilities have been struggling to maintain sufficient water for the residents and other users. The *Handbook of Irrigation Hydrology and Management: Irrigation Methods* examines and analyzes irrigated ecosystems in which water storage, applications, or drainage volumes are artificially controlled in the landscape and the spatial domain of processes varies from micrometers to tens of kilometers, while the temporal domain spans from seconds to centuries. The continuum science of irrigation hydrology includes the surface, subsurface (unsaturated and groundwater systems), atmospheric, and plant subsystems. Further, the book addresses the best practices for various types of irrigation methods including pressure, smart, surface and subsurface, and presents solutions for water scarcity and soil salinity in irrigation.

Features:

- Offers water-saving strategies to increase the judicious use of scarce water resources
- Presents strategies to maximize agricultural yield per unit of water used for different regions
- Compares irrigation methods to offset changing weather patterns and impacts of climate change

Handbook of Irrigation Hydrology and Management
Irrigation Methods

Edited by
Saeid Eslamian and Faezeh Eslamian

CRC Press
Taylor & Francis Group
Boca Raton London New York

CRC Press is an imprint of the
Taylor & Francis Group, an **informa** business

Designed cover image: Shutterstock

First edition published 2023
by CRC Press
6000 Broken Sound Parkway NW, Suite 300, Boca Raton, FL 33487-2742

and by CRC Press
4 Park Square, Milton Park, Abingdon, Oxon, OX14 4RN

CRC Press is an imprint of Taylor & Francis Group, LLC

© 2023 selection and editorial matter, Saeid Eslamian and Faezeh Eslamian; individual chapters, the contributors

Reasonable efforts have been made to publish reliable data and information, but the author and publisher cannot assume responsibility for the validity of all materials or the consequences of their use. The authors and publishers have attempted to trace the copyright holders of all material reproduced in this publication and apologize to copyright holders if permission to publish in this form has not been obtained. If any copyright material has not been acknowledged please write and let us know so we may rectify in any future reprint.

Except as permitted under U.S. Copyright Law, no part of this book may be reprinted, reproduced, transmitted, or utilized in any form by any electronic, mechanical, or other means, now known or hereafter invented, including photocopying, microfilming, and recording, or in any information storage or retrieval system, without written permission from the publishers.

For permission to photocopy or use material electronically from this work, access www.copyright.com or contact the Copyright Clearance Center, Inc. (CCC), 222 Rosewood Drive, Danvers, MA 01923, 978-750-8400. For works that are not available on CCC please contact mpkbookspermissions@tandf.co.uk

Trademark notice: Product or corporate names may be trademarks or registered trademarks and are used only for identification and explanation without intent to infringe.

Library of Congress Cataloging-in-Publication Data
Names: Eslamian, Saeid, editor. | Eslamian, Faezeh A., editor.
Title: Handbook of irrigation hydrology and management : irrigation
management and optimization / Edited by Saeid Eslamian and Faezeh Eslamian.
Description: Boca Raton, FL : CRC Press, 2023. | Includes bibliographical
references and index. | Also available online. | Description based on
print version record and CIP data provided by publisher; resource not viewed.
Identifiers: LCCN 2022050740 (print) | LCCN 2022050741 (ebook) |
ISBN 9780429290152 (ebook) | ISBN 9780367258306 (hardback) |
ISBN 9781032457468 (paperback) | ISBN 9780367258191(v. 1 ;hardback) |
ISBN 9781032457451(v. 1 ;paperback) | ISBN 9781032457468(v. 2;paperback) |
ISBN 9780429290152(v. 2 ;ebook) | ISBN 9781032406077(v. 3 ;hardback)|
ISBN 9781032429106(v. 3 ;paperback) | ISBN 9781003353928(v. 3 ;ebook)
Subjects: LCSH: Irrigation. | Irrigation engineering.
Classification: LCC TC805 (ebook) | LCC TC805 .H36 2023 (print) |
DDC 627/.52 23/eng/20221–dc16
LC record available at https://lccn.loc.gov/2022050740

ISBN: 978-0-367-25830-6 (hbk)
ISBN: 978-1-032-45746-8 (pbk)
ISBN: 978-0-429-29015-2 (ebk)

DOI: 10.1201/9780429290152

Typeset in Minion
by codeMantra

To

Leonardo di ser Piero da Vinci (1452–1519),

an Italian polymath of the High Renaissance who was active as a painter, draughtsman, engineer, scientist, theorist, sculptor, architect, and a Hydrodynamic Man.

"Water is the driving force of all nature"

"All the branches of a water at every stage of its course, if they are of equal rapidity, are equal to the body of the main stream"

He studied the motion of water, in particular the forms taken by fast-flowing water on striking different surfaces.

Contents

Preface ix
Editors..xi
Contributors xv

SECTION I Solutions for Water Scarcity and Soil Salinity in Irrigation

1. Deficit Irrigation: A Solution for Water Scarcity Conditions? 3
   Jie Liu, Zhenhua Wei, and Fulai Liu

2. Irrigation Assessment: Efficiency and Uniformity 13
   Patricia Angélica Alves Marques, Verônica Gaspar Martins Leite de Melo, José Antônio Frizzone, and Antonio Pires de Camargo

SECTION II Surface Irrigation

3. Canal Irrigation.. 33
   Qin Qian and Saeid Eslamian

4. Ditch and Furrow Irrigation 51
   Abebech A. Beyene and Saeid Eslamian

5. Level-Basin Irrigation 65
   Nasrin Azad, Javad Behmanesh, Vahid Rezaverdinejad, and Saeid Eslamian

SECTION III Pressure Irrigation

6. Sprinkler Irrigation Systems.. 113
   Neil A. Coles, Mark R. Rivers, and Saeid Eslamian

7. Mini-Bubbler Irrigation 141
   K. Y. Raneesh and Saeid Eslamian

8. Measured Irrigation 149
   Bernie Omodei

SECTION IV Subsurface Irrigation

9 Qanat Irrigation Systems 203
Naser Valizadeh, Dariush Hayati, Hamid Karimi, Saeid Eslamian, and Mohammad Mohammadzadeh

10 Subsurface Drainage and Irrigation System in Paddy Fields 219
Shuichiro Yoshida

11 Pitcher Irrigation as a Viable Tool of Enhancing Nutrition and Livelihood of Small-Scale Farmers: The Case Study of the Drier Areas of Katsina State, Nigeria 239
S. Jari, T. A. Rimi, R. T. Nabinta, and Saeid Eslamian

12 Pot Irrigation 247
Hamideh Faridi and Ghasem Zarei

SECTION V Smart Irrigation

13 Automation and Smart Irrigation 275
Sajjad Roshandel and Saeid Eslamian

14 Intelligent Irrigation and Automation 295
Hamideh Faridi, Babak Ghoreishi, and Hamidreza Faridi

15 Smart Irrigation in Urban Development Using Treated Wastewater: Irrigation Systems and Management 321
Leonor Rodríguez-Sinobas, Freddy Canales-Ide, and Sergio Zubelzu

SECTION VI Water Pumps for Irrigation

16 Water Pumps for Irrigation: An Introduction 339
Saeid Eslamian, Mousa Maleki, and Asal Soltani

17 Pumps for Irrigation Systems 355
Mahbub Hasan, Aschalew Kassu, and Saeid Eslamian

18 Inverted Siphon Implementation Method in Karun River for Farm Water ... 387
Faramarz Ghalambaz and Saeid Eslamian

Index 397

Preface

Water is known as the most important input required for plant growth for agricultural production. Irrigation can be defined as the completion of soil water storage in the root zone of the plant by methods other than natural rainfall. Irrigation seems to have the roots in human history from the beginning. This makes it possible to reduce uncertainties in agricultural practices, particularly climatic ones.

Archaeological research has found evidence of irrigation where natural rainfall was not sufficient to support farming. The ancient Egyptians, for example, practiced basin irrigation by using the Nile flood to flood land surrounded by dykes. Another example is the aqueduct, which was built in ancient Persia around 800 BC, and is one of the oldest known methods of irrigation still in use.

Irrigation hydrology is constrained to analysis of irrigated ecosystems in which water storage, applications, or drainage volumes are artificially controlled in the landscape and the spatial domain of processes varies from micrometers to tens of kilometers, while the temporal domain spans from seconds to centuries. The continuum science of irrigation hydrology includes the surface, subsurface (unsaturated and groundwater systems), atmospheric, and plant subsystems.

Irrigation management involves many different decisions: selection of economically viable cropping patterns, land allocation by crop, water resource allocation by crop, irrigation scheduling, deficit management irrigation, etc. Plants need adequate amounts of water, and its distribution throughout the growth cycle has a huge influence on the final yield of the crop. This means that the management of soil water content is crucial to obtain an optimal allocation of water resources, provided that the other production factors are adequate.

Our book has several merits on the previous published books as follows:

- A comprehensive book on majority methods of Irrigation
- A new focus on Irrigation Hydrology, Landscape, Scales, and Social Context
- A robust tool for computational analysis of Irrigation Methods
- A updated book including global warming, adaptation, resilience, and sustainability associated with Irrigation
- Deficit and Over-irrigation Merits and Disadvantages
- Inclusion of Smart and Precision Irrigation
- Offering Solutions for Water Scarcity and Soil Salinity in Irrigation
- Satellite Measurements for Irrigation Management
- Selected case studies from different climates across world.

Volume 2 of this handbook entitles *Irrigation Methods*. It includes the following topics:

- Pressure Irrigation
- Smart Irrigation
- Solutions for Water Scarcity and Soil Salinity in Irrigation
- Subsurface Irrigation

- Surface Irrigation
- Water Pumps for Irrigation.

The information contained in this *Handbook of Irrigation Hydrology and Management* can be beneficial to students at the following levels of the study: Undergraduate, Postgraduate, Research Students, and Short Courses Programs, and could also be a useful resource for courses such as Surface Hydrology, Water Resources, Climatology, Agrometeorology, Irrigation Principals, Surface Irrigation, Irrigation System Design, Drip Irrigation, Sprinkler Irrigation, Water, Soils and Plants Relationships, Evapotranspiration and Water Requirement, Drainage Engineering, Irrigation and Drainage Networks, Irrigation Hydraulic Structures, and Special Problems in Irrigation.

All the scholars and students of Applied Geography, Geosciences, Environmental Engineering, Environmental Health, Natural Resources, Agricultural Engineering, Irrigation Engineering and the related courses, as well as professionals, will find this handbook of great value.

Three-volume *Handbook of Irrigation Hydrology and Management* could be recommended not only for universities and colleges, but also for research centers, governmental departments, policy makers, engineering consultants, federal emergency management agencies and the related bodies.

Saeid Eslamian and Faezeh Eslamian
Isfahan University of Technology and McGill University

Editors

Saeid Eslamian is a full professor of environmental hydrology and water resources engineering in the Department of Water Engineering at Isfahan University of Technology (IUT), where he has been serving since 1995. His research focuses mainly on statistical and environmental hydrology in a changing climate. In recent years, he has worked on modeling natural hazards, including floods, severe storms, wind, drought, pollution, water reuses, sustainable development and resiliency, etc. Formerly, he was a visiting professor at Princeton University, New Jersey, and the University of ETH Zurich, Switzerland. On the research side, he started a research partnership in 2014 with McGill University, Canada. He has contributed to more than 1K publications in journals, books, and technical reports. He is the founder and chief editor of the *International Journal of Hydrology Science and Technology (IJHST-Web of Sxience)*. Eslamian is now associate editor of three important publications: *Eco-Hydrology and Hydrobiology* (Elsevier), *Water Reuse* (IWA), and *Journal of the Saudi Society of Agricultural Sciences* (Elsevier). Professor Eslamian is the author of approximately 65 books and Special Issues and also 350 chapter books.

Dr. Eslamian's professional experience includes membership on editorial boards, and he is a reviewer of approximately 100 Web of Science (ISI) journals, including the *ASCE Journal of Hydrologic Engineering, ASCE Journal of Water Resources Planning and Management, ASCE Journal of Irrigation and Drainage Engineering, Advances in Water Resources, Groundwater, Hydrological Processes, Hydrological Sciences Journal, Global Planetary Changes, Water Resources Management, Water Science and Technology, Eco-Hydrology, Journal of American Water Resources Association, American Water Works Association Journal*, etc. UNESCO has also nominated him for a special issue of the *Eco-Hydrology and Hydrobiology Journal* in 2015.

Professor Eslamian was selected as an outstanding reviewer for the *Journal of Hydrologic Engineering* in 2009 and received the EWRI/ASCE Visiting International Fellowship in Rhode Island (2010). He was also awarded outstanding prizes from the Iranian Hydraulics Association in 2005 and Iranian Petroleum and Oil Industry in 2011. Professor Eslamian has been chosen as a distinguished researcher of Isfahan University of Technology (IUT) and Isfahan Province in 2012 and 2014, respectively. In 2016, he was a candidate for national distinguished researcher in Iran.

He has also been the referee of many international organizations and universities. Some examples include the U.S. Civilian Research and Development Foundation (USCRDF), the Swiss Network for International Studies, the Majesty Research Trust Fund of Sultan Qaboos University of Oman, the Royal Jordanian Geography Center College, and the Research Department of Swinburne University of Technology of Australia. He is also a member of the following associations: American Society of Civil

Engineers (ASCE), International Association of Hydrologic Science (IAHS), World Conservation Union (IUCN), GC Network for Drylands Research and Development (NDRD), International Association for Urban Climate (IAUC), International Society for Agricultural Meteorology (ISAM), Association of Water and Environment Modeling (AWEM), International Hydrological Association (STAHS), and UK Drought National Center (UKDNC).

Professor Eslamian finished Hakimsanaei High School in Isfahan in 1979. He was admitted to IUT for a BS in water engineering and graduated in 1986. After graduation, he was offered a scholarship for a master's degree program at Tarbiat Modares University, Tehran. He finished his studies in hydrology and water resources engineering in 1989. In 1991, he was awarded a scholarship for a PhD in civil engineering at the University of New South Wales, Australia. His supervisor was Professor David H. Pilgrim, who encouraged him to work on "Regional Flood Frequency Analysis Using a New Region of Influence Approach." He earned a PhD in 1995 and returned to his home country and IUT. In 2001, he was promoted to associate professor and in 2014 to full professor. For the past 28 years, he has been nominated for different positions at IUT, including university president consultant, faculty deputy of education, and head of department. Eslamian is now director for center of excellence in Risk Management and Natural Hazards (RiMaNaH).

Professor Eslamian has made three scientific visits to the United States, Switzerland, and Canada in 2006, 2008, and 2015, respectively. In the first, he was offered the position of visiting professor by Princeton University and worked jointly with Late Professor Eric F. Wood at the School of Engineering and Applied Sciences for one year. The outcome was a contribution in hydrological and agricultural drought interaction knowledge by developing multivariate L-moments between soil moisture and low flows for northeastern U.S. streams.

Recently, Professor Eslamian has published the 14 handbooks by Taylor & Francis (CRC Press): the three-volume *Handbook of Engineering Hydrology* (2014), *Urban Water Reuse Handbook* (2016), *Underground Aqueducts Handbook* (2017), the three-volume *Handbook of Drought and Water Scarcity* (2017), *Constructed Wetlands: Hydraulic Design* (2019), *Handbook of Irrigation System Selection for Semi-Arid Regions* (2020), *Urban and Industrial Water Conservation Methods* (2020), and the three-volume *Flood Handbook* (2022).

An Evaluation of Groundwater Storage Potentials in a Semiarid Climate and *Advances in Hydrogeochemistry Research* by Nova Science Publishers are also his book publications in 2019 and 2020, respectively. Two-volume *Handbook of Water Harvesting and Conservation* (Wiley) and *Handbook of Disaster Risk Reduction and Resilience (New Frameworks for Building Resilience to Disasters)* are early 2021 book publications of Professor Eslamian. *Handbook of Disaster Risk Reduction and Resilience (Disaster Risk Management Strategies)* and two-volume *Earth Systems Protection and Sustainability* are early 2022 handbooks of Professor Eslamian. The 3-volume Handbook of Hydroinformatics (Elsevier) has been the latest book publication of him in early 2023.

Professor Eslamian has been appointed as World Top 2-Percent Researcher by Stanford University, USA, in 2019 and 2020. He has also been a Grant Assessor, Report Referee, Award Jury, and Invited Researcher for international organizations such as United States Civilian Research and Development Foundation (2006), Intergovernmental Panel on Climate Change (2012), World Bank Policy and Human Resources Development Fund (2021), and Stockholm International Peace Research Institute (2022), respectively.

Editors

Faezeh Eslamian is a PhD holder of bioresource engineering from McGill University. Her research focuses on the development of a novel lime-based product to mitigate phosphorus loss from agricultural fields. Faezeh completed her bachelor's and master's degrees in civil and environmental engineering from Isfahan University of Technology, Iran, where she evaluated natural and low-cost absorbents for the removal of pollutants such as textile dyes and heavy metals. Furthermore, she has conducted research on the worldwide water quality standards and wastewater reuse guidelines. Faezeh is an experienced multidisciplinary researcher with research interests in soil and water quality, environmental remediation, water reuse, and drought management.

Contributors

Nasrin Azad
Urmia University
Urmia, Iran

Javad Behmanesh
Urmia University
Urmia, Iran

Abebech A. Beyene
Bahir Dar University
Bahir Dar, Ethiopia

Freddy Canales-Ide
Universidad Politécnica de Madrid
Madrid, Spain

Neil A. Coles
University of Leeds
Leeds, United Kingdom
and
The University of Western Australia
Perth, Australia
and
Verdant Earth
Cambridge, United Kingdom

Saeid Eslamian
College of Agriculture, Isfahan University of Technology
Isfahan, Iran
and
Excellence Center on Risk Management and Natural Hazards, Isfahan University of Technology
Isfahan, Iran

Hamideh Faridi
University of Manitoba
Winnipeg, Canada

Hamidreza Faridi
University of Tehran
Tehran, Iran

José Antônio Frizzone
University of São Paulo
Piracicaba, Brazil

Faramarz Ghalambaz
Khuzestan Water and Power Authority
Ahvaz, Iran

Babak Ghoreishi
Forest, Range and Watershed Management Organization
Hamedan, Iran

Mahbub Hasan
Alabama A&M University
Normal, Alabama

Dariush Hayati
Shiraz University
Shiraz, Iran

S. Jari
Federal University Dutsin-Ma
Dutsin-Ma, Nigeria

Hamid Karimi
University of Zabol
Zabol, Iran

Aschalew Kassu
Alabama A&M University
Normal, Alabama

Verônica Gaspar Martins Leite de Melo
University of São Paulo
Piracicaba, Brazil

Fulai Liu
University of Copenhagen
Taastrup, Denmark

Jie Liu
Northwest A&F University
Yangling, China

Mousa Maleki
Illinois Institute of Technology Chicago, Illinois

Patricia Angélica Alves Marques
University of São Paulo
Piracicaba, Brazil

Mohammad Mohammadzadeh
Shiraz University
Shiraz, Iran

R. T. Nabinta
Federal University Dutsin-Ma
Dutsin-Ma, Nigeria

Bernie Omodei
Measured Irrigation
Woodville Park, Australia

Antonio Pires de Camargo
State University of Campinas
Campinas, Brazil

Qin Qian
Lamar University
Beaumont, Texas

K. Y. Raneesh
Sri Shakthi Institute of Engineering and Technology
Coimbatore, India

Vahid Rezaverdinejad
Urmia University
Urmia, Iran

T. A. Rimi
Federal University Dutsin-Ma
Dutsin-Ma, Nigeria

Mark R. Rivers
ClearWater Research and Management Pty. Ltd.
Perth, Australia

Leonor Rodríguez-Sinobas
Universidad Politécnica de Madrid
Madrid, Spain

Sajjad Roshandel
Xiamen University
Xiamen, China

Asal Soltani
The University of Toledo
Toledo, Ohio

Naser Valizadeh
Shiraz University
Shiraz, Iran

Zhenhua Wei
Northwest A&F University
Yangling, China

Shuichiro Yoshida
The University of Tokyo
Tokyo, Japan

Ghasem Zarei
Agricultural Engineering Research Institute, Agricultural Research, Education and Extension Organization
Karaj, Iran

Sergio Zubelzu
Universidad Politécnica de Madrid
Madrid, Spain

Solutions for Water Scarcity and Soil Salinity in Irrigation

I

1

Deficit Irrigation: A Solution for Water Scarcity Conditions?

Jie Liu and
Zhenhua Wei
Northwest A&F University

Fulai Liu
University of Copenhagen

1.1	Introduction	3
1.2	Deficit Irrigation Techniques	4
1.3	Mechanisms for Saving Water	4
1.4	Effect on Product Quality	5
1.5	Water-Saving Potential	6
1.6	Perspectives and Future Focus	7
1.7	Conclusions	8
	References	8

1.1 Introduction

Water is an increasingly scarce resource globally. Water shortage threatens humans' health and survival, as nearly 800 million people have no access to safe water and 2.5 billion have no suitable sanitation (Schiermeier 2014). The situation may be exacerbated by the middle of this century as the population is expected to increase by 30%; meanwhile, regional and global drought induced by climate change will occur frequently. To meet the increasing food demand due to the population growth, the irrigated arable land area for agriculture is continuously increasing. Agriculture is the most water-consuming sector, using 70% of total freshwater withdrawals in the world, and the percentage will increase in the future (Clothier et al. 2010). However, agricultural water use efficiency (WUE) is low, and the competition for water resources with other sectors (i.e. industry) is increasing. In many areas of the planet, irrigation water has been overexploited, leading to serious environmental issues, such as deepened groundwater table and desertification. Therefore, the application of water management strategies to improve agriculture water productivity is the key to ensure global water and food security, and environment sustainable development.

Drought results in soil water deficit and subsequently plant water deficit. Plants respond to soil water deficit by avoiding the occurrence of leaf water deficit (drought avoidance) or tolerating low cellular water contents (drought tolerance) (Dodd & Ryan 2016). Drought affects plant growth by affecting leaf stomatal behaviors. The prime water loss of plants is via pores termed stomata, which are bound with guard cells. Stomata are sensitive to environmental factors, such as atmospheric CO_2 concentration (Ainsworth & Rogers 2007), leaf to air vapor pressure deficit (VPD) (Yong et al. 1997; Du et al. 2019) and soil water status (i.e. Liu et al. 2015). Though crops can resist water deficit under different mechanisms, water stress usually has a negative effect on crop growth and production (Andrade et al. 2005; Zhou et al. 2018), mainly due to reduced carbon assimilation, cell division and enlargement. However, moderate drought stress is a flexible way to improve plant WUE with acceptable yield.

Many irrigation strategies have been developed to save water, including biological water-saving techniques, agronomy water-saving techniques and engineering water-saving techniques. Among these, the biological water-saving techniques are based on plant physiology that plants are not sensitive to water stress at certain growth period; thus, plants can be supplied with insufficient water. Deficit irrigation (DI) has been considered as a valuable and sustainable strategy to maximize the water productivity in agriculture (Kang et al. 1998). DI is an optimal strategy that deliberately allows crops to experience some degree of water deficit either during a special period or throughout the whole growing season with slight decrease in yield and significant reduction of irrigation water, thereby improving plant WUE (Banihabib et al. 2016). The effects of DI on crop yield, quality and WUE were highly studied; however, the systematic theory and the underlying mechanisms were rarely reviewed. What's more, whether DI is a promising solution for future water scarcity still needs to be reconsidered since eco-systematical problems such as soil salinization occurred in areas of DI application. Therefore, this chapter mainly focuses on DI strategies, its water-saving potential and mechanism and its effect on WUE and crop quality, and the disadvantages and future perspectives are also discussed.

1.2 Deficit Irrigation Techniques

Faced with water scarcity, more efficient technological innovation and irrigation strategies are urgently needed to develop. The purpose of irrigation has been shifted from satisfying crop evapotranspiration requirement for more crop yield per unit of land to decreasing inefficient luxury transpiration for maximizing crop yield per drop of water. DI is an optimal strategy that deliberately allows crops to experience some degree of water deficit either during a special period or throughout the whole growing season with slight decrease in yield and significant reduction of irrigation water, thereby improving plant WUE. Though DI strategy leaves crops in a state of water deficit, especially for plants grown in soil with low water shortage capacity resulting in a declined harvestable yield, it maximizes the water productivity and improves crop quality (Kang & Zhang 2004; Du et al. 2008). Additionally, DI may reduce the risk of fungal diseases due to the less humid environment around the roots than full irrigation.

There are three main DI strategies based on the physiological knowledge of crops response to water deficit, named as sustained deficit irrigation (SDI, water deficit is uniformly distributed over the whole stage), regulated deficit irrigation (RDI, water deficit is applied at certain developmental stages) and partial root-zone drying (PRD, two parts of the root system are wetted or dried alternatively) (FAO 2002). SDI is the application of less than 100% crop evapotranspiration to the whole growing crop, thus avoiding the occurrence of serious plant water deficit at any growth stage. RDI is based on phenological knowledge of crop that transpiration is more sensitive to water deficit than photosynthesis. It is a method that irrigates the entire root zone with less amount of water than the potential evapotranspiration for optimal plant growth in particular periods, especially avoiding drought-sensitive phenological stage as slight stress depresses the yield. The PRD technique, to some extent, is a further development of RDI, which divides the root system into two parts, one exposed to drying soil while the other irrigated with decreased water. The dried and wetted areas of root system are alternated with a frequency according to soil moisture and crop water requirement. PRD can improve the plant WUE without significant crop yield depletion. Researchers reported that crop under PRD harvested more yield than RDI when the same amount of irrigation water was applied (Antolín et al. 2008). DI could be applied in several ways in the field and has been largely developed, such as surface irrigation, furrow irrigation, drip irrigation and so on (Kang & Zhang 2004).

1.3 Mechanisms for Saving Water

The sensitivity of plant to water deficit is related to phenological period; thus, DI strategies can be applied when plants are less sensitive to water. For plants during seedling stage, the leaf area is small and transpiration is low; thus, plants have little demand on water and can suffer water deficit with declined stomatal conductance (g_s) and photosynthetic rate (A_n). Once irrigated, photosynthetic capacity could largely recover from water stress at this stage. From the flowering to fruit maturity stage, the water demand is increasing. Particularly, during fruit development stage, soil moisture has been a limit factor depressing tomato yield (Kuşcu et al. 2014). The effect of DI on crop not only depends on the period of water stress exposed, but also depends on the degree of stress (Conesa et al. 2014).

One of the physiological mechanisms by which DI technique saves water and improves WUE was that root can sense soil water deficit and produce chemical signal, mainly abscisic acid (ABA), which can be transported to shoots, inducing stomata partial closure and g_s reduction, thereby reducing water loss. Transpiration was more sensitive than photosynthesis to soil water deficit; thus, plant could maintain A_n when exposed under soil water stress. For the PRD strategy, the underlying mechanism by which water is saved could be easier understood, that half part of root system is exposed to soil water stress, and thus chemical signals like ABA are produced and transported to leaves, resulting in reduced stomata aperture; and at the same time, the other part of root system is irrigated to have access to water, and thus plant has intensified and constant ABA level and water status, and maintains plant A_n, contributing to a more efficient control of plant transpiration, thereby improving WUE (Dood 2009; Wang et al. 2010; Kang & Zhang 2004; Liu et al. 2006). Besides, PRD decreases soil evaporation; meanwhile, it improves the plant transpiration to evaporation ratio (Liu 2011). Additionally, PRD at vegetative growth stage could stimulate the development of roots, including root numbers, total root dry weight and root density in an arid environment, which allows plants to have access to water and nutrient in larger soil volumes and potentially benefit for water and nutrient uptake (Kang et al. 1998, 2000; Liu et al. 2006).

On the other hand, plant can respond to soil water stress through osmotic adjustment to maintain cell turgor and enable plant grow under reduced water potential in leaves and roots (Blum 2017). Under water stress, the increased osmotic active solutes (mainly including sugars, organic acids, proline and betaine) will decrease osmotic potential, leading to the improved ability of cells to absorb and retain water. Sucrose and glucose participate as substrates for cellular respiration or as osmolytes to maintain cell homeostasis. Sucrose plays an important role in adjusting osmotic potential as it can be broken down into fructose and glucose by invertase, doubling the osmotic concentration of sucrose, thereby increasing water flux into fruit cells (Beckles et al. 2012).

1.4 Effect on Product Quality

In recent years, consumers and growers pay increasing attention to product quality (Samoticha et al. 2016). Crop quality largely depends on agronomical (i.e. intensity and duration of water deficit) factors except for genetic and seasonal factors. Fruit quality is determined by primary metabolites (sucrose, fructose, glucose, malic acid and citric acid), secondary metabolites (vitamin C, protein, phenolics and other bioactive compounds) and mineral nutrients (macro- and micro-mineral nutrients). The sugar to acid ratio is responsible for fruit flavor quality (Dorais et al. 2001), and secondary metabolites and mineral nutrients are main indicators to assess fruit nutritional quality (Wei et al. 2018b). Total soluble solids (TSS) is an aggregative indicator of all soluble components in fruit, such as soluble sugar, acid and vitamin C. Apart from being a strategy improving WUE, DI has become an important agronomic measure to modulate many fruit quality parameters (Du et al. 2015). DI has been successfully applied to crops, fruit trees and vegetables, optimizing the vegetative and reproductive growth with acceptable yield and improved fruit quality (Kang & Zhang 2004; Du et al. 2008; Patanè & Saita 2015; Malejane et al. 2018). For example, compared to full irrigation, DI strategy with 25%–30% water deficit significantly improved protein content in maize with acceptable yield (Ertek & Kara 2013; Oktem 2008; Kresović et al. 2018). The improvement of protein content in sweet corn under DI could be a result of increased N uptake by root as DI stimulates the 'Birch effect' (Birch 1958), which contributes to decomposition and mineralization of organic matter.

Tomato (*Solanum lycopersicum* L.) is considered as one of the most popular vegetables or fruits because it is an important source of secondary metabolites such as antioxidant substances including lycopene, phenolics and vitamin C in human diet (Toor et al. 2006). The effects of DI on tomato quality were intensively studied (Chen et al. 2013; Wei et al. 2018b; Patanè & Saita 2015; Mattar et al. 2019). DI improved TSS and sugar concentrations in tomato fruit, and this was due to (1) the increase of carbohydrate accumulation toward fruit maturation with reduction of side shoots and reproductive growth, (2) the increase of starch decomposition in fruit maturation, and (3) the enhanced activities of enzymes involving in carbohydrate metabolism triggered by ABA (Wei et al. 2018b). Ripoll et al. (2014) have reviewed the effects of various drought forms on fleshy fruit quality (primary and secondary metabolites), and proposed two potential mechanisms to reveal the change in secondary metabolites. On the one hand, water deficit reduced leaf g_s and A_n, thus influencing the transport of primary metabolites to the fruits for secondary metabolites biosynthesis. On the other hand, water deficit may induce oxidative stress which stimulates the antioxidant response in plant leaves. Recently, Hou et al. (2019b) further supplemented the mechanisms of vitamin C, lycopene and β-carotene metabolisms in tomato fruits under DI, and proposed that DI decreased leaf area and increased fruit sunlight exposure and temperatures that influence fruit vitamin C and lycopene metabolisms.

Mineral nutrient (macro-nutrients and micro-nutrients) accumulation is both affected by water management and species specific. The effect of water deficit on mineral nutrient was studied in many crop species, i.e. maize (Kresović et al. 2018), grape (Centofanti et al. 2019) and almond (Lipan et al. 2019). The decreased uptake of nutrients from the soil can be ascribed to the water stress induced by DI, given that the mobilization of nutritional elements was closely related to water availability. Furthermore, the PRD strategy also intensified the 'Birch effect'. This was mainly due to microbial physiology and soil physical processes regulated by repeated soil drying and rewetting cycles, enriching bioavailable N or P in soil (Wang et al. 2016). Moreover, nutrient uptake, particularly for less mobile nutrient (i.e. P), is largely determined by plant root systems. The PRD leaves plant root zone in two water status, and this could potentially affect root morphological and architectural traits. Thus, the nutrient uptake from the soil to root surface of plant will be potentially improved, and some studies have reported that PRD improves plant N acquisition (Liu et al. 2015; Wei et al. 2018a) and P uptake (Wang et al. 2012; Liu et al. 2015), further being helpful for improving fruit quality.

1.5 Water-Saving Potential

WUE is a ratio between two physiological (i.e. photosynthesis and transpiration) or agronomic (i.e. yield and water consumption) indices (Blum 2005). It can be defined at different scales, such as stomatal, leaf, plant and field. Physiologically, WUE indicates the intrinsic trade-off between carbon fixation and water loss, as water evaporates from the interstitial tissues (stomata) of leaves, where CO_2 enters for photosynthesis (Bramley et al. 2013). At stomatal level, WUE is expressed as the ratio of A_n to g_s, also named as intrinsic water use efficiency (WUE_i). At a stable atmospheric CO_2 environment, WUE_i is less disturbed by environmental factors, and mainly related to crop species and varieties (Taylor et al. 2018), e.g. WUE_i in C4 plant (i.e. ca.160 μmol/mol in sorghum) is higher than that in C3 plant (i.e. peak value of ca. 70 μmol/mol in potato) (Schulze & Hall 1982; Liu et al. 2005); thus, C4 crops have greater water-saving potential than C3 crops. Additionally, WUE_i could be improved by moderate soil water stress, as moderate soil water stress slightly affected A_n while significantly decreased g_s (Liu et al. 2005). At leaf level, WUE is the ratio of A_n to transpiration rate (T_r), also named as instantaneous water use efficiency (WUE_T), indicating carbon fixation per unit of water loss through transpiration in a certain area of leaf. WUE_T is affected by both plant species and the environment, for example, WUE_T decreases with increasing VPD, and WUE_T was higher for plants treated with PRD strategy (Liu et al. 2009). In plant scale, WUE (WUE_P) is the ratio of plant biomass to water consumption, characterizing the biomass accumulation and water consumption of whole plant in certain period, which is an important parameter to study the physiological and ecological mechanism of plant CO_2 fixation. WUE_P is significantly affected by climate, i.e. WUE_P in southern and northern hemisphere is different and varies with months (Turner 2004). WUE_P can be modulated by irrigation strategies, such as DI; especially PRD strategy significantly improves WUE_P (Wang et al. 2010), indicating PRD had greater potential in saving water. WUE at field scope (WUE_Y) is defined as total biomass or yield produced by water consumption (precipitation or irrigation) by the whole field ecosystem. WUE acts as a key variable in the assessment of plant response to the DI-induced water stress.

The regulation of stomatal aperture is a central process to influence plant WUE. Stomatal behavior is closely related to environmental factors, e.g. atmospheric CO_2 concentration (Azoulay-Shemer et al. 2015), VPD (Yong et al. 1997; Du et al. 2019) and plant water status (Liu et al. 2005). WUE_i and WUE_T can be determined by measuring leaf gas exchange and be used to investigate the short-term response of plant water use to environmental factors. For example, the response of g_s and A_n in tomato plant to progressive soil drying was retarded in elevated CO_2 environment compared to that in ambient CO_2 environment, and WUE_i and WUE_T were improved by elevated CO_2 (Liu et al. 2019). This suggests in future CO_2 enrichment climate, plant may adapt water deficit and strengthen the drought resistance or tolerance to keep higher WUE_i and WUE_T to perform great water-saving potential (Kang et al. 2002; Liu et al. 2019). In practice, PRD plays an important role in the efficient regulation of stomata aperture to decrease Tr and water loss, thereby improving WUE_P. What's more, micro-sprinkler irrigation could also contribute to higher WUE_P by adjusting atmospheric humidity and temperature to lower VPD. WUE_Y can reflect water use status and crop production in field, while it is difficult to measure, and stable isotope technique provided a quick and stable way to determine WUE_Y as carbon isotope discrimination of leaf and fruit was negatively correlated with WUE_Y (Wei et al. 2016). Immense potential of water saving (up to 30%) by DI strategies has been reported in literature (Aragüés et al. 2014; Ertek & Kara 2013; Golzardi et al. 2017). This was closely due to the reduced water losses through evaporation and deep percolation with the adaption of DI (Jia et al. 2014).

1.6 Perspectives and Future Focus

Intensive studies focus on the effect of DI on crop yield, quality and WUE, but few studies emphasized its environmental impacts. As it is known, CO_2 is the major component of greenhouse gases driving global warming, and agricultural soils are major sources of greenhouse gas emission. A recent study by Hou et al. (2019a) reported that DI reduced the cumulative soil CO_2 emissions by 10.2%–25.5% than full irrigation during most of the growing season. Thus, promoting the DI strategies can make positive effect on slowing down global warming.

Irrigation strategies contribute to water saving at farm scale, while they may increase water consumption at a watershed or basin scale (Perry et al. 2017; Grafton et al. 2018). The improvement of irrigation efficiency by reduced water supplied per hectare for field crops could increase net income, and thus may stimulate the expansion of irrigated area, resulting in over-extraction of groundwater and increase of total water consumption at watershed or basin scale, which threatens the survival of other species and ecological stability. Thus, the extension of DI strategy should be carried out within a certain growing area to avoid the decreased marginal interest at a watershed or basin scale.

Another potential disadvantage of DI is that it may induce soil salinization, especially in areas where irrigation water contains relatively high concentration of salt. In a cucumber study, salt accumulation occurs in the center of the plant root, 30 cm depth from the soil surface (Kaman & Özbek 2012). The salt is originally from irrigation water, and its concentration will be intensified by crop evapotranspiration and by limited absorption via root; thus, salt accumulates in the soil (Liu 2011). DI supplies insufficient water, resulting in reduced leaching (Geerts & Raes 2009; Aragüés et al. 2014). Soil salinity inhibits plant nutrient acquisition and increases the risk of nutrient imbalances (Cheng et al. 2019), and thus depresses the development of the plant and thereby productivity. However, PRD irrigated half part of roots with more water amount each time, which may facilitate to enlarge the soil wetted volume and reduce the soil salinization.

Combined effect of DI and soil amendments (i.e. biochar or plant growth-promoting bacteria, PGPR) should also be taken into consideration in future. PGPR colonize plant roots system and promote plant growth by the various mechanisms, i.e. nitrogen fixation (Glick 2012). The application of PGPR could not only reduce or alleviate the contamination of soil and water induced by the extensive use of chemical fertilizers (Gouda et al. 2018), but may also synergistically improve WUE combined with DI, and thereby beneficial to the sustainable development of agriculture.

1.7 Conclusions

DI, as an efficient strategy to improve agricultural water productivity, has been highly studied and applied in agricultural water management nowadays. In relation to traditional irrigation, DI maximizes the economical yield and crop quality with the advantage of less water input during plant growth period. DI could induce stomata closure by chemical signaling synthesis to depress luxury transpiration; meanwhile, DI could stimulate osmotic adjustment to maintain cell turgor, thus enabling plant grow under reduced water potential in leaves and roots. Moreover, DI promotes root development to improve the ratio of root to shoot, further enhancing root ability of water and nutrient uptake. In terms of improving crop quality, DI plays an efficient role by increasing the allocation of carbohydrates toward fruit, the decomposition of starch in fruit maturation and the enhancement activities of carbohydrate metabolism enzymes. On the other hand, limited water availability depressed the mobilization of nutritional elements, resulting in decrease in accumulation of mineral nutrition under DI. However, PRD could provide a solution because the PRD could intensify the 'Birch effect' to improve plant N acquisition. What's more, the PRD strategy possesses greater potential in saving water with improved plant WUE. In the perspective of environmental impacts, DI strategy shows a positive effect in reducing greenhouse gas emissions and water use at farm scale, while it may induce soil salinization. DI was suggested to apply within a certain growing area, as it threatens the ecological stability by over-extraction of groundwater at watershed or basin scale with reduced marginal interest.

References

Ainsworth, E., Rogers, A. 2007. The response of photosynthesis and stomatal conductance to rising [CO_2]: mechanisms and environmental interactions. *Plant, Cell and Environment*, 30, 258–270.

Andrade, F., Sadras, V., Vega, C., Echarte, L. 2005. Physiological determinants of crop growth and yield in maize, sunflower and soybean: their application to crop management, modeling and breeding. *Journal of Crop Improvement*, 14, 51–101.

Antolín, M., Santesteban, H., María, E., Aguirreolea, J. 2008. Involvement of abscisic acid and polyamines in berry ripening of *Vitis vinifera* (L.) subjected to water deficit irrigation. *Australian Journal of Grape and Wine Research*, 14 (2), 123–133.

Aragüés, R., Medina, E., Martínez-Cob, A., Faci, J. 2014. Effects of deficit irrigation strategies on soil salinization and sodification in a semiarid drip-irrigated peach orchard. *Agricultural Water Management*, 142, 1–9.

Azoulay-Shemer, T., Palomares, A., Bagheri, A., Israelsson-Nordstrom, M., Engineer, C., Bargmann, B., Stephan, A., Schroeder, J. 2015. Guard cell photosynthesis is critical for stomatal turgor production, yet does not directly mediate CO_2- and ABA-induced stomatal closing. *Plant Journal*, 83 (4), 567–581.

Banihabib, M. E., Zahraei, A., Eslamian, S. 2016. Dynamic programming model for the system of a non-uniform deficit irrigation and a reservoir. *Irrigation and Drainage*, 66 (1), 71–81.

Beckles, D., Hong, N., Stamova, L., Luengwilai, K. 2012. Biochemical factors contributing to tomato fruit sugar content: a review. *Fruits*, 67, 49–64.

Birch, H. 1958. The effect of soil drying on humus decomposition and nitrogen. *Plant and Soil*, 10, 9–31.

Blum, A. 2005. Drought resistance, water-use efficiency, and yield potential—are they compatible, dissonant, or mutually exclusive? *Australian Journal of Agricultural Research*, 56, 1159–1168.

Blum, A. 2017. Osmotic adjustment is a prime drought stress adaptive engine in support of plant production. *Plant, Cell and Environment*, 40, 4–10.

Bramley, H., Turner, N., Siddique, K. 2013. Water use efficiency. In: Kole, C. (ed) *Genomics and Breeding for Climate-Resilient Crops*, Vol. 2, pp. 225–268. Springer, Berlin, Heidelberg.

Centofanti, T., Bañuelos, G., Ayars, J. 2019. Fruit nutritional quality under deficit irrigation: the case of table fruits in California. doi:10.1002/jsfa.9415

Chen, J., Kang, S., Du, T., Qiu, R., Guo, P., Chen, R. 2013. Quantitative response of greenhouse tomato yield and quality to water deficit at different growth stages. *Agricultural Water Management*, 129, 152–162.

Cheng, Z., Chen, Y., Zhang, F. 2019. Effect of cropping systems after abandoned salinized farmland reclamation on soil bacterial communities in arid northwest China. *Soil & Tillage Research*, 187, 204–213.

Clothier, B., Green, S., Deurer, M. 2010. Green, blue and grey waters: minimizing the footprint using soil physics. 19th World Congress of Soil Sciences, Soil Solution for a Changing World, Brisbane, Australia.

Conesa, M., García-Salinas, M., Rosa, J., Fernandez-Trujillo, J., Domingo, R., Perez-Pastor, A. 2014. Effects of deficit irrigation applied during fruit growth period of late mandarin trees on harvest quality, cold storage and subsequent shelf-life. *Scientia Horticulturae*, 165: 344–351.

Dodd, I., Ryan, A. 2016. *Whole-Plant Physiological Responses to Water-Deficit Stress*. John Wiley & Sons, Ltd, Chichester, UK.

Dorais, M., Papadopoulos, A., Gosselin, A. 2001. Greenhouse tomato fruit quality. *Horticultural Reviews*, 26, 239–319.

Du, Q., Liu, T., Jiao, X., Song, X., Zhang, J., Li, J. 2019. Leaf anatomical adaptations have central roles in photosynthetic acclimation to humidity. *Journal of Experimental Botany*, 70 (18), 494–4962.

Du, T., Kang, S., Zhang, J., Davies, W. 2015. Deficit irrigation and sustainable water-resource strategies in agriculture for China's food security. *Journal of Experimental Botany*, 66, 2253–2269.

Du, T., Kang, S., Zhang, J., Li, F., Yan, B. 2008. Water use efficiency and fruit quality of table grape under alternate partial root-zone drip irrigation. *Agricultural Water Management*, 95 (6), 659–668.

Ertek, A., Kara, B. 2013. Yield and quality of sweet corn under deficit irrigation. *Agricultural Water Management*, 129, 138–144.

FAO. 2002. Deficit irrigation practices. FAO Water Reports 22, Rome, Italy.

Geerts, A., Raes, D. 2009. Deficit irrigation as an on-farm strategy to maximize crop water productivity in dry areas. *Agricultural Water Management*, 96, 1275–1284.

Glick, B. 2012. Plant growth-promoting bacteria: mechanisms and applications. *Scientifica*. doi:10.6064/2012/963401

Golzardi, F., Baghdadi, A., Afshar, R. 2017. Alternate furrow irrigation affects yield and water-use efficiency of maize under deficit irrigation. *Crop and Pasture Science*, 68, 726–734.

Gouda, S., Kerry, R., Das, G., Paramithiotis, S., Shin, H., Patra, J. 2018. Revitalization of plant growth promoting rhizobacteria for sustainable development in agriculture. *Microbiological Research*, 206, 131–140.

Grafton, R., Williams, J., Perry, C., Molle, F., Ringler, C., Steduto, P., Udall, B., Wheeler, S., Wang, Y., Garrick, D., Allen, R. 2018. The paradox of irrigation efficiency. *Science*, 361 (6404), 748–750.

Hou, H., Yang, Y., Han, Z., Cai, H., Li, Z. 2019a. Deficit irrigation effectively reduces soil carbon dioxide emissions from wheat fields in northwest China. *Journal of the Science of Food and Agriculture*, 99 (12), 5401–5408.

Hou, X., Zhang, W., Du, T., Kang, S., Davies, W. 2019b. Responses of water accumulation and solute metabolism in tomato fruit to water scarcity and implications for main fruit quality variables. *Journal of Experimental Botany*, 71 (4): 1249–1264.

Jia, D., Dai, X., Men, H., He, M. 2014. Assessment of winter wheat (*Triticum aestivum* L.) grown under alternate furrow irrigation in northern China: grain yield and water use efficiency. *Canadian Journal of Plant Science*, 94, 349–359.

Kaman, H., Özbek, Ö. 2012. Salt and water distributions in the plant root zone under deficit irrigation. *Journal of Food, Agriculture and Environment*, 10 (3&4), 496–500.

Kang, S., Liang, Z., Hu, W., Zhang, J. 1998. Water use efficiency of controlled alternate irrigation on root divided maize plants. *Agricultural Water Management*, 38, 69–76.

Kang, S., Liang, Z., Pan, Y., Shi, P., Zhang, J. 2000. Alternate furrow irrigation for maize production in an arid area. *Agricultural Water Management*, 45, 267–274.

Kang, S., Zhang, F., Hu, X., Zhang, J. 2002. Benefits of CO_2 enrichment on crop plants are modified by soil water status. *Plant and Soil*, 238 (1), 69–77.

Kang, S., Zhang, J. 2004. Controlled alternate partial root-zone irrigation: its physiological consequences and impact on water use efficiency. *Journal of Experimental Botany*, 55 (407), 2437–2446.

Kresović, B., Gajić, B., Tapanarova, A., Dugalić, G. 2018. How irrigation water affects the yield and nutritional quality of maize (*Zea mays* L.) in a temperate climate. *Polish Journal of Environmental Studies*, 27 (3), 1123–1131.

Kuşcu, H., Turhan, A., Demir, A. 2014. The response of processing tomato to deficit irrigation at various phenological stages in a sub-humid environment. *Agricultural Water Management*, 133, 92–103.

Lipan, L., Martín-Palomo, M., Sánchez-Rodríguez, L., Cano-Lamadrid, M., Sendra, E., Hernández, F., Burló, F., Vázquez-Araújo, L., Andreu, L., Carbonell-Barrachina, Á. 2019. Almond fruit quality can be improved by means of deficit irrigation strategies. *Agricultural Water Management*, 217, 236–242.

Liu, C., Rubœk, G., Liu, F., Andersen, M., 2015. Effect of partial root zone drying and deficit irrigation on nitrogen and phosphorus uptake in potato. *Agricultural Water Management*, 159, 66–76.

Liu, F. 2011. Irrigation strategies for sustainable environmental and influence on human health. In: Nriagu, J. (ed) *Encyclopedia of Environmental Health*, pp. 297–303. Elsevier, Amsterdam.

Liu, F., Andersen, M., Jacobsen, S., Jensen, C. 2005. Stomatal control and water use efficiency of soybean (Glycine max, L. Merr.) during progressive soil drying. *Environmental and Experimental Botany*, 54 (1), 33–40.

Liu, F., Andersen, M., Jensen, C. 2009. Capability of the 'Ball - Berry' model for predicting stomatal conductance and water use efficiency of potato leaves under different irrigation regimes. *Scientia Horticulturae*, 122, 346–354.

Liu, F., Shahanzari, A., Andersen, M., Jacobsen, S., Jensen, C. 2006. Effects of deficit irrigation (DI) and partial root drying (PRD) on gas exchange biomass partitioning, and water use efficiency in potato. *Scientia Horticulturae*, 109, 113–117.

Liu, J., Hu, T., Fang, L., Peng, X., Liu, F. 2019. CO_2 elevation modulates the response of leaf gas exchange to progressive soil drying in tomato plants. *Agricultural and Forest Meteorology*, 268, 181–188.

Malejane, D., Tinyani, P., Soundy, P., Sultanbawa, Y., Sivakumar, D. 2018. Deficit irrigation improves phenolic content and antioxidant activity in leafy lettuce varieties. *Food Science & Nutrition*, 6 (2), 334–341.

Mattar, M., Zin El-Abedin, T., Alazba, A., Al-Ghobari, H. 2019. Soil water status and growth of tomato with partial root-zone drying and deficit drip irrigation techniques. *Irrigation Science*. doi:10.1007/s00271-019-00658-y

Oktem, A. 2008. Effect of water shortage on yield, and protein and mineral compositions of drip-irrigated sweet corn in sustainable agricultural systems. *Agricultural Water Management*, 95, 1003–1010.

Patanè, C., Saita, A. 2015. Biomass, fruit yield, water productivity and quality response of processing tomato to plant density and deficit irrigation under a semi-arid Mediterranean climate. *Crop and Pasture Science*, 66 (2), 224–234.

Perry, C., Steduto, P., Karajeh, F. 2017. Does improved irrigation technology save water? A view of the evidence. Food and Agricultural Organization of the United Nations, Cairo, Egypt.

Ripoll, J., Urban, L., Staudt, M., Lopez-Lauri, F., Bidel, L., Bertin, N. 2014. Water shortage and quality of fleshy fruits – making the most of the unavoidable. *Journal of Experimental Botany*, 65, 4097–4117.

Samoticha, J., Wojdyło, A., Golis, T. 2016. Phenolic composition, physicochemical properties and antioxidant activity of interspecific hybrids of grapes growing in Poland. *Food Chemistry*, 215, 263–273.

Schiermeier, Q. 2014. The parched planet: water on tap. *Nature*, 510, 326–328.

Schulze, E., Hall, A. 1982. Stomatal responses, water loss and CO_2 assimilation rates of plants in contrasting environments. In Lange, O. L., Nobel, P. S., Osmond, C. B., Ziegler, H. (eds) *Physiological Plant Ecology, II. Encyclopedia of Plant Physiology*, pp 181–230, 12B. Springer-Verlag, Berlin, Germany.

Taylor, S., Aspinwall, M., Blackman, C., Choat, B., Tissue, D., Ghannoum, O. 2018. CO_2 availability influences hydraulic function of C_3 and C_4 grass leaves. *Journal of Experimental Botany*, 69 (10), 2731–2741.

Toor, R., Savage, G., Heeb, A. 2006. Influence of different types of fertilizers on the major antioxidant components of tomatoes. *Journal of Food Composition and Analysis*, 19, 20–27.

Turner, N. 2004. Agronomic options for improving rainfall-use efficiency of crops in dryland farming system. *Journal of Experimental Botany*, 55 (407), 2413–2425.

Wang, Y., Jensen, C., Liu, F. 2016. Nutritional responses to soil drying and rewetting cycles under partial root-zone drying irrigation. *Agricultural Water Management*, 179, 254–259.

Wang, Y., Liu, F., Andersen, M., Jensen, C. 2010. Improved plant nitrogen nutrition contributes to higher water use efficiency in tomatoes under alternate partial root-zone irrigation. *Functional Plant Biology*, 37, 175–182.

Wang, Y., Liu, F., Jensen, C. 2012. Comparative effects of partial root-zone irrigation and deficit irrigation on phosphorus uptake in tomato plants. *Journal of Horticultural Science & Biotechnology*, 87 (6), 600–604.

Wei, Z., Du, T., Li, X., Fang, L., Liu, F. 2018a. Interactive effects of CO_2 concentration elevation and nitrogen fertilization on water and nitrogen use efficiency of tomato grown under reduced irrigation regimes. *Agricultural Water Management*, 202, 174–182.

Wei, Z., Du, T., Li, X., Fang, L., Liu, F. 2018b. Interactive effects of elevated CO_2 and N fertilization on yield and quality of tomato grown under reduced irrigation regimes. *Front in Plant Science*, 9, 328.

Wei, Z., Du, T., Zhang, J., Xu, S., Cambre, P., Davies, W. 2016. Carbon isotope discrimination shows a higher water use efficiency under alternate partial root-zone irrigation of field-grown tomato. *Agricultural Water Management*, 165, 33–43.

Yong, J., Wong, S., Farquhar, G. 1997. Stomatal responses to changes in vapour pressure difference between leaf and air. *Plant, Cell and Environment*, 20, 1213–1216.

Zhou, J., Liu, D., Deng, X., Zhen, S., Wang, Z., Yan, Y. 2018. Effects of water deficit on breadmaking quality and storage protein compositions in bread wheat (*Triticum aestivum* L.). *Journal of the Science of Food and Agriculture*, 98 (11), 4357–4368.

2
Irrigation Assessment: Efficiency and Uniformity

Patricia Angélica
Alves Marques,
Verônica Gaspar
Martins Leite de
Melo, and José
Antônio Frizzone
University of São Paulo

Antonio Pires
de Camargo
*State University
of Campinas*

2.1	Introduction	13
2.2	Use and Consumption of Water	14
2.3	Partitioning of the Applied Irrigation Water	15
2.4	Irrigation Efficiency Indicators	16
	Irrigation Efficiency (*IE*) • Irrigation Consumptive Use Coefficient • Irrigation Sagacity (*IS*) • Application Efficiency • Distribution Efficiency • Effective Application Rate • Storage Efficiency	
2.5	Irrigation Uniformity Indicators	22
	Christiansen's Uniformity Coefficient (%) • Uniformity Coefficient of Wilcox–Swailes (*UCW*) • Uniformity Coefficient of Hart (*UCH*) • Distribution Uniformity	
2.6	Adequacy of Application	24
2.7	Relation between Distribution Uniformity and Deep Percolation	26
2.8	Relation between Adequacy of Application and Distribution Efficiency	27
2.9	Conclusions	28
References		28

2.1 Introduction

Knowing the objectives of system evaluation is essential for the planning and operation of irrigation systems [7]. One of the objectives is related to irrigation scheduling strategies followed by evaluation of the corresponding effects on the irrigation uniformity. Another one is to quantify the system performance in order to benefit from incentives offered by water regulatory agencies and to demonstrate proper use of water resources. Technical and financial assessment based on the results of systems evaluation might support a decision about upgrading an irrigation system in order to make it more efficient. In a systematic way, the main goals of irrigation system evaluation are: (1) to determine the efficiency of the system as it has been operated; (2) to determine how effectively the system can be operated and improved; (3) to obtain information to assist engineers in the design of other systems; and (4) to obtain information for the comparison of various methods, systems and strategies of operation as a basis for economic decision-making.

Improvement in irrigation efficiency must comply with a holistic analysis of the process stages. The holistic approach necessary to understand irrigated agriculture considers all factors that may influence

DOI: 10.1201/9780429290152-3

how water is used, including technical, economic, political, social, or environmental aspects. The analysis can be further improved by weighting the impact of each factor on the water use in a given situation.

Investigation of the reasons why an irrigation system has not been used successfully requires cause-and-effect analyses. A challenging aspect of irrigated agriculture refers to the complexity and peculiarities of places, problems, and potential solutions. For instance, inefficient water distribution can be caused by an inadequate irrigation schedule and results in low crop yields and water wastage. An inadequate irrigation schedule may be caused by lack of training, low motivation among users, unsuitable data collection, and so on. The identification of interfering factors is the basis of an adequate diagnosis and the feasibility of potential solutions must always be assessed. Improvements in irrigation scheduling and operation by training require commitment by staff.

The assessment of irrigation system performance must consider three aspects: (1) an understanding of what is happening; (2) how evaluations might contribute to irrigation scheduling practices; and (3) where the adoption of technologies, practices, and procedures is required to improve water use efficiency. The irrigation scheduling strategies and their potential to improve the overall performance must be evaluated in order to reduce water waste and to increase the yield and incomes of the agricultural activity.

The quality of irrigation is expressed by performance indicators named uniformity, efficiency, and adequacy of application. Uniformity refers to the indicators associated with the variability of the irrigation depth applied over the area, which may interfere with the irrigation system design. Regarding agronomic aspects, uniformity affects the calculation of the amount of water required for irrigation, while regarding hydraulics, it affects the spacing of emitters, the system flow rate, and the duration of irrigation.

Efficiency is quantified by indicators calculated by dividing water amounts related to irrigation processes. Typically, the quantities of water (1) derived from the water source, (2) applied to the plot, (3) stored in soil at the roots' effective depth, (4) evaporated and drifted by the wind, (5) percolated and drained superficially outside the boundaries of the cultivated area, and (6) beneficially or reasonably used in the cultivated area are used to obtain efficiency indicators.

Adequacy of application has been proposed to express how much the irrigation system fulfills the crop water requirements in order to maintain the product quality and crop yield at economic levels. The adequacy of application is usually defined in relation to the percentage of the area receiving at least the water depth required to supply the water deficit. This definition requires the specification of the crop, soil, and market conditions.

In irrigation, the applied water must meet the requirements of beneficial uses. High levels of water distribution uniformity (DU) and application efficiency (AE) are necessary for the irrigation system to achieve satisfactory performance.

In an irrigation community, competition for water use and shortage of supply highlight the need for efficient water use. Irrigation equipment, systems, and practices compete in terms of water consumption, cost of capital, and operating and maintenance costs. Irrigation performance indicators may serve as the basis for these comparisons and for the selection of competing activities.

2.2 Use and Consumption of Water

Water use is any intentional application of water for a specific goal [21]. The term does not distinguish between uses that remove water from a place (e.g., evaporation, transpiration, percolation) and uses that have a small impact on the water availability (e.g., navigation, hydroelectric power plants, domestic uses). Therefore, water is use applied to:

  i. Consumed fraction (evaporation and transpiration):
      a. Beneficial use: water evaporated or transpired to achieve the intended purpose (e.g., evaporation from a cooling tower, transpiration of an irrigated crop).
      b. Non-beneficial use: water evaporated or transpired for purposes other than the intended one (e.g., evaporation from a free water surface or from wet soil, transpiration of weeds or plants not related to the economic activity of interest).

Irrigation Assessment

ii. Non-consumed fraction:
   a. Recoverable fraction: the water that can be captured and reused, for example flow into drains returning to the river system and percolation into irrigated fields for aquifers; return flow of sewage systems (wastewater).
   b. Non-recoverable fraction: water that is lost, making further use impossible, for example, flow of saline water into groundwater aquifers; deep aquifers that are not economically exploitable; flow toward the sea.

The sum of the consumed and non-consumed fractions must equal one. The use of such terms to describe water use and the impacts of changes in water management has several benefits [21], such as: (1) compatibility with hydrology: water percolation from an open channel to an aquifer makes this water retrievable, while draining flow in an estuary near the sea is likely to be non-recoverable; (2) the terminology is consistent in several sectors and enables unambiguous discussions about hydrological impacts of technologies and alternative interventions; (3) the terminology can be applied at any scale without any modifications; (4) the terminology has a "neutral value", meaning that a beneficial use is one defined as such by society, whatever that is; and (5) attention can be focused on beneficial use (what we want), non-beneficial use (what we do not want), and retrievable flows (aspects of secondary interest).

Irrigation systems are often managed to maximize production of irrigated crops. This provides benefits but also means that more transpiration is favored, and the amount of water lost becomes unavailable for other purposes. The withdrawn volumes of unconsumed water flow over the surface and to underground aquifers. Although such amounts of water may be considered losses from the perspective of irrigation, in fact the water is normally recovered and reused elsewhere in the watershed [7].

2.3 Partitioning of the Applied Irrigation Water

Water applied by irrigation may be partitioned as for retrievable use:

i. *Consumptive use*: the total amount of water absorbed by the plants for transpiration, tissue growth, the evaporation of water from the soil, and the water intercepted by the canopy. Thus, all water in transit to the atmosphere by evaporation and transpiration (i.e., evapotranspiration processes) or that remains in plant tissues and in products harvested is irrecoverable, and this is called consumptive use. Some of the consumptive uses are evapotranspiration of plants (crops or weeds), evaporation of water applied by sprinkling emitters, evaporation from water reservoirs, evaporation from wet soils, and water exported with harvested products.
ii. *Non-consumptive use*: water that moves out the cultivated area but is not lost to the atmosphere in the form of steam. Non-consumptive water can be reused in the area itself or downstream in the river basin or returned to the water source. The term is related to the water that percolates beyond the root zone, the water that flows superficially, and water which infiltrates open channels. Some of the non-consumptive uses are water depth for leaching excess salts in a soil profile, deep seepage, surface runoff, infiltration into open channels, and leakages along pipelines.

Defining these various flow components that occur during the management of water resources is important. Efficient domestic supply systems practically involve non-consumptive use, since the outflows are captured, treated, and returned to the water-resource system. Efficient irrigation systems result in substantial consumptive use, because about 85% or more of the withdrawn water is used by the evapotranspiration process.

Partitioning of water applied by irrigation for agronomic benefits occurs as follows:

i. *Beneficial uses*: water used for crop production. All water used to achieve agronomic goals is considered beneficial. Examples include water consumed in the crop evapotranspiration process; water consumed in the evapotranspiration process of beneficial plants, like windbreaks, soil cover and protection, and habitat for beneficial insects; water within the harvested product and in plant tissues; water used to leach excess salts from the root zone; water used for weather control; water used for seedbed preparation; water for softening the soil crust for seedling emergencies; and water used for application of pesticides and fertilizers.

  ii. *Non-beneficial uses*: the amount of water applied for purposes not related to agronomic ones. Examples include excessive percolation when leaching of salts is required, excess runoff, weed evapotranspiration, unnecessary evaporation of water on the moist soil surface of areas near to the cultivated area, and water drift due to wind effects. Water application cannot be perfectly uniform; hence, sometimes there is excess water and percolation may occur, which characterizes non-beneficial use.

Non-beneficial uses can be further grouped into reasonable and unreasonable uses, while all beneficial uses are reasonable. Reasonable non-beneficial uses are those that can be justified in specific conditions, places, and moments. Unreasonable non-beneficial uses are those which have no technical, economic, social, or environmental justification.

Examples of reasonable non-beneficial uses are evaporation from reservoirs and open channels, evaporation from soil, evaporation of water droplets applied by sprinkler systems, water used to clean filters, water to satisfy some environmental purpose, water for maintenance of drainage systems, water percolated due to uncertainties in irrigation management, water percolated due to non-uniform distribution of water by the irrigation system, and losses that cannot be avoided due to economic reasons.

Examples of unreasonable non-beneficial uses are deep seepage, excessive surface runoff, evaporation of irrigated soil outside the cultivated area, leakage from pipes, and leakage or excessive infiltration from open channels.

2.4 Irrigation Efficiency Indicators

Partitioning of water applied by irrigation is fundamental to determine certain performance parameters, such as the efficiency indicators. Efficiency indicators represent fractions of water volume intended for certain purposes and may be expressed as percentages. Some indicators are easily obtained, while others cannot be defined without the time period and boundaries of the region analyzed.

2.4.1 Irrigation Efficiency (*IE*)

Irrigation efficiency (*IE*) is the ratio between water consumed by the evapotranspiration process of the crop (ET_c) and water withdrawn from a river or other source [14]. Jensen [15] pointed out that the term may be inappropriate because the water not consumed by evapotranspiration is considered to be wasted. Losses at the scale of the individual field or irrigation project are not necessarily losses in the hydrological sense. The "lost" water may be available for use elsewhere in the watershed or aquifer [21].

A system or irrigation project can serve other agricultural needs beyond satisfying the ET_c [12]. Irrigation efficiency represents the ratio between the beneficial irrigation volume and the total volume of irrigation water [4] (beneficial plus non-beneficial use) (Eq. 2.1):

$$IE = \frac{\text{Volume of irrigation water beneficially used}}{\text{Volume of irrigation water applied} - \Delta \text{storage of irrigation water}} \times 100 \quad (2.1)$$

The numerator of Eq. (2.1) represents the beneficial use of irrigation water for multiple purposes, including ET_c, salt leaching, frost protection, temperature control, application of fertilizers and pesticides, seedbed preparation, softening the soil crust for seedling emergencies, and so on. Deep seepage, surface runoff, spray drift beyond field boundaries, and water used by weeds will tend to reduce irrigation efficiency and are non-beneficial.

The denominator of Eq. (2.1) corresponds to the total volume of irrigation water (beneficial plus non-beneficial uses) that leaves field boundaries (output equals volume applied minus variation in irrigation water storage in the soil). These volumes of water leave the area within a specified time interval, for example, the period between two successive irrigation events or during a growing season. If, at the end of a period, the volume of water stored in the soil is the same as at the beginning, then water storage variation is zero (Δ storage of irrigation water $= 0$). So, all applied water has left the field by ET_c, drainage, deep seepage, and so on. Thus, water stored temporarily in the root zone for use outside the specified time interval is not accounted for nor added nor subtracted from beneficial use.

Irrigation efficiency is defined between two dates, usually corresponding to the whole crop cycle. It is determined only a posteriori and does not presuppose future beneficial uses. For the defined time interval, it requires an accurate assessment of the fraction of irrigation water that has been beneficially used. Uniform application of water over the area is not a requirement. The values of efficiency rely on the choice of time interval.

The expression "irrigation water" excludes water applied to the crop by natural precipitation or capillary rise from the water table. Also, volumes can be replaced with irrigation water depths in the numerator and denominator of Eq. (2.1). The ratio between *IE*, beneficial use, and non-beneficial use of irrigation water can be found by Eqs. (2.2) and (2.3).

$$\text{Beneficial use} + \text{Non beneficial use} = 100\% \qquad (2.2)$$

$$\text{Nonbeneficial use} = (100 - IE)\% \qquad (2.3)$$

2.4.2 Irrigation Consumptive Use Coefficient

The irrigation consumptive use coefficient (*ICUC*) is defined as the ratio of the volume of irrigation water used consumptively to the total volume of irrigation water that has left the region, both in a specified period of time, expressed as a percentage (Eq. 2.4). Therefore, it quantifies the water not recovered [4].

$$ICUC = \frac{\text{Volume of irrigation water used consumptively}}{\text{Volume of irrigation water applied} - \Delta \text{ storage of irrigation water}} \times 100 \qquad (2.4)$$

Like the *IE*, the *ICUC* can be used at field, farm, project, irrigation district, or watershed scale. At project scale, for example, the total outflow of liquid water from the project area (surface and subsurface) originated from irrigation over a specified time period is $(100 - ICUC)\%$ of the irrigation water supplied to the project minus the storage variation of irrigation water in the soil. The *ICUC* quantifies unrecovered irrigation water.

2.4.3 Irrigation Sagacity (*IS*)

Irrigation efficiency (*IE*) is a performance indicator that has been used to quantify the beneficial use of irrigation water and to compare beneficial and non-beneficial uses. Water is used beneficially when it contributes directly to the agronomic production of the crop. However, due to physical, economic, or administrative constraints and various environmental demands, some degree of non-beneficial use is reasonable. Other benefits to society may arise from irrigation water; for example, a fraction not used by plants can supply the rural population. Therefore, an additional indicator that incorporates quantification of reasonable non-beneficial use (uses which may not contribute to agronomic production but are justified under particular circumstances) is necessary. Authors of Ref. [22] proposed an irrigation performance index which considers reasonable beneficial and non-beneficial water uses, called irrigation sagacity (*IS*). The term means wise or prudent use of water and is defined by Eq. (2.5). It is not suggested that *IS* replace *IE*. With clear goals, both can offer useful information.

$$IS = \frac{\text{Volume of irrigation water beneficially and/or reasonably used}}{\text{Volume of irrigation water applied} - \Delta \text{ storage of irrigation water}} \times 100 \qquad (2.5)$$

Conceptually, irrigation efficiency has often been poorly applied or misinterpreted by failures in the differentiation between consumptive use and beneficial use. It should not be understood that $(100-IE)\%$ of the water applied by irrigation represents the amount that is wasted and therefore the potential for conservation or reallocation. An $IE=75\%$ does not mean that 25% of applied irrigation water could be retained and redirected for any other purpose. Some non-beneficial uses may be reasonable, so the potential for conservation and relocation consists only of unprofitable and unreasonable uses. IS incorporates the beneficial and non-beneficial uses and makes it possible to define the potential for conservation and reallocation, limiting it to $(100-IS)\%$ of the water applied.

Reasonable non-beneficial uses are those which, while not directly benefiting agronomic production, can be justified under some physical and economic conditions. For example, a small loss of water through channels that does not economically justify their impermeabilization constitutes a reasonable but not beneficial use. An irrigation system cannot be designed to apply water with perfect uniformity, and therefore some deep seepage due to non-uniformity is inevitable and reasonable, though not beneficial. Likewise, losses of water by evaporation in the air during sprinkler irrigation, water used to clean filters, and so on can be cited as reasonable but not beneficial uses.

2.4.4 Application Efficiency

In irrigation, the target is to store enough water in the soil profile to supply crop demand (to meet the demands of evapotranspiration, ET_c, or soil moisture depletion, SMD) and other beneficial uses, such as leaching of salts. The definition of AE [4] is given by Eq. (2.6).

$$AE = \frac{\text{Average depth of irrigation water contributing to target}}{\text{Average depth of irrigation water applied}} \times 100 \qquad (2.6)$$

The numerator of Eq. (2.6) is the average depth of irrigation water stored in the soil profile, which contributes to meeting the crop water requirements (ET_c or SMD), leaching of excess salts, and other beneficial uses. The target quantity may be smaller when effective rainfall is factored in.

The AE is an indicator of performance applied at field or smaller scales and refers to an irrigation event. Normally, the irrigation depth required is assumed to be uniform over the field and all water for beneficial use will in fact be beneficially used. The AE quantifies the extent to which the irrigation system satisfies the irrigation requirements and is used to estimate what happens during an irrigation event. The choice of the target irrigation quantity may be ET_c, SMD, or another quantity that considers an amount of water for leaching or other beneficial uses.

The definition presented by ASCE On-Farm Irrigation Committee [1] and reproduced by [23] and [20] mentions that AE considers only the depth required to replenish the SMD but does not include any water for leaching or other perceived beneficial uses. For this special case, Eq. (2.7) is applied.

$$AE_{SMD} = \frac{\text{Average depth for replenishing } SMD}{\text{Average depth of irrigation water applied}} \times 100 \qquad (2.7)$$

Figure 2.1 shows the cumulative frequency distribution of water applied by a sprinkler system. A closed normal distribution in $[\mu \pm 3\sigma]$ is considered, representing a 99% confidence interval, which is appropriate for the intended analyses [2]. When the probability density function equals zero, 0% of the total area receives more than $\mu \pm 3\sigma$; when it equals one, 100% of the area receives more than $\mu \pm 3\sigma$. Based on that distribution, the AE is calculated by Eq. (2.8) and may be related to the total losses of water in the plot $(100-AE)$. Although AE is an indicator of the excessive application of water to the plot, it cannot be related to the adequacy of application or to water DU.

$$AE = \frac{A+C}{A+B+C+E} \times 100 \qquad (2.8)$$

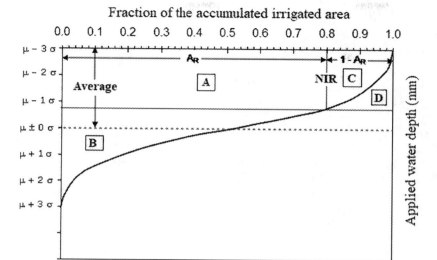

FIGURE 2.1 Cumulative probability density or cumulative frequency curve of the normal distribution ($\mu + k\sigma$ required irrigation depth).

In Eq. (2.8), assuming the surface runoff is insignificant, E is the quantity of water lost by evaporation and wind drift. In sprinkler irrigation, losses due to evaporation and wind drift are relevant in regions of high atmospheric demand and winds of velocity higher than 4 km/h. Losses may reach about 10% in normal conditions and up to 20% in severe weather conditions.

Four areas are shown in Figure 2.1 (A, B, C, and D), representing relative quantities of water. $A+B+C$ represents the total amount of water infiltrated in the soil. $A+C$ is the quantity that contributes to the target *SMD*. The amounts of deep seepage (B) and deficit (D) are calculated based on a normal distribution of irrigation depths [2] (Eqs. 2.9–2.15):

$$A+B = \frac{S}{\sqrt{2\pi}} e^{-T_R^2/2} + \overline{W} \, A_R \qquad (2.9)$$

$$A = A_R \left(\overline{W} + I_R \, S \right) \qquad (2.10)$$

$$B = \frac{S}{\sqrt{2\pi}} e^{-T_R^2/2} - A_R \, I_R \, S \qquad (2.11)$$

$$A+C = \overline{W}_{\text{arm}} + A_R \, I_R \, S \frac{S}{\sqrt{2\pi}} e^{-T_R^2/2} \qquad (2.12)$$

$$C = \overline{W} \left(1 - A_R \right) - \frac{S}{\sqrt{2\pi}} e^{-T_R^2/2} \qquad (2.13)$$

$$C+D = \left(1 - A_R \right) \left(\overline{W} + I_R S \right) \qquad (2.14)$$

$$D = \left(1 - A_R \right) I_R \, S + \frac{S}{\sqrt{2\pi}} e^{-I_R^2/2} \qquad (2.15)$$

where:

$$\overline{W} = A + B + C;$$

$$I_R = \frac{RAW - \overline{W}}{S};$$

I_R = Irrigation depth required;
$\overline{W}$ = Average water depth infiltrated;
RAW = Readily available water, which is a fraction of total available soil water (TAW) that can be depleted from the root zone before moisture stress occurs;
A_R = Fraction of the area receiving at least the irrigation depth required;
S = Standard deviation of the average water depth infiltrated;
$\overline{W}_{arm}$ = Average useful depth stored.

The AE is 100% after an irrigation event when the whole soil profile is in deficit and no water has been lost due to evaporation or wind drift. In this case, all the infiltrated water is available only to meet the crop needs (ET_c). When the soil moisture is below a threshold value (i.e., a critical soil moisture) or when water application is not fully uniform, the crop may exhibit symptoms of water stress both in the over-irrigated fraction of the plot, due to hypoxia effects, as well as in under-irrigated areas, due to deficit. The excess water in relation to the RAW may cause surface runoff and deep seepage. An additional irrigation depth is sometimes required to leach excess salts from the root zone. Drifting of water outside the cultivated area due to wind effects, surface runoff, and evaporation of water applied by sprinklers tends to reduce the AE. These volumes of water represent non-beneficial uses and are difficult to quantify.

A simple approach to estimate the AE of a sprinkler irrigation system may be proposed. As an example, consider a solid-set sprinkler irrigation system for which Christiansen's uniformity coefficient (CUC)=88%; the average water depth applied over the area $\overline{W}_a$=27 mm, obtained by dividing the sprinkler discharge by an area surrounded by four sprinklers (assuming a rectangular arrangement of sprinklers); and the average water depth infiltrated $\overline{W}$=23.4 mm to replenish water in the soil profile and obtain the target RAW=20 mm ($\overline{W}$ is assumed to be the average water depth collected in the area surrounded by four sprinklers). Based on the water distribution, in 87% of the area the infiltrated depth is higher than the target (over-irrigated area), while 13% of the area is under-irrigated. The minimum water depth infiltrated is 15 mm. The average useful depth stored in the soil, $\overline{W}_{arm}$, is estimated to be $0.87 \times 20 + 0.13[(20+15)/2] = 19.7$ mm. In this situation, $AE = 100\,(19.7/27) = 73\%$. This means that 27% of the applied depth does not contribute to reaching the target; that is, 27% is lost through deep seepage, evaporation, and wind drift.

2.4.5 Distribution Efficiency

Water distribution by irrigation systems is not fully uniform, and thus part of the irrigated area receives more water than the target RAW (over-irrigated area) and part receives less than the target (under-irrigated area) (Figure 2.1). If the depth infiltrated in the whole area is less than RAW, the whole of the area is in deficit and deep seepage will not occur. However, if the depth infiltrated in the whole area is higher than RAW, the irrigation is excessive and deep seepage is expected to occur. Balanced irrigation is obtained when the infiltrated depth is higher than RAW in only part of the area. The distribution efficiency (DE) (Eq. 2.16) describes the ratio between the average depth that contributes to reach the target and the average water depth infiltrated during an irrigation event [8, 9].

$$DE = \frac{\text{Average depth that contributes to the target}}{\text{Average water depth infiltrated}} \times 100 = \frac{A+C}{A+B+C} \times 100 \quad (2.16)$$

The DE is 100% if the whole of the area is under-irrigated and DE is very low if over-irrigation occurs. The following reference values of DE were proposed by authors of Ref. [8]: excellent ≥ 0.8; $0.5 \leq$ satisfactory < 0.8; unsatisfactory < 0.5. Therefore, DE is an indicator sensitive to over-irrigation, and hence low values of DE indicate deep percolation. The DE relies on the water DU and strategies of irrigation scheduling. In the previous example, the target depth is 20 mm (RAW), the average depth that contributes to the target is 19.7 mm, and the average depth infiltrated is 23.4 mm. Finally, $DE = 100 \, (19.7/23.4) = 84.2\%$, which means 15.8% of water has percolated.

2.4.6 Effective Application Rate

The Effective Application Rate (EAR) defined by authors of Ref. [8] is also called the potential AE. EAR is used as a performance indicator of a single irrigation event. Assuming there is no surface runoff, EAR is obtained as the ratio between the irrigation depth available for infiltration and the applied irrigation depth (Eq. 2.17).

$$EAR = \frac{\text{Irrigation depth available for infiltration}}{\text{Applied irrigation depth}} \times 100 = \frac{A+B+C}{A+B+C+E} \times 100 \quad (2.17)$$

The complement of EAR indicates the amount of water lost through wind drift and evaporation before the water reaches the soil surface. The numerator of Eq. (2.17) represents the water depth collected over a surface irrigated by sprinklers, and thus the amount is available for infiltration ($A+B+C$). The denominator represents the irrigation depth applied by the sprinklers; assuming the surface runoff is insignificant, the applied irrigation depth is the sum of the infiltrated depth and the depth lost through evaporation and wind drift ($A+B+C+E$), where E represents water that moves outside the boundaries due to evaporation and wind drift.

Based on the previous example for sprinkler irrigation, $EAR = (23.4/27) \times 100 = 86.7\%$, which means that 13.3% of the applied water depth is lost due to wind drift and evaporation before reaching the soil surface. In addition, when surface runoff is neglected, $AE = \left[(DE/100) \times (EAR/100) \right] \times 100$.

2.4.7 Storage Efficiency

Assuming surface runoff does not occur, the AE is high when the infiltrated water depth is not excessive and RAW is not fully replenished. Consequently, deep percolation is not expected to occur. However, the average infiltrated depth stored in the root zone may not suffice for the irrigation requirements (beneficial use), reducing crop yield. Therefore, an irrigation deficit may lead to high values of AE. The storage efficiency (SE) is the proper indicator to satisfy adequate application and crop water requirements. SE is the ratio between the average depth that contributes to reaching the target and the required irrigation depth [8, 9] (Eq. 2.18).

$$SE = \frac{\text{Average depth that contributes to reach the target}}{\text{Required irrigation depth}} \times 100 = \frac{A+C}{A+C+D} \times 100 \quad (2.18)$$

Based on the previous example for sprinkler irrigation, the required irrigation depth is 20 mm (RAW) and the average depth stored in the soil that contributes to reaching the target is 19.7 mm. Therefore, $SE = 98.5\%$. Assuming there are no other beneficial uses, the required irrigation depth is defined as the RAW, the SMD, or the evapotranspiration (ET_c). The RAW represents the fraction of TAW that can be depleted from the root zone before moisture stress occurs. The SMD considers the current soil moisture and represents the water shortage relative to field capacity. The following are reference values of SE [8]: excellent, ≥ 0.8; satisfactory, ≤ 0.5 to <0.8; unsatisfactory, <0.5.

2.5 Irrigation Uniformity Indicators

The irrigation uniformity indicators express the variability of the irrigation depth applied over the soil surface in relation to an average value. Usually, such indicators are obtained by dimensionless dispersion measures. Several uniformity coefficients have been proposed for sprinkler systems since Christiansen [5].

2.5.1 Christiansen's Uniformity Coefficient (%)

Christiansen [5] used the mean absolute deviation as the dispersion measure. The mean absolute deviation corresponds to the arithmetic average of absolute deviations between the applied irrigation depth and its average value (Eq. 2.19).

$$CUC = \left(1 - \frac{\sum_{i=1}^{N} |W_i - \overline{W}|}{N\,\overline{W}}\right) \times 100, \qquad (2.19)$$

where N is the number of observations, W_i is the irrigation depth applied on the soil surface at point i, and $\overline{W}$ is the average irrigation depth applied.

CUC is the most frequently used uniformity coefficient for sprinkler systems and generally values higher than 80% are acceptable. Lower values may be admissible if precipitation contributes significantly during the crop cycle or if a relevant decrease in the system costs is obtained due to a reduction in uniformity and it is expected that economic feasibility will be enhanced in this situation.

In center-pivot irrigation systems, the water depths must be weighted according to the position of the water collector (i.e., the catch-can), since the area represented by each collector increases proportionally to the distance from the lateral inlet. In the literature, authors of Ref. [11] proposed an approach to weight the depths based on the distance R_i of a collector i from the lateral inlet (Eq. 2.20).

$$CUC = \left(1 - \frac{\sum_{i=1}^{N} R_i |W_i - \overline{W}_p|}{\sum_{i=1}^{N} W_i\,R_i}\right) \times 100, \qquad (2.20)$$

where the weighted mean depth ($\overline{W}_p$) is obtained by Eq. (2.21):

$$\overline{W}_p = \frac{\sum_{i=1}^{N} W_i\,R_i}{\sum_{i=1}^{N} R_i}. \qquad (2.21)$$

2.5.2 Uniformity Coefficient of Wilcox–Swailes (UCW)

In 1947, it was proposed the sample standard deviation as the dispersion measure [25]. Their indicator of performance is usually called the statistical uniformity coefficient (UCW) and is commonly used for sprinkler and micro-irrigation systems (Eq. 2.22).

$$UCW = \left(1 - \frac{S}{\overline{W}}\right) \times 100, \qquad (2.22)$$

where S is the standard deviation of irrigation depths applied on the soil surface and $\overline{W}$ is the average irrigation depth applied. Satisfactory values of UCW are higher than 75% for sprinkler systems and 85% for micro-irrigation systems.

The *CUC* is not sensitive to the effect of irrigation depths that are too dispersed from the average value because its dispersion measure is based on the mean absolute deviation [17]. However, *UCW* is a more sensitive uniformity coefficient and is especially useful to evaluate center-pivot irrigation systems and micro-irrigation systems, in which the emitter discharge may be affected by clogging or mechanical damage. For center pivots, *UCW* must be calculated using a weighted mean depth (Eq. 2.21) and a weighted standard deviation, S_w (Eq. 2.23).

$$S_w = \sqrt{\frac{\sum_{i=1}^{N} R_i (W_i - \overline{W}_p)^2}{\sum_{i=1}^{N} R_i}} \qquad (2.23)$$

2.5.3 Uniformity Coefficient of Hart (*UCH*)

Hart [9] proposed a uniformity coefficient for solid-set sprinkler systems that also use the standard deviation as the dispersion measure. In addition, *UCH* assumes that the applied irrigation depths comply with the normal distribution (Eq. 2.24). Acceptable values of *UCH* are higher than 75%.

$$UCH = \left(1 - \sqrt{\frac{2}{\pi}} \times \frac{S}{\overline{W}}\right) \times 100 \qquad (2.24)$$

If the applied irrigation depths follow the normal distribution, $UCH = CUC$ to meet the following property of the normal distribution: $\sum_{i=1}^{N} |W_i - \overline{W}|/N = S\sqrt{2/\pi}$. Based on that, the standard deviation is obtained as a function of CUC (dimensionless) and $\overline{W}$ using Eq. (2.25).

$$S = \frac{\overline{W}(1 - CUC)}{0.8} \qquad (2.25)$$

From the average and standard deviation values, the percentage of the area that receives at least a given quantity of water can be estimated. Such information may be combined in models of crop yield as a function of water depth to estimate crop production. The gross revenue related to production may be compared with irrigation costs in order to define optimal strategies for operating the irrigation system [19]. However, the hypothesis that irrigation depths follow the normal distribution is partially valid. Even when CUC values are relatively high (higher than 85%), sometimes irrigation depths are not normally distributed [3]. For center pivots, UCH is calculated using the average computed by Eq. (2.21) and the standard deviation by Eq. (2.23).

2.5.4 Distribution Uniformity

DU was proposed by Soil Conservation Service – United States Department of Agriculture (SCS-USDA) and is obtained as the ratio between the average low-quarter depth (W_{lq}) and the average depth (Eq. 2.26). The average low-quarter depth represents the average of the 25% smallest depths over the irrigated field. *DU* has been used for sprinkler, surface, and micro-irrigation systems. For surface irrigation systems, infiltrated depths are considered, while emitter discharge is required for micro-irrigation systems.

$$DU = \frac{W_{lq}}{\overline{W}} \times 100 \qquad (2.26)$$

Low values of *DU* indicate excessive deep percolation, which is expected to occur when the average infiltrated depth in 25% of the area is higher than the SMD. Assuming a normal distribution of W_i, the values of DU may be related to CUC by Eq. (2.27) [16].

$$DU = 100 - 1.59(100 - CUC) \qquad (2.27)$$

TABLE 2.1 Performance Indicators

Indicator	Excellent	Satisfactory	Unsatisfactory
SE	≥0.80	0.50–0.80	<0.50
DE	≥0.80	0.50–0.80	<0.50
DU	≥0.90	0.70–0.90	<0.70
EAR	≥0.90	0.80–0.90	<0.80
AE	≥0.72	0.40–0.72	<0.40

Source: Adapted from Grafton et al. [8] and Heermann et al. [13].

Assuming a normal distribution of W_i, authors of Ref. [8] proposed DU_H (Eq. 2.28).

$$DU_H = \left(1 - 1.27 \frac{S}{\overline{W}}\right) \times 100, \qquad (2.28)$$

where $DU = DU_H$, when the applied irrigation depths follow a normal distribution. The coefficient 1.27 means that for the normal distribution, the low-quarter average is 1.27 S lower than the dataset arithmetic average. From this value, 10.22% of the field receives irrigation depths lower than W_{1q}. For center pivots, weighted values of the standard deviation and average must be used. Authors of Refs. [8] and [13] suggest the criteria for interpreting the results of performance indicators as listed in Table 2.1.

For high-value crops with shallow roots, the most economic irrigation system usually is the one that provides high DU: $DU > 80\%$ and $CUC > 88\%$. For fruit trees with deep roots and supplementary irrigation, satisfactory results are obtained with DU between 50% and 70% or CUC between 70% and 82% [18]. For crops with intermediate root depth, the recommended values are $70\% < DU < 80\%$ and $82\% < CUC < 88\%$.

2.6 Adequacy of Application

The adequacy of application (A_R) refers to the fraction of the area that receives the amount of water required to assure the product quality and the productivity at a target economic level. Thus, A_R indicates the irrigated area complying with the needs of the agricultural production system. Since this definition is too wide and requires specification of the crop, soil, and market conditions, the adequacy of application is usually defined in relation to the percentage of the area receiving at least the water depth required to supply the water deficit.

The evaluation of A_R is based on a cumulative frequency curve (Figure 2.1), which shows the percentage of the area having a deficit or excess of water. The dashed horizontal line represents the average water depth infiltrated into the soil. The solid horizontal line represents the RAW. For each RAW, there is a corresponding value of A_R. The cumulative frequency curve is plotted from the irrigation depths collected during in-field water distribution tests and considering the fraction of the area receiving a minimum depth.

It was proposed [6], the following CUC values based on A_R: for annual crops, $CUC = 80\%$ and $A_R = 75\%$; for fruit trees, $CUC = 70\%$ and $A_R = 50\%$; and for vegetables, $CUC = 85\%$ and $A_R = 90\%$. That criterion aims to consider the crop economic value, the irrigation system costs, and soil water redistribution. The efficiency levels expected in irrigation systems designed to fully satisfy the crop water requirements will provide a minimum adequacy of application of 90% for shallow roots and high-value crops, 80% for crops of intermediate root depth and moderate value, and 75% for deep root crops. Thus, irrigation is defined in terms of the amount of water applied to avoid water deficit in 90%, 80%, or 75% of the area. In conventional irrigation scheduling, efficiencies correspond to specific values of adequacy of application and DU.

Irrigation Assessment

Assuming the quantities of applied water follow the normal distribution, the DU may be related to Christiansen's uniformity coefficient (CUC) by Eq. (2.27). Authors of Ref. [10] related measures of water application uniformity (CUC) with adequacy of application (A_R) to estimate the DE and SE and for the design and operation of irrigation systems (Table 2.2). For a sprinkler irrigation system with $DU=84\%$ (i.e., $CUC=90\%$) and $A_R=80\%$, the obtained value in Table 2.2 is $DE=89.4\%$. In 80% of the area, the

TABLE 2.2 Dimensionless Values of Distribution Efficiency (DE) and Storage Efficiency (SE) as a Function of Adequacy of Application (A_R) and Indicators of Water Distribution Uniformity

DU (%)	CUC (%)	Effic.	\multicolumn{10}{c}{Adequacy of Irrigation – A_R (%)}
			95	90	85	80	75	70	65	60	55	50
97	98	DE	0.959	0.968	0.974	0.979	0.983	0.987	0.990	0.994	0.997	1.000
		SE	0.958	0.967	0.972	0.976	0.979	0.982	0.984	0.986	0.988	0.990
94	96	DE	0.917	0.936	0.948	0.958	0.966	0.974	0.981	0.987	0.994	1.000
		SE	0.916	0.933	0.944	0.952	0.959	0.964	0.969	0.973	0.977	0.980
91	94	DE	0.876	0.903	0.922	0.937	0.949	0.961	0.971	0.981	0.991	1.000
		SE	0.875	0.900	0.916	0.928	0.938	0.946	0.953	0.959	0.965	0.970
84	90	DE	0.794	0.839	0.870	0.894	0.915	0.934	0.952	0.968	0.984	1.000
		SE	0.791	0.833	0.860	0.880	0.897	0.910	0.922	0.932	0.941	0.950
81	88	DE	0.753	0.807	0.844	0.873	0.899	0.921	0.942	0.962	0.981	1.000
		SE	0.749	0.800	0.832	0.856	0.876	0.892	0.906	0.919	0.930	0.940
78	86	DE	0.711	0.775	0.818	0.852	0.882	0.908	0.932	0.956	0.978	1.000
		SE	0.707	0.767	0.804	0.832	0.855	0.874	0.891	0.905	0.918	0.930
75	84	DE	0.670	0.743	0.792	0.831	0.865	0.895	0.923	0.949	0.975	1.000
		SE	0.666	0.733	0.776	0.809	0.835	0.856	0.875	0.892	0.906	0.920
71	82	DE	0.629	0.711	0.766	0.810	0.848	0.882	0.913	0.943	0.971	1.000
		SE	0.624	0.700	0.749	0.785	0.814	0.839	0.860	0.878	0.895	0.910
68	80	DE	0.588	0.679	0.740	0.789	0.831	0.869	0.903	0.937	0.968	1.000
		SE	0.582	0.667	0.721	0.761	0.793	0.821	0.844	0.865	0.883	0.900
65	78	DE	0.546	0.647	0.714	0.768	0.814	0.855	0.894	0.930	0.965	1.000
		SE	0.541	0.633	0.693	0.737	0.773	0.803	0.829	0.851	0.871	0.890
62	76	DE	0.505	0.614	0.688	0.747	0.797	0.842	0.884	0.924	0.969	1.000
		SE	0.499	0.600	0.665	0.713	0.752	0.785	0.813	0.838	0.860	0.880
59	74	DE	0.464	0.582	0.662	0.726	0.780	0.829	0.875	0.917	0.959	1.000
		SE	0.457	0.567	0.637	0.689	0.731	0.767	0.797	0.824	0.848	0.869
56	72	DE	0.423	0.550	0.636	0.704	0.763	0.816	0.865	0.911	0.956	1.000
		SE	0.415	0.533	0.609	0.665	0.711	0.749	0.782	0.811	0.836	0.859
52	70	DE	0.381	0.518	0.610	0.683	0.747	0.802	0.855	0.905	0.953	1.000
		SE	0.373	0.500	0.581	0.641	0.690	0.731	0.766	0.797	0.825	0.849
49	68	DE	0.340	0.486	0.585	0.662	0.730	0.790	0.845	0.899	0.949	1.000
		SE	0.332	0.467	0.553	0.617	0.669	0.713	0.751	0.784	0.813	0.839
46	66	DE	0.299	0.454	0.559	0.641	0.713	0.777	0.836	0.892	0.946	1.000
		SE	0.290	0.433	0.525	0.593	0.649	0.695	0.735	0.770	0.801	0.829
43	64	DE	0.258	0.421	0.533	0.620	0.696	0.763	0.826	0.886	0.943	1.000
		SE	0.248	0.400	0.497	0.569	0.628	0.677	0.719	0.757	0.789	0.819
40	62	DE	0.217	0.389	0.507	0.599	0.679	0.750	0.817	0.879	0.940	1.000
		SE	0.206	0.367	0.469	0.545	0.607	0.659	0.704	0.743	0.778	0.809
36	60	DE	: –	0.357	0.401	0.578	0.662	0.737	0.807	0.873	0.937	1.000
		SE	: –	0.333	0.441	0.521	0.587	0.641	0.688	0.730	0.766	0.799

Source: Adapted from Hart [10].

infiltrated amount will be equal to or higher than 89.4% of the average required amount of water. In addition, in 20% of the area, the infiltrated water will be less than 89.4% of the required water (Eq. 2.29).

$$DU_{AR} = \frac{W_{\min AR\%}}{\overline{W}}, \qquad (2.29)$$

where $W_{\min AR\%}$ is the minimum water depth required at the area properly irrigated and $\overline{W}$ is the average water depth infiltrated.

Based on Eq. (2.29), to infiltrate 1 mm of water in at least 80% of the irrigated area (numerator of Eq. 2.29) with $DU=84\%$, the expected DE is 89.4% (Table 2.2) and the average water depth applied over the soil surface (denominator of Eq. 2.29) will be $1/0.894=1.12$ unities of water (not considering water lost due to evaporation and wind drift and neglecting surface runoff). For a system with $DU=90\%$ ($CUC \cong 94\%$) and $A_R=80\%$, the obtained DE is 84%; that is, to infiltrate 1 unit of water useful to plants in at least 80% of the area, we must apply 1.06 units of water over the surface. For a center pivot operating with $CUC=90\%$, to apply $RAW=12$ mm with adequacy of application of 80%, the DE expected is 0.894. Therefore, the infiltrated depth must be 13.4 mm. Assuming an Effective Application Rate (EAR) of 90%, the average water depth applied by the system will be 14.9 mm.

According to Table 2.2, for $A_R=80\%$, $DE_{80} \cong CUC$. Therefore, when CUC is used as the DU, 80% of the area will receive at least the required amount of water while 20% of the area will be in water deficit. Table 2.2 also shows that $DU \cong DE_{90}$. According to Eq. (2.29), when $CUC=94\%$, DU will be approximately 90% and consequently $DE_{90}=0.90$ (Table 2.2).

2.7 Relation between Distribution Uniformity and Deep Percolation

The importance of uniformity in the determination of the optimal depth has been widely recognized, but a universal quantitative analysis has not yet been developed. The optimal depth is influenced by the cost of water, the cost of capital invested in the system, the DU, the AE, and the economic value of the crop.

The relation between DU and water percolation in the plot for $A_R=50\%$ is shown in Figure 2.2. For $CUC=75\%$, the excess and deficit volumes are higher than for $CUC=90\%$. When the DU is high, even

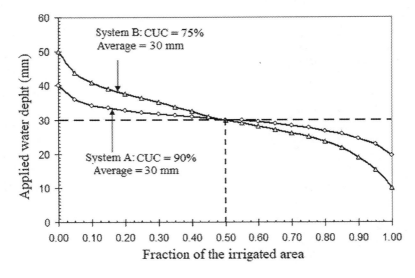

FIGURE 2.2 Water distribution for adequacy of application of 50% and two uniformity levels (CUC) with an infiltrated average depth of 30 mm.

if the adequacy of application is only 50%, the excess and deficit volumes are not high, resulting in less adverse effects on crop production and irrigation economics.

In irrigation with a high adequacy of application, losses by percolation are higher for the system with $CUC=75\%$, resulting in lower DE, compared to $CUC=90\%$. Thus, design criteria and irrigation strategies that aim at high uniformity rates might achieve high DE even when irrigation is performed with a high adequacy of application. However, even for highly uniform irrigation, DE depends on the amount of water applied.

2.8 Relation between Adequacy of Application and Distribution Efficiency

The relation between A_R and DE is outlined in Figure 2.3. It is observed that if the system applies water according to curve A and the required irrigation depth is 27 mm, the adequacy of application is 76%. If this same system operates according to curve B, with the same uniformity and depth, the adequacy of application is 38%. With an adequacy of application of 76%, the system B satisfies only the irrigation requirement of 23 mm. The area $a+b$ represents the amount of percolation if the system operates under condition A. Area b represents percolation considering operation condition B. Note that the area representing the excess water in operation condition A is greater than that in condition B. Therefore, operation condition A will provide a lower DE.

The reduction of adequacy of application allows an increase in DE. This increase is due to the increase of the poorly irrigated area. Deficit irrigation generally results in decreased productivity but may be more economical if there are no problems concerning accumulation of salts in the soil. However, a high degree of deficit, besides reducing productivity, can affect the quality of the product, making it uneconomical. Therefore, in order to maximize income, a balance between the benefits associated with high DU and the reduction of gross revenue due to lower productivity and product quality must be achieved.

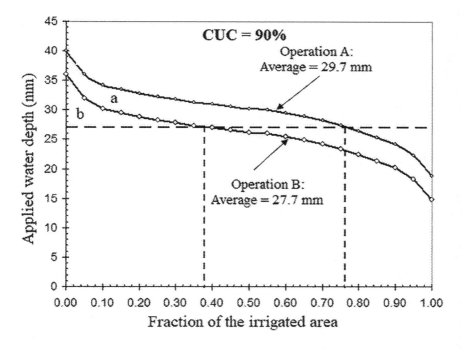

FIGURE 2.3 Relation between adequacy of application and distribution efficiency.

2.9 Conclusions

To improve irrigation efficiency, a holistic analysis of the stages of the process should be carried out. They should be considered all factors that can influence how water is used, including technical, economic, political, social, or environmental aspects. The evaluation of the performance of the irrigation system, through the indicators studied, allows you to assist decision-making to improve the overall performance of the irrigation system, reduce waste of water, and increase yield and income from agricultural activity.

References

1. American Society of Civil Engineering, 1978. Describing irrigation efficiency and uniformity. *Journal of the Irrigation and Drainage Division*, V.104, N.1, pp. 35–41.
2. Anyoji, H.; Wu, I.P., 1994. Normal distribution water application for drip irrigation schedules. *Transactions of the ASAE*, V.37, N.1, pp. 159–164.
3. Benami, A.; Ofen, A., 1984. *Irrigation engineering: sprinkler, trickle, surface irrigation principles, design and agricultural practice*. Haifa, Israel Irrigation Scientific Publications. 257p.
4. Burt, C.M.; Clemmens, A.J.; Strelkoff, T.S.; Solomon, K.H.; Bliesner, R.D.; Hardy, L.A.; Howell, T.A.; Eisenhauer, D.E., 1997. Irrigation performance measures: efficiency and uniformity. *Journal of Irrigation and Drainage Engineering*, V.123, N.6, pp. 423–442.
5. Christiansen, J.E., 1942. *Irrigation by sprinkler*. Berkeley: California Agricultural Station. 124p.
6. Cuenca, R.H., 1989. *Irrigation systems design: an engineering approach*. Englewood Cliffs: Prentice Hall. 551p.
7. Fathian, F., Dehghan, Z., Eslamian, S., Adamowski, J., 2016. Assessing irrigation network performance based on different climate change and water supply scenarios: a case study in Northern Iran. *International Journal of Water*, V.11, N.3, pp. 191–208.
8. Grafton, R.Q.; Williams, J.; Perry, C.J.; Molle, F.; Ringler, C.; Steduto, B.; Udall, S.A.; Wheeler, S.A; Wang, Y.; Garrick, D.; Allen, R.G., 2018. The paradox of irrigation efficiency. *Science*, V.361, N.6404, pp. 748–750.
9. Hart, W.E., Pery, G.; Skogerboe, G.V., 1979. Irrigation performance: an evaluation. *Journal of the Irrigation and Drainage Division*, V.105, N.3, pp. 275–288.
10. Hart, W.E., 1961. Overhead irrigation pattern parameters. *Agricultural Engineering*, V.42, N.7, pp. 354–255.
11. Hart, W.E.; Reynolds, W.N., 1965. Analytical design of sprinkler systems. *Transactions of the ASAE*, V.8, N.1, pp. 83–85, 89.
12. Heermann, D.F.; Hein, P.R. 1968. Performance characteristics of self-propelled center pivot sprinkler irrigation system. *Transactions of the ASAE*, V.11, N.1, pp. 11–15.
13. Heermann, D.F.; Wallender, W.W.; Bos, M.G., 1992. Irrigation efficiency and uniformity. In: Hoffman, G.J.; Howell, T.A.; Solomon, K.H. (Eds.). *Management of farm irrigation systems*. St. Joseph: American Society of Agricultural Engineers. pp. 125–149.
14. Hsiao, T.C.; Steduto, P.; Fereres, E., 2007. A systematic and quantitative approach to improve water use efficiency in agriculture. *Irrigation Science*, V.25, N.2, pp. 209–231.
15. Israelsen, O.W., 1950. *Irrigation principles and practices*. New York: Wiley. 471p.
16. Jensen, M.E., 2007. Beyond irrigation efficiency. *Irrigation Science*, V.25, N.4, pp. 233–245.
17. Keller, J.; Bliesner, R.D., 1990. *Sprinkle and trickle irrigation*. New York: Van Nostrand Reinhold. 651p.
18. Marek, T.U.; Undesander, D.J.; Ebeling, L.L., 1986. An aerial weighted uniformity coefficient for pivot irrigation systems. *Transactions of the ASAE*, V.29, N.6, pp. 1665–1670.
19. Merriam, J.L.; Keller, J., 1978. *Farm irrigation system evaluation: a guide for management*. Logan: Utah State University, 271p.

20. Paz, V.P.S.; Frizzone, J.A.; Botrel, T.A.; Folegatti, M.V., 2002. Otimização do uso da água em sistemas de irrigação por aspersão. *Revista Brasileira de Engenharia Agrícola e Ambiental*, V.6, N.3, pp. 404–408.
21. Pereira, L. S. 1999. Higher performance through combined improvements in irrigation methods and scheduling: a discussion. *Agricultural Water Management*, V.40, N.2, pp. 153–169.
22. Perry, C.; Steduto, P.; Allen, R.G.; Burt, C.M., 2009 Increasing productivity in irrigated agriculture: Agronomic constraints and hydrological realities. *Agricultural Water Management*, V.96, pp. 1517–1524.
23. Solomon, K.H.; Burt, C.M., 1999. Irrigation sagacity: a measure of prudent water use. *Irrigation Science*, V.18, N.3, pp. 135–140.
24. Wang, Z.; Zerihum, D., Feyen, J., 1996. General irrigation efficiency for field water management. *Agricultural Water Management*, V.30, N.2, pp. 123–132.
25. Wilcox, J.C.; Swailes, G.E., 1947. Uniformity of water distribution by some undertree orchard sprinklers. *Scientific Agriculture*, V.27, N.11, pp. 565–583.

Surface Irrigation

II

3
Canal Irrigation

Qin Qian
Lamar University

Saeid Eslamian
Isfahan University of Technology

3.1	Introduction	33
3.2	Irrigation Canal Design	34
	Canal Structure Design • Seepage Reduction Design	
3.3	Canal Model	36
	Hydraulic Model • Hydrological Model • Integrated Model	
3.4	Canal Control	40
	Control Methods • Canal Control Algorithm • Automatic Control	
3.5	Optimal Irrigation Canal Management	44
	Machine Learning Approaches • Automatic Control Optimization	
3.6	Conclusions	46
References		47

3.1 Introduction

Irrigation canals are the constructed channels to convey water from sources of supply to farmers and other uses according to a specific schedule. Canals are used because they can be constructed more inexpensively than other available methods with available local labor and materials. The objectives of the irrigation canal design are to reduce erosion, prevent degradation of water quality, and improve efficient use of water by minimizing conveyance losses from seepage, or structural failure. A typical irrigation canal system (Figure 3.1) is composed of related structures such as a network of channels, turnouts, checks, crossings, and other facilities needed for successful operations as an efficient system. The design and installation of all related structures are required to follow the local quality standards, e.g., USDA Natural Resource Conservation Service – practice code 320 (USDA, 2010 https://efotg.sc.egov.usda.gov/references/public/AR/320SOW.pdf).

The simplest operation method of distributing water in a canal network is to supply a continuous flow of water to each farmer, who is then responsible for distributing this water over different parts of his farm (Clemmens, 1986). However, this method would supply too much water early in the season and perhaps not enough water later in the season since plants use water at different rates over the growing season. To achieve a reasonable efficiency, an available irrigation canal must be rotated between different areas of land. Such irrigation canal management involves operating gates, pumps, and valves in order to satisfy user's demands and minimize costs and water losses. In addition, a set of constraints imposed by the physical system and management policies have to be considered, e.g., maximum and minimum water level and flow in the canal. Different integrated models have been developed to reduce the canal cost (Imam et al., 1991), manage the water use effectively in the irrigation canal by integrating hydrological, hydraulic, and crop water use models (Islam et al., 2008; Bhadra et al., 2009a, b).

FIGURE 3.1 A typical irrigation canal from India.

Source: http://india-wris.nrsc.gov.in/wrpinfo/index.php?title=File%3AGhataprabha_Canal.JPG.

Recently, automatic control techniques are widely used in irrigation canals to improve the water management and ensure an improvement of crop production per unit of water consumed and simultaneous protection of the environment. Although a traditional manual operation to monitor system conditions, supervisory manual control to manipulate control structures, and some feedback-control algorithms, e.g., the Proportional-Integral (PI) controller based on current measurements (Burt et al., 1998; Schuurmans et al., 1995, 1999; Litrico et al., 2003), are widely applied to manage water delivery, these control methods cannot meet the increasing standard due to water losses. The more advanced control technique and model predictive control (MPC) have been successfully applied to control the water quantity and water quality in irrigation canals, rivers, and reservoirs (Camacho and Bordons, 2004; Brian and Clemmens, 2006; Xu et al., 2010).

In summary, this chapter will cover the design of the irrigation canal structures, integrated canal hydrological and hydraulic models, canal operation and automatic controls, and optimal irrigation canal management.

3.2 Irrigation Canal Design

An irrigation canal network consists of a group of canals with varying orders and a number of control structures. It is crucial that the conveyance canal and its related structures can be designed to perform their functions efficiently and competently with minimum maintenance, ease of operation, and minimum water loss.

3.2.1 Canal Structure Design

The design of the physical canal infrastructure is standardized with little variation based on either U.S. Bureau of Reclamation design manuals (USBR, 1967, 1999, 2000, 2001) or French Hydraulic Designs (Wahlin and Zimbelman, 2014). The Design of Small Canal Structures (USBR, 1978) (https://www.usbr.gov/tsc/techreferences/mands/mands-pdfs/SmallCanals.pdf) demonstrates the design requirements and general considerations on conveyance structures, regulating structures, water measurement structures, protective structures, and pipe components and appurtenances. The typical canal structures are summarized in Table 3.1.

TABLE 3.1 Canal Structures

Conveyance structure	(1) Canals and pipelines, (2) inverted siphons, (3) road crossings, (4) beach flumes, and (5) drops or chutes.
Regulating structure	(1) Checks, (2) check-drops, (3) turnouts, (4) division structures, (5) check inlets, and (6) control inlets.
Water measurement structure	(1) Parshall flumes, weirs, weir boxes, and open-flow meters weirs, open-flow meters, constant-head orifices; (2) acoustic velocity meters and magnetic flow meters are expected to offer a reliable method of flow measurement in irrigation systems.
Protective structure	(1) Controlled entrance into the canal through a drain inlet; (2) controlled conveyance over the canal in an overchute; (3) controlled conveyance under the canal through a culvert; or (4) the canal must be routed under the cross-drainage channel in a siphon.
Pipe system	(1) Pipes may be manufactured from many different materials such as reinforced concrete, corrugated metal, asbestos cement, reinforced plastic mortar, and welded steel; (2) appurtenances to a pipe are those structural elements necessary to provide efficient hydraulics, structural integrity, effective water tightness, adequate percolation path, easy access for inspection and maintenance, and effective safety.

Design capacity for an irrigation canal is determined by the maximum water demand, which is dependent on irrigation area, corps, rotation, and demand system for turnout deliveries, water losses from evaporation and seepage, and anticipated efficiency of water application to the corps. The canal shall be designed to convey surface runoff, meet delivery demands, cover the estimated conveyance losses, and convey the available water supply in water-short areas (USDA, 2010). The design process for a canal system comprises of the alignment (layout), the determination of the water levels in the network (synoptic diagram), the determination of the design flow, and the design of the canal cross and longitudinal sections considering freeboard, linings, longitudinal slope, side slopes, and top width (Imam et al., 1991). The alignment of the canals ensures to distribute irrigation water to all parts of the command area. The required farm size and the method of irrigation determine the length and spacing of canals. The water levels of the different canal orders are relative to each other to make sure that water flows from the lower order (bigger canals) to the higher order (smaller canals). The design of the longitudinal and cross sections of a given canal is usually carried out simultaneously with the water surface profile because the water surface (or bed) slopes are tentatively selected before the cross section can be figured out to carry the design flow. The major design variables for the cross section are the bed width, the maximum water depth, the freeboard, the side slope, and the longitudinal bed slope. The side slope is usually selected based on the stability requirements of the material forming the canal. Adequate freeboard is chosen based on the canal capacity. The bed width and the maximum water depth are determined by Manning's equation (Imam et al., 1991). For unlined canals, the velocity limits for specific soils shall be considered (USDA, 2007) and the Manning's coefficient should be less than 0.025 (USDA, 2010). The design of erodible, unlined canals is a complex process because the canal instability is dependent on not only just the properties of the materials composing the canal boundaries, but also scours, depositions, or their combination. The lined or piped canal should be considered where erosive water velocities occur (NRCS, 2001). The detailed design requirements and general considerations on unlined and lined canals can be found in The Design Standards No. 3 (USBR, 1967).

An irrigation canal is often used to convey water not only for farmland irrigation, but also sometimes transporting water to meet requirements for municipal, industrial, and outdoor recreational uses. A successful canal network design should ensure that the alignment, cross and longitudinal sections of all its components are worked out to suit the functional, hydraulic, and maintenance requirements. To obtain the least-cost design of irrigation canals, an integrated approach to estimate the total cost (Imam et al., 1991) has been developed to incorporate elements of the water section and the above-water section, which is governed by maintenance and/or transportation requirement, to generate design alternatives covering the solution domain.

3.2.2 Seepage Reduction Design

Canal seepage is the common problem in irrigation canals. It can lead to reduce water deliveries to farmers, increase pumping costs when pumps are used in the system, cause crop yield and health problems due to increasing drainage, etc. The most common Best Management Practices (BMPs) are lining canals, and replacement of them with pipeline (TWDB, 2013). Lining canals with a fixed impervious material can minimalize the seepage loss. The three most commonly used liners are Ethylene-Propylene-Diene Monomer (EPDM), urethane, and concrete. The EPDM is the least expensive and can result in minimal or no seepage, but large animals or other traffics might tear the liner. Reinforced concrete liners have the longest durability but they are the most expensive and may have the largest seepage rate due to cracks and expansion joints (Abedi-Koupai et al., 2011). Urethane has low seepage rates but uses hazardous chemicals during the installation (USBR, 2000). A detailed description of these and other liners are provided in the U.S. Bureau of Reclamation report titled "Canal Lining Demonstration Project Year 7 Durability Report" (USBR, 2000).

The buried pipelines and appurtenances to convey water from the source (well, river, reservoir) to a farm or irrigation turnout can be used to replace most types of small canals or lateral canals to reduce loss from evaporation and prevents soil erosion or loss of water quality (Kannan et al., 2011). In general, district irrigation pipelines use either PVC Plastic Irrigation Pipe (PIP) or Reinforced Concrete Pipe (RCP) with gasketed joints, and have size of 72 inches in diameter or less, with 12 inches through 48-inch diameter pipes being common (USBR, 1988). The PIP is available in diameters from 6 to 27 inches with pressure ratings from 80 to 200 psi and can result in minimal or no seepage losses. The RCP is typically available in diameters between 24- and 72 inches. The cost for low-pressure PVC PIP pipe is based on the pipe diameter and the distance between the pipe factory and the installation site. The RCP is usually manufactured in the area in which the pipe is being installed. The cost for pipeline design, site preparation, trenching, bedding materials, backfill, compaction, and finish work are depended on the specific site and project (USBR, 1999).

These two options are expensive compared to the soil compaction for seepage reduction (Burt et al., 2010). The Irrigation Training and Research Center, with support from the USBR and the California State University Agricultural Research Initiative, has conducted "in-situ" compaction with a vibratory roller of canal banks and canal bottoms for reducing canal seepage. The test results show that the seepage reduction is significant and reached 86%–90% when both the sides and bottom were compacted (Burt et al., 2010). Since silt contents in the 20% and 50% had high percent reductions, it appears that within the range observed between 20% and 50%, silt content of the soil does not significantly affect the seepage rate reduction due to compaction (Burt et al., 2010). Therefore, the soil compaction can be a consideration to reduce seepage when the silt content is below 20% and above 50%.

3.3 Canal Model

Improving the performance of the major irrigation project is one of the economically viable options in meeting the growing water demands and sustaining the productivity of irrigated agriculture under present financial, environmental, and physical constraints. A good understanding of all components of water movement and balance is essential to manage water use effectively in an irrigation canal system. Therefore, the hydraulic and hydrologic model of the system and their integrated model are the basic tools.

3.3.1 Hydraulic Model

The unsteady hydraulic model is based on the conservation of mass and momentum/energy equations. The partial differential equations of Saint-Venant equations (Strelkoff, 1969; Cunge et al., 1980) express gradually varied one-dimensional unsteady open-channel flows as follows:

Canal Irrigation

$$\frac{\partial Q}{\partial x} + \frac{\partial A}{\partial t} + I = 0. \tag{3.1}$$

$$\frac{\partial Q}{\partial t} + \frac{\partial}{\partial x}\left(\frac{\beta Q^2}{A}\right) + (gA)\frac{\partial z}{\partial x} + gAS_f = 0. \tag{3.2}$$

Equations (3.1) and (3.2) are continuity and momentum equations, respectively, where Q = flow rate, A = flow cross-section area, z = water surface elevation, I = net seepage outflow, β = momentum correction factor, x = distance, t = time, g = ratio of weight to mass, and S_f = friction slope. The friction slope term S_f can be estimated using Manning's equation as follows:

$$S_f = \frac{n^2 Q |Q|}{A^2 R^{4/3}}, \tag{3.3}$$

where n = Manning's roughness coefficient, and R = hydraulic radius A/P, and P = wetted perimeter.

An analytical solution for the nonlinear partial differential equation is restricted to problems of simple geometry, subject to very simple initial and boundary conditions. It is required to apply accurate and efficient numerical techniques to solve Eqs. (3.1) and (3.2) for practical problems. Several techniques such as finite element (Szymkiewicz, 1991; Tavakoli and Zarmehi, 2011), finite volume (Mangeney-Castelnau et al., 2003), and upwind conservation scheme (Ying et al., 2004) have been used to numerically solve the Saint-Venant equations. The most common method for solving the Saint-Venant equations in irrigation canal system (Gichuki et al., 1990; Islam et al., 2008) is the weighted four-point Preissmann finite-difference scheme (Cunge et al., 1980), which approximates the dependent variables and their corresponding space and time derivatives using weighted means in a computational grid of space and time domain as described below (Islam et al., 2008).

$$f(x,t) = \frac{\theta}{2}\left(f_{j+1}^{n+1} + f_j^{n+1}\right) + \frac{1-\theta}{2}\left(f_{j+1}^n + f_j^n\right) \tag{3.4}$$

$$\frac{\partial f(x,t)}{\partial x} = \theta\left(\frac{f_{j+1}^{n+1} - f_j^{n+1}}{\Delta x}\right) + (1-\theta)\left(\frac{f_{j+1}^n - f_j^n}{\Delta x}\right) \tag{3.5}$$

$$\frac{\partial f(x,t)}{\partial t} = \frac{\left(f_{j+1}^{n+1} - f_{j+1}^n + f_j^{n+1} - f_j^n\right)}{2\Delta t} \tag{3.6}$$

Applying Eqs. (3.4–3.6) to Eqs. (3.1) and (3.2), then we obtain:

$$\theta\left(\frac{Q_{j+}^{n+1} - Q_j^{n+1}}{\Delta x}\right) + (1-\theta)\left(\frac{Q_{j+1}^n - Q_j^n}{\Delta x}\right) + \frac{\left(A_{j+1}^{n+1} - A_{j+1}^n + A_j^{n+1} - A_j^n\right)}{2\Delta t} + I_j^n = 0 \tag{3.7}$$

$$\frac{(Q_{j+1}^{n+1} - Q_{j+1}^n + Q_j^{n+1} - Q_j^n)}{2\Delta t} + \theta\left[\frac{\frac{(Q_{j+1}^{n+1})^2}{A_{j+1}^{n+1}} - \frac{(Q_j^{n+1})^2}{A_j^{n+1}}}{\Delta x}\right] + (1-\theta)\frac{\frac{(Q_{j+1}^n)^2}{A_{j+1}^n} - \frac{(Q_j^n)^2}{A_j^n}}{\Delta x}$$

$$+ qA_j^{n+1}\left[\theta\left(\frac{z_{j+}^{n+1} - z_j^{n+1}}{\Delta x}\right) + (1-\theta)\left(\frac{z_{j+1}^n - z_j^n}{\Delta x}\right)\right] + g\frac{n^2 Q_j^{n+1}|Q_j^{n+1}|}{(R_j^{n+1})^{4/3} A_j^{n+1}} = 0 \tag{3.8}$$

The area (A) are calculated using elevation (z) and canal cross-section data. Equations (3.7) and (3.8) can yield the following linearized equations with two unknowns (Q and z) (Islam et al., 2008):

$$Cu_1 \Delta Q_j + Du_1 \Delta z_j + Eu_1 \Delta Q_{j+1} + Fu_1 \Delta z_{j+1} = Gu_1, \quad (3.9)$$

$$Cu_2 \Delta Q_j + Du_2 \Delta z_j + Eu_2 \Delta Q_{j+1} + Fu_2 \Delta z_{j+1} = Gu_2, \quad (3.10)$$

where Cu_1, Du_1, Eu_1, Fu_1, and Cu_2, Du_2, Eu_2, Fu_2 are coefficients of the Jacobian matrix corresponding to Eqs. (3.7) and (3.8), respectively; and Gu_1 and Gu_2 represent residuals of the Eqs. (3.7) and (3.8), respectively.

The initial condition and boundary conditions at upstream and downstream are required to solve Eqs. (3.9) and (3.10). The most common used boundary conditions in the subcritical flow region are discharge hydrograph $Q(t)$ at the upstream boundary, and stage hydrograph $Z(t)$, rating curve $Q = f(z)$, or constant depth/stage at the downstream boundary.

The Saint-Venant Eqs. (3.1) and (3.2) are applicable within canal reaches only. For other elements of a canal network such as junctions or hydraulic structures, the hydraulic conditions at the junctions may be expressed by continuity and energy equations:

$$\sum Q_i = \sum Q_o, \quad (3.11)$$

$$h_i + z_{bi} + \frac{v_i^2}{2g} = h_o + z_{bo} + \frac{v_o^2}{2g} + k_h \frac{v_o^2}{2g}, \quad (3.12)$$

where h = depth of water, z_b = bed elevation, V = flow velocity, k_h = head loss coefficient at the junction, and the subscripts i and o represent inflow and outflow, respectively. When the junction losses and the differences in velocity heads at the junctions are negligible, the energy conservation equation at the junction points can be approximated as $z_i = z_o$.

An irrigation canal system also consists of different control/regulating structures for regulating the flow in various sections of the system. The most common hydraulic structures used for regulating and distributing water in an irrigation canal network are checks, turnouts, weirs, and orifice gates as listed in Table 3.1. Equations for calculating the flow rates are included in the Design of Small Canal Structures (USBR, 1978) and used as internal boundary conditions. An irrigation canal network model developed using Saint-Venant equations provides fast and accurate prediction to simulate the filling and operation of canals, and can handle bulk lateral outflow from turnouts and waste way weirs (Gichuki et al., 1990). A numerical "CanalMod" (Islam et al., 2008) has been developed using Saint-Venant equations combined with the flow control internal boundary conditions to simulate the Right Bank Main Canal of Kangsabati Irrigation Project in India for steady and unsteady flow conditions to manage the canal water effectively. Therefore, Hydraulic Simulation Models (HSMs) offer unlimited opportunities for improving the performance of the irrigation systems by studying the flow behavior in a large and complex canal network under a variety of design and management scenarios.

3.3.2 Hydrological Model

The mathematical models of canal operation and distribution (Islam, 2008) were developed over the years exclusively concentrate on hydraulic aspects of a canal system and do not take into account the hydrology of irrigation system. A clear understanding of all the components of the water balance is essential to manage water use effectively in an irrigation canal system. When modeling the hydrology of irrigated watershed, the soil and water assessment tool (SWAT) (Arnold et al., 1993) is a popular tool for understanding the hydrological processes in the watershed and the irrigated field.

The SWAT is a physically based conceptual continuous simulation model developed to quantify the impact of land management practices on surface water quality in large watersheds (Gassman et al., 2007; Neitsch et al., 2004) (http://www.brc.tamus.edu/swat). It provides a continuous simulation of processes such as evapotranspiration, surface runoff, percolation, return flow, groundwater flow, channel transmission losses, pond and reservoir storage, channel routing, field drainage, crop growth, and material transfers (soil erosion, nutrient and organic chemical transport, and fate) (https://swat.tamu.edu/media/99192/swat2009-theory.pdf). It incorporates the combined and interacting effects of weather and land management (e.g., irrigation, planting, and harvesting operations, and the application of fertilizers, pesticides, or other inputs). The SWAT divides the watershed into subwatersheds/subbasins using topography. Each subwatershed/subbasin is divided into hydrological response units (HRUs), which are unique combinations of soil and land cover. Although individual HRUs are simulated independently from one and another, predicted water flows are routed within the channel network, which allows for large watersheds with hundreds or even thousands of HRUs to be simulated.

The SWAT+ is a completely revised version of the SWAT model to provide a more flexible spatial representation of interactions and processes within a watershed released on April 26, 2019 (https://swat.tamu.edu). Two interface for setting up the application are QSWAT+ and SWAT+ Editor. The QSWAT+ is a QGIS interface (https://swat.tamu.edu/media/116301/qswat-manual_v18.pdf) to process the geographic information system (GIS) inputs for soils, land cover and land uses, climatic data including precipitation, temperature, solar radiation, wind speed, and relative humidity. The SWAT Editor reads the project database generated by ArcSWAT or QSWAT, allowing the user to edit SWAT input files, execute SWAT run, and perform sensitivity, auto-calibration, and uncertainty analysis. The previous versions of SWAT 2009 (Kannan et al., 2011), SWAT 2005 (Dechmi et al., 2012), and SWAT 2000 (Liu et al., 2013) have been used to model the different irrigation systems.

Development of SWAT model is a very complex process and requires a lot of input files. Three basic input files are required for delineating the basin into subbasins and HRUs are a digital elevation model (DEM), a soil map, and a land use/land cover map. The topographic parameters including slope, slope length, drainage network, watershed delimitation, and number of subbasins can be obtained from the DEM. The weather input data including maximum and minimum daily air temperature, solar radiation, wind speed, and relative humidity can be obtained from the weather stations. Tentative quantity, timing, and frequency of irrigation required for major crops are also needed. To model canal irrigation, land cover map, soil map, and subbasin map are overlaid using GIS tools, and a comprehensive map is prepared that has all three types of information (HRU information). A HRU under agriculture land cover can be either irrigated or not irrigated. If irrigated, the model followed the canal irrigation procedure to estimate the tentative quantity of water that could have been diverted from the source for irrigating the crop. The stream flow at different outlets of the irrigation canal needs to be measured to calibrate and validate the model.

3.3.3 Integrated Model

The irrigation canal system with a reservoir can be challenging to manage because rainfall–runoff relationship is an extremely complex and difficult problem involving many variables (Bhadra et al., 2009a, b). On the other hand, the correct estimation of conveyance water losses due to seepage from an irrigation system is vital for the proper management for the system. Therefore, it is very important to develop an integrated model for better understanding the hydrology and hydraulics of the irrigation canal system.

For the irrigation canal system with a reservoir, the development of an integrated reservoir-based canal irrigation model (IRCIM) includes catchment hydrologic modeling, reservoir water balance, crop water demand modeling, and a rotational canal irrigation management system (Bhadra et al., 2009a). The catchment module predicts daily runoff from the catchment that inflows to the reservoir, which provides with the flexibility of choosing between the Soil Conservation Service (SCS) curve number method combined with the Muskingum routing technique, and an ANN-based model depending on the data availability; the reservoir module calculates the daily reservoir storage based on conservation of mass approach; the crop water demand module is comprised of water-balance models for both paddy and field crops; and irrigation management system serves as the program flow controller for the model and runs the required module when needed (Bhadra et al. 2009a). The IRCIM has been applied to the Kangsabati Irrigation Project, West Bengal, India, as a case study (Bhadra et al., 2009b). The model has successfully simulated the operation of the test reservoir after proper calibration and determined better delivery schedules to improving the performance of the test irrigation project considerably over the actual delivery schedule for most of the simulation years.

Canal seepage losses are affected by many factors, which include the permeability of the canal bed and lower soil layers, water depth of the canal, form of the canal lining, form of the canal section, and groundwater depth (Swamee et al., 2000; Kahlown and Kemper, 2004, 2005; Ghazaw, 2011). Field experiments, the Kostiakov formula, empirical formulas, and numerical methods are common methods to qualify field canal seepage losses (Akkuzu, 2012; Luo et al., 2005). Field tests include the inflow–outflow measurement method, ponding tests, point measurement method, double-ring infiltration test, and permeameter measurement method (Alam and Bhutta, 2004), and these methods have advantage and disadvantage under different irrigation canal systems and require heavy lab work. With the development of science and technology, several new methods have been introduced to measure canal seepage. Acoustic Doppler devices are useful for determination of seepage rates (Kinzli et al., 2010) and have been utilized to estimate the uncertainty of canal seepage losses (Martin and Gates, 2014). Hotchkiss et al. (2001) have proposed techniques to determine irrigation canal seepage through electrical resistivity, which can easily pinpoint seepage zones more precisely than the stream gauging approach. Although the Kostiakov formula is widely utilized in the planning and design of canals and has also been applied to seepage loss estimation during canal water conveyance and distribution under normal canal operating conditions, the value of each coefficient in the formulas exhibits a large fluctuation and the formulas involve numerous constraints when used; thus, they are limited and only suitable for specific situations (Zhang et al., 2016). The empirical formulas are relatively accurate in determining the mechanism of seepage loss and expressing results when their initial and boundary conditions are applied and restricted to make simplifications and assumptions (Yussuff et al., 1994). When the appropriate equation and boundary conditions are selected consistent seepage condition, the numerical method can be established to predict seepage losses changing with time and space under various complicated conditions. Although most of the existing canal seepage models have certain limitations, several models have been applied to predict accurate seepage losses at different irrigation canals. The saturated–unsaturated flow model HYDRUS-2D has been applied to predict reliable canal seepage under complex soil structures and different canal liners (Yao et al., 2012). The finite-element software AutoBank (Du, 2012) has been employed to calculate the seepage losses in Hetao Irrigation District, China, and the reasonable predication has been obtained (Zhang et al., 2016). The finite-volume MODFLOW model has been utilized to predict the groundwater in a canal-well irrigation district in lower Yellow River Basin, China (Liu et al., 2013). Therefore, these numerical models can be integrated with hydrological and hydraulic models to predict water use effectively in an irrigation system. Liu et al. (2013) establish the integrated surface water and groundwater model using modified soil and water assessment tool (SWAT2000) and modular three-dimensional groundwater flow (MODFLOW) models to analyze sustainable management strategies in surface–groundwater conjunctive use irrigation districts in the lower Yellow River Basin.

3.4 Canal Control

The main intent of canal networks is to supply irrigation water to farmers to grow crops, although some canals also supply water to urban areas or transport of drainage water through them. Experience has shown that farmers can be more productive when they have better control on frequency, flow rate, and duration of their water supply by developing improved canal operations (Wahlin and Zimbelman,

2014). Supply-oriented, demand-oriented, or their combinations are common irrigation control methods. A government agency or central authority makes the decisions regarding the amount of water diverted and delivered to users in supply-oriented systems. Demand-oriented systems allow farmers to order water when required for their farm management. However, supply-and-demand mismatches occur all the time because canal and outlet capacities limit flows, water supply itself often is limited, as well as it is not possible or reasonable to adjust the supply in a canal spontaneously due to a significant time lag between releasing of water at the source (e.g., reservoir) and delivering of water to users. Therefore, an improved operation strategy to deal with these supply-and-demand mismatches is utilized by canal control.

3.4.1 Control Methods

The canal control variable can be water level, flow rate, pressure, volume, or others in the canal pool, which is a canal section between check structures. Comparing between the actual value of the variable and the setpoint, the error in that value is determined to make a control decision, and a control action is made to the structure (gate, actuator, or pump). The control decision is to move the structure at a certain amount, and the control action physically makes that change. Canal control can be classified into five categories (Malaterre et al., 1998):

- *Manual control and automatic control:* Canals or parts of canals can be controlled by utilizing manual control or automatic control. In case of manual control, a local operator does a visual reading of the water level that needs to be controlled and takes action by changing the setting of the structure in place or remotely with a properly designed supervisory control and data acquisition (SCADA) system. For automatic control, electronic sensors measure the water variable and send a signal to a microprocessor or programmable logic controller (PLC). An analog-to-digital convertor makes the signal readable for the computer. An algorithm calculates the required control action and sends this through a digital-to-analog convertor or a digital on-off signal to the motor of the structure, such as AVIS, AVIO, or AMIL gates (Wahlin and Zimbelman, 2014).
- *Feedback control and feedforward control:* Feedback control compares an observed water variable value with a target value. A controller calculates a required change for the input of the canal pool, based on a deviation between the observed and desired value. The required change is sent to the structure that implements the required change, resulting in an adjusted structure flow that influences the hydraulic conditions of the canal pool. On the other hand, feedforward control does not react to an error between the target and observed water level but to a disturbance that could create a future error (e.g., a future water delivery). To counteract the effect that the disturbance has on the water level, the feedforward controller uses a model of the response from which control actions are determined that minimize the effects of this disturbance (Wahlin and Zimbelman, 2014).
- *Upstream control and downstream control:* In case of upstream control, when an increase in flow arrives from upstream, a change in water level is detected and the controller passes this disturbance (change in flow) to the downstream pool. In opposite, with downstream control, the observation is located downstream from the control structure (Wahlin and Zimbelman, 2014).
- *Local control, distant control, and remote control:* Local control refers to the concept where the observation is made in close proximity to the control structure; distant control refers to the observation located at a significant distance away from the control structure, usually at the other end of the canal pool; remote control is the concept whereby all actions are decided and sent from a central control center (Wahlin and Zimbelman, 2014).

3.4.2 Canal Control Algorithm

In a canal network, control actions are often changes in gate position, which is a nonlinear function of upstream and downstream water level for a given flow rate. For given upstream and downstream levels at certain location, it is approximately linear. For a desired change in gate discharge, ΔQ, the change in gate opening Δhg can be determined as:

$$\Delta h_g = \Delta Q \frac{\delta h_g}{\delta Q}, \qquad (3.13)$$

where $\delta hg/\delta Q$ may change over the range of upstream and downstream levels. The ratio $\delta hg/\delta Q$ can be determined from the head discharge equations for different control structures as in USBR (1978).

The delay times from the head gate to turnouts are used to count routing water through a canal.

Bautista and Clemmens (2005) have developed a simple scheme for routing flow changes based on the volume change resulting from canal flow change. The delay time ΔT is found from:

$$\Delta T = \frac{\Delta V}{\Delta Q}, \qquad (3.14)$$

where ΔV is the change in canal pool volume in going from one steady-state flow to another, and ΔQ is the change in steady-state flow. The canal pool volume and the delay time are determined using fitted relationships for a series of flow rate data as:

$$V = aQ_{in}^b + c; \quad \Delta T = abQ_{in}^{b-1}. \qquad (3.15)$$

This approach requires knowledge of the flow rate in all canal pools at all times and effort to develop the volume relationship. The flow rates at all locations along a canal for steady and unsteady flow are best described by the Saint-Venant equations and can be predicted using various hydraulic modeling packages. However, such modeling packages are sometimes considered to be too complex for controller design. Therefore, simplified first-order models described by a limited number of pool properties have been developed to capture the relevant dynamics:

- *Integrator-delay (ID) model* (Schuurmans et al., 1995): calculates the downstream water depth h_d as a function of upstream flow Q_{in}, downstream flow Q_{out}, storage area or surface area A_s, and delay time τ.

$$h_d(t+\Delta t) = h_d(t) + \frac{Q_{in}(t-\tau)\Delta t}{A_s} - \frac{Q_{in}(t)\Delta t}{A_s} \qquad (3.16)$$

- *Integrator-delay zero (IDZ) model* (Litrico and Fromion, 2004): describes the canal with a sudden step in structure flow:

$$h_d(t+\Delta t) = h_d(t) + \frac{Q_{in}(t-\tau)\Delta t}{A_s} - \frac{Q_{in}(t)\Delta t}{A_s} + c_{z1}\Delta Q_{in}(t-\tau) - c_{z2}\Delta Q_{out}(t), \qquad (3.17)$$

where c_{z1} and c_{z2} are the positive parameters that represent the immediate change in water level from a change in the delayed upstream flow and a change in the downstream flow, respectively.

The tail-end structures such as weirs and orifices impose a relationship between level variation and flow variation. A delay model has been developed by Munier et al. (2010) as:

$$\Delta Q_{out}(t) = \Delta Q_{in}(t)\left(1 - e^{-(t-\tau)/K}\right), \qquad (3.18)$$

where $K = A_s/k_d$, and k_d is the rate of change of check gate discharge with water level, namely $k_d = dQout,d/dy$ for the check structure.

The routing/operation schedule is developed by summing the changes in flow from downstream to upstream using delay times for each pool from the point of delivery to the head. Each change can be routed separately. Canal pool volume generally is based on measurements of water level; thus, volume control can be thought of as a special version of water level control and not treated separately. To control a water level, the automatic control must change the flow rate into or out of the pool. A separate calculation in a local controller/algorithm is made to determine the change in gate or valve position to implement this control.

A discussion of flow control routing in irrigation canals has been included in ASCE Manuals and Reports on Engineering Practice No. 131 (Wahlin and Zimbelman, 2014). Feedforward (Bruggers, 2004) and feedforward-decoupling (Li et al., 2005) control methods adjust the target water level/flow setpoint changes to speed up the delivery of a flow rate change from the head of the canal down to the end of the canal. Few of such control options have been successfully applied in irrigation districts over long periods of time, and under multiple conditions because the control method requires not only to move the proper flow change down a canal for the proper duration, but to simultaneously maintain fairly constant water levels in the canal pools (Burt et al., 2018). Typically, a volume compensation method of feedforward control in conjunction with feedback control (Clemmens et al., 2010) can handle multiple flow changes simultaneously for the downstream water level. The different mathematical procedures have been developed to tune controllers for the control of water levels. Clemmens et al. (2005) implemented a *linear-quadratic regulator control* (LQR) to model the entire canal consisting of all water level deviations and control interactions. Negenborn et al. (2009) described the use of a serial, iteration-based, distributed MPC scheme for a 7-pool open irrigation canal system and presented the simulated graphs of water-level setpoints and water-level deviations. MPC is a structural methodology to determine the control actions that maintain water levels after disturbances to obtain the system-wide multiple variable optimum performance. As with LQR, it uses an objective function with penalties on different, conflicting objectives.

3.4.3 Automatic Control

Automatic control is defined as a procedure or method used to regulate mechanical or electrical equipment without human observation, effort, or decision (Wahlin and Zimbelman, 2014). The canal automatic control system is useful for operators to have better control of water levels, improvement of irrigator service levels, water savings, operational cost savings, etc. As indicated on Fig. 8-2 of ASCE Manuals and Reports on Engineering Practice No. 131 (Wahlin and Zimbelman, 2014), the implementing canal control process includes:

- Selection, installation, configuration, and testing of instrumentation or field equipment such as regulating gates, sensors, actuators, power supply, access to the gates, Remote Terminal Unit (RTU)/Programmable Logical Controller (PLC), Human-machine interface (HMI) electronics for the RTU/PLC, and the software within the field equipment/RTU;
- Selection, installation, configuration, and testing of the IT infrastructure hardware such as server and storage, virtualization, networking (WAN, LAN, etc.), enterprise management system; infrastructure software including SCADA software and associated databases, network visualization, and potentially for collecting water orders, managing demand, and planning deliveries; and communications infrastructure and protocols such as optical fiber cable VSAT, radio, and protocols (e.g., TCP/IP, DNP3, MDLC/IP, and MODBUS/IP);
- Design, rollout, and performance assessment of the control system and its operation.

Once every element of the canal control system is implemented in the field, a system performance acceptance test is needed to conduct to ensure that the client's operational and management expectations have been met and that the vendor that delivered the project is paid. Tests have been found useful in Australia are (Wahlin and Zimbelman, 2014):

- Water-level alarm performance metric;
- Outfall volume performance metric or administrative spill volume performance metric;
- Successful irrigation index.

Post-acceptance operations and maintenance is an important part of canal automation. The irrigation district can expect manufacturers' warranties or negotiate a comprehensive warranty on the field instrumentation, communications infrastructure, and computer hardware. Irrigation authority operators generally need to be trained and learn a number of new computer skills and adapt to a style of operation that involves reacting to operational problems rather than simply implementing routines.

3.5 Optimal Irrigation Canal Management

The success of an irrigation water delivery system can be measured by how well it meets the objectives of delivering an adequate and dependable supply of water in an equitable, efficient manner to users served by the system (Molden and Gates, 1990). The innovation of the last century modernized today's irrigation systems to improve system efficiency and management effectiveness of every component of the system (reservoirs, canals, and gates) using automation technologies, along with HSMs. Various control optimizations for irrigation canals have been developed to integrate the automatic control techniques with hydraulic/hydrological simulation models.

3.5.1 Machine Learning Approaches

The canal flow control scheme resulting from the coupling of the system automation and the simulation models has proven to be an excellent irrigation water management instrument around the world (Torres-Rua, 2012). However, uncertainties or errors in the components of canal flow control induced by the harsh environment of irrigation systems can mislead or confuse both human and computer controllers. To minimize these errors, new tools from recent Learning Machine (LM) can perform relationship analysis and exploit the statistical characteristics of time-series data in irrigation and water resources (Pulido-Calvo et al., 2003, 2007). To explore the use of statistical LMs as a technique to reduce the impact of the aggregate error occurring in a canal flow control scheme, two LM algorithms, the relevance vector machine (RVM) and multilayer perceptron (MLP), have been tested separately by Torres-Rua (2012). A HSM based on Saint-Venant equation has been validated using data from SCADA system at the Lower Sevier River Basin in Utah, and then the coupled HSM–RVM and the HSM–MLP provided the required aggregate error correction to reduce local biases and structure in the model predication error (Torres-Rua, 2012).

3.5.2 Automatic Control Optimization

Automatic control techniques based on local control of gates using classic approaches as PI controllers (Malaterre et al., 1998) do not take into account the coupling effect among the different local controllers. The use of a single global controller for the control of the whole system with MPC approaches has been widely and successfully applied in water systems. MPC is a setup of a constrained optimization problem that uses a certain optimization algorithm to find optimal solutions of an objective function over a finite prediction horizon. Both equality and inequality constraints can be the objective function (Camacho and Bordons, 2004). Most of the irrigation system with MPC used lumped models and fixed targets for controllers when the water delivery is on demand. The first optimal solution over the prediction horizon is implemented in practice and the optimization is executed at each control step as time proceeds forward. An optimization problem setting can be formulated as (Camacho and Bordons, 2004):

$$j = \min_{U}(X'QX + U'RU) \quad \text{subject to} \begin{cases} x(k+1) = Ax(k) + B_u u(k) + B_d d(k) \\ u_{\min} \leq u(k) \leq u_{\max}, \ k = 1,\ldots,n \end{cases}, \quad (3.19)$$

where X = state vector$[x(1), x(2), ..., x(n)]$, U = control input vector $[u(1), u(2), ..., u(n)]$, Q=weighing matrix on states, R = weighing matrix on control inputs, A = state matrix, Bu = control input matrix, B_d = disturbance matrix, d = disturbance, u_{min} and u_{max} = upper and lower bounds of control inputs, n = number of prediction steps, and k = time step index. The equality constraints of Eq. (5.1) can be represented by an integrator-delay model in canal control; assuming a constant control target, water level deviation from the target e at $k + 1$ time step can be written as:

$$e(k+1) = e(k) + \frac{T_c}{A_s}\{Q_{in}(k-k_d) - [Q_c(k) + Q_{off\text{-}take}(k)]\}, \tag{3.20}$$

where k_d = number of delay steps, T_c = control time step, A_s = water surface area, Q_{in} = inflow, Q_c = control flow, and $Q_{off\text{-}take}$ = flow of farm off-take.

However, when a fixed control target could meet the request due to on-supply delivery, a dynamic target trajectory approach is proposed by Xu (2017) to calculate changes of control targets that are used by MPC. The dynamic target trajectory calculates the percentage change of each setpoint based on the total volume mismatch spreading over the available capacity in each canal pool. Defining SP = dynamic time-variant setpoint in the canal pool, Eq. (5.2) can be expressed as (Xu, 2017):

$$e(k+1) = e(k) + \frac{T_c}{A_s}\{Q_{in}(k-k_d) - [Q_c(k) + Q_{offtake}(k)]\} - \Delta SP(k+1). \tag{3.21}$$

The setpoint change is considered as an extra disturbance vector D_e to the system; thus, the state-space equation using dynamic setpoints over the prediction horizon can be rewritten as:

$$x(k+1) = Ax(k) + B_u u(k) + B_d d(k) + D_e. \tag{3.22}$$

The dynamic setpoint is updated for each time step before it comes into the MPC algorithm. When changes of setpoint act as a corrector to soften constraints and facilitate problem solving, the objective function becomes (Xu, 2017):

$$j = \min_U [X'QX + U'RU + (B_d D_e)'Q(2X + B_d D_e)]. \tag{3.23}$$

The approach is applied to the Central Main Canal in Arizona and the results demonstrate that the dynamic trajectory can help the optimization find the optimum more quickly and facilitate tuning of the controller.

MPC techniques with decentralized and centralized approaches have been utilized. Gómez et al. (2002) present a decentralized predictive control for an irrigation canal composed of a series of pools to estimate future discharges and the hypothesis of being linearly approaching the reference, to finally reach it at the end of the prediction horizon. Sawadogo et al. (1998, 2000) demonstrate a similar decentralized adaptive predictive control using the reach's head gate opening as a controllable variable, and the reach's tail gate opening and irrigation off-take discharge as known disturbances. Several centralized MPC approaches have also been proposed. Malaterre and Rodellar (1997) performed a multivariable predictive control of a two-reach canal using a state-space model to vary the control perspective form a local to a global problem.

Wahlin (2004) tests a multivariable constrained predictive controller using a state-space model to perform tests where the controller either knew or did not know the canal parameters and with and without the minimum gate movement restriction. A higher-level, centralized controller with local PI controller has been applied to shorten the time necessary to stabilize the new flow rate at the buffer reservoir in the Central California Irrigation District (CCID) (Burt et al., 2018). The main channel of CCID is split into the upper main canal with nine check structures and the lower main channel with eight check structures, as well as a regulating reservoir between them. Each check structure is equipped with a district-standard RTU that runs modified proportional-integral-filter (PIF) control algorithms in a SCADA system. To supply the water from head to the downstream, automated upstream feedback control has been applied along with distributed individual check-structure PLCs operated independently. An unsteady flow simulation model plus a proprietary program to characterize resonance and storage, and another optimization program to determine the original check-structure PIF control constants were coupled together to simulate typical flow rate adjustment (Burt et al., 2018). The flow rate control has been implemented by switching all gates to the manual model to adjust the gate positions and switching to the automation model when the steady water depth reaches to adjust the gate with the simulation results of the extra flow (Burt et al., 2018).

The MPC is a technique with strong computational requirements that hinder its application to large-scale systems due to the communication difficulties in the automatic system and/or operated by different systems and/or organizations. Distributed model predictive control (DMPC) techniques to optimize the management of water in irrigation canals provide a reasonable trade-off between complexity and performance and have been a focus of research during the last few years. Basically, the idea is to provide communication among local controllers in such a way that agents can exchange information or even negotiate and reach agreements. Alvarez et al. (2013) present the application of a DMPC algorithm which has been implemented in Matlab and interfaced to the hydraulic model, which is developed with the Saint-Venant equations and implemented using the well-known Simulation of Irrigation Canals' modeling software for irrigation canals in Spain. The local control policies are included in the distributed-control algorithms to obtain better control performance than a more-conventional decentralized control scheme without information exchange.

3.6 Conclusions

The main objective of irrigation canals is to supply water to farmers according to a specific schedule. An irrigation canal is composed of several order canal reaches, connected by gates checks, pipes etc., and usually following a tree structure of off-take points. This chapter summarizes the design of an irrigation canal, hydraulic and hydrological model for a canal system, along with the canal controls to better manage the canal water by optimized operation in order to satisfy user demands and minimize costs and water loses.

The process for the design of an irrigation canal comprises alignment, cross and longitudinal sections, and of all its components, that are worked out to suit the functional, hydraulic, and maintenance requirements. The design standards for an irrigation canal infrastructure are summarized and an integrated approach to estimate the total cost has been developed to incorporate cost elements of structures, maintenance, and operation. The BMPs to deal with seepage losses are lining canals or replacement of them with pipeline. In addition, soil compaction for seepage reduction is also introduced as a cheaper alternative.

Better understanding of water movement and balance in all the components of the canal is essential to manage water use effectively in an irrigation canal system. Therefore, the hydraulic and hydrologic model of the system and their integrated model are introduced. An irrigation canal hydraulic model can be developed using Saint-Venant equations combined with the flow control internal boundary conditions to provide fast and accurate prediction to simulate the filling and operation of canals, and can handle bulk lateral outflow from turnouts. SWAT is a popular tool for understanding the hydrological processes in the irrigation canal watershed and the irrigated field. The integrated hydraulic and hydrological models have been established to deal with the reservoir-based irrigation system and estimate the seepage losses for the irrigation canal system.

The main intent of canal networks is to supply irrigation water to farmers/other users with better controls on frequency, flow rate, and duration of the water supply by developing improved canal operation schemes. Canal control methods, canal flow routing algorithm, and automatic canal control are introduced to reduce the supply–demand mismatches. A volume compensation method of feedforward control in conjunction with feedback control has been demonstrated to successfully handle multiple flow changes simultaneously for the downstream water level. The different mathematical flow routing procedures have been developed to tune controllers for the control of water levels. Automatic control techniques are widely used in irrigation canals to control the canal system more accurate and effectively.

With the technology development, today's irrigation systems have been modernized with automatic technology to improve system efficiency and management effectiveness of every component of the system (reservoirs, canals, and gates). Machine learning algorithms, the RVM, and MLP have been coupled with the hydraulic model along with data from the SCADA system to reduce the impact of the aggregate error occurring in a canal flow control scheme. A set of MPC approaches of automatic control techniques have been summarized for different applications optimization to optimize the canal operation. The MPC has been a power tool to determine the control actions that maintain water levels after disturbances to obtain the system-wide multiple variable optimum performance.

References

Abedi-Koupai, J., Eslamian, S. S., Gohari, S. A., and Khodadadi, R. 2011. Evaluation of mechanical properties of water conveyance concrete canals incorporating nano pozzolan of wheat ash sheath. *Water Soil Sci.*, 14(54): 39–52 (in Persian).

Akkuzu, E. 2012. Usefulness of empirical equations in assessing canal losses through seepage in concrete-lined canal. *J. Irrig. Drain Eng.*, 138(5): 455–460. doi: 10.1061/(ASCE)IR.1943-4774.0000414

Alam, M. M., and Bhutta, M. N. 2004. Comparative evaluation of canal seepage investigation techniques. *Agric. Water Manage.*, 66(1): 65–76.

Alvarez, A., Ridao, M. A., Ramirez, D. R., and Scanchez, L. 2013. Constrained predictive control of an irrigation cannel. *J. Irrig. Drain Eng.*, 139(10): 841–854. doi: 10.1061/(ASCE)IR.1943-4774.0000619

Arnold, J. G., Allen, P. M., and Bernhardt, G. 1993. A comprehensive surface-groundwater flow model. *J. Hydrol. (Amsterdam)*, 142(1–4): 47–69.

Bautista, E., and Clemmens, A. J. 2005. Volume compensation method for routing irrigation canal demand changes. *J. Irrig. Drain Eng.*, DOI: 10.1061/(ASCE)0733-9437(2005)131:6(494)

Bhadra, A., Bandyopadhyay, A., Raghuwanshi, N. S., and Singh, R. 2009a. Integrated reservoir-based canal irrigation model. II: description. *J. Irrig. Drain Eng.*, 135 (2): 149–157.

Bhadra, A., Bandyopadhyay, A., Raghuwanshi, N. S., and Singh, R. 2009b. Integrated reservoir-based canal irrigation model. II: application. *J. Irrig. Drain Eng.*, 135 (2): 158–168.

Brian, W. T., and Clemmens, A. J. 2006. Automatic downstream waterlevel feedback control of branching canal networks: theory. *J. Irrig. Drain. Eng.*, 3(198), 198–207. doi: 10.1061/(ASCE)0733-9437(2006)132:3(198)

Bruggers, M. 2004. Application feed forward controller on Delta-Mendota Canal. M.Sc. Thesis, Delft Univ. of Technology, Delft, Netherlands.

Burt, C. M., Mills, R. S., and Khalsa, R. D. 1998. Improved proportional-integral (PI) logic for canal automation. *J. Irrig. Drain. Eng.*, 1(53), 53–57. doi: 10.1061/(ASCE)0733-9437124:1(53)

Burt, C. M., Orvis, S., and Alexander, N. 2000. Canal seepage reduction by soil compaction. *J. Irrig. Drain. Eng.*, 136(7): 479–485. doi: 10.1061/(ASCE)IR.1943-4774.0000205

Burt, C. M., Feist, K. E., and Piao, X. 2018. Accelerated irrigation canal flow change routing. *J. Irrig. Drain Eng.*, 144(6): 04018006. doi: 10.1061/(ASCE)IR.1943-4774.0001307

Camacho, E. F., and Bordons, C. 2004. *Model predictive control*, 2nd Ed., Springer, New York.

CanalCAD [Computer software]. Cal Poly Irrigation Training & Research Center, California Polytechnic State Univ., San Luis Obispo, CA, USA.

Clemmens, A. J. 1986. Canal capacities for demand under surface irrigation. *J. Irrig. Drain Eng.*, 112(4): 331–347.

Clemmens, A. J., Bautista, E., Wahlin, B. T., and Strand, R. J. 2005. Simulation of automatic canal control systems. *J. Irrig. Drain. Eng.*, 4(324), 324–335. doi: 10.1061/(ASCE)0733-9437(2005)131:4(324)

Clemmens, A. J., Strand, R. J., and Bautista, E. 2010. Routing demand changes to users on theWMlateral canal with SacMan. *J. Irrig. Drain. Eng.* doi: 10.1061/(ASCE)IR.1943-4774.0000226

Cunge, J. A., Holly, F. M., Jr., and Verwey, A. 1980. *Practical aspects of computational river hydraulics*, Pitman, London, UK.

Dechmi, F., Burguete, J., and Skhiri, A. 2012. SWAT application in intensive irrigation systems: model modification, calibration and validation. *J. Hydrol.*, 470–471, 227–238. doi: doi.org/10.1016/j.jhydrol.2012.08.055

Du, S.-L. 2012. Application of AutoBank software in earth dam seepage flow stability computation. *Mod. Agric. Sci. Technol.*, 6, 252–253 (in Chinese).

Gassman, P. W., Reyes, M. R., Green, C. H., and Arnold, J. G. 2007. The soil and water assessment tool: historical development, applications and future research directions. *Trans. ASABE*, 50(4), 1211–1250.

Ghazaw, Y. M. 2011. Design and analysis of a canal section for minimum water loss. *Alexandria Eng. J.*, 50(4), 337–344.

Gichuki, F. N., Walker, W. R., and Merkley, G. P. 1990. Transient hydraulic model for simulating canal-network operation. *J. Irrig. Drain Eng.*, 116(1), 67–82.

Gómez, M., Rodellar, J., and Mantecón, J. A. 2002. Predictive control method for decentralized operation of irrigation canals. *Appl. Math. Model.*, 26(11), 1039–1056.

Hotchkiss, R. H., Wingert, C. B., and Kelly, W. E. 2001. Determining irrigation canal seepage with electrical resistivity. *J. Irrig. Drain.*, 1(20), 20–26. doi: 10.1061/(ASCE)0733-9437(2001)127:1(20)

HYDRUS-2D [Computer software]. International Groundwater Modeling Center, Golden, CO, USA.

Imam, E. H., Bazaraa, A. S., and Zaghlool, A. S. 1991. Design of irrigation canals: integrated approach. *J. Irrig. Drain Eng.*, 117(6): 852–869.

Islam, A., Raghuwanshi, N. S., and Singh, R. 2008. Development and application of hydraulic simulation model for irrigation canal network. *J. Irrig. Drain Eng.*, 134(1): 49–59.

Kahlown, M. A., and Kemper, W. D. 2004. Seepage losses as affected by condition and composition of channel banks. *Agric. Water Manage.*, 65(2), 145–153.

Kahlown, M. A., and Kemper, W. D. 2005. Reducing water losses from canals using linings: costs and benefits in Pakistan. *Agric. Water Manage.*, 74(1), 57–76.

Kannan, N., Jeong, J., and Srinivasan, R. 2011. Hydrologic modeling of a canal-irrigated agricultural watershed with irrigation best management practices: case Study. *J. Hydrol. Eng.*, 16(9): 746–757.

Kinzli, K.-D., Martinez, M., Oad, R., Prior, A., and Gensler, D. 2010. Using an ADCP to determine canal seepage loss in an irrigation district. *Agric. Water Manage.*, 97(6), 801–810.

Li, Y., Cantoni, M., and Weyer, E. 2005. On water-level error propagation in controlled irrigation channels. Proc., Combined 44th IEEE CDC and ECC 2005, Seville, Spain, 1502–1519.

Litrico, X., and Fromion, V. 2004. Simplified modeling of irrigation canals for controller design. *J. Irrig. Drain. Eng.*, 130(5), 373–383.

Litrico, X., Fromion, V., Baume, J. P., and Rijo, M. 2003. Modelling and PI control of an irrigation canal. Proc., European Control Conf., IEEE, Piscataway, NJ, 850–855.

Liu, L., Cui, Y., and Luo, Y. 2013. Intergeted modeling of conjunctive water use in a canal-well irrigation district in the lower yellow river basin, China. *J. Irrig. Drain Eng.*, 139(9): 775–784. doi: 10.1061/(ASCE)IR.1943-4774.0000620

Luo, Y. F., Cui, Y. L., and Zheng, Z. J. 2005. Research progress on methods of quantifying seepage from rivers and canals. *Adv. Water Sci.*, 16(3), 444–449 (in Chinese).

Malaterre, P., and Rodellar, J. 1997. Multivariable predictive control of irrigation canals. Design and evaluation on a 2-pool model. Proc. of the Int. Workshop on Regulation of Irrigation Canals, Marrakech, Morocco.

Malaterre, P., Rogers, D., and Schuurmans, J. 1998. Classification of canal control algorithms. *J. Irrig. Drain. Eng.*, 124 (1), 3–10. doi: 10.1061/(ASCE)0733-9437(1998)124:1(3)

Mangeney-Castelnau, A., Vilotte, J.-P., Bristeau, M. O., Perthame, B., Bouchut, F., Simeoni, C. and Yerneni, S. 2003. Numerical modeling of avalanches based on Saint Venant equations using a kinetic scheme. *J. Geophys. Res.*, 108(B11), 2527. doi: 10.1029/2002JB002024

Martin, C. A., and Gates, T. K. 2014. Uncertainty of canal seepage losses estimated using flowing water balance with acoustic Doppler devices. *J. Hydrol.*, 517, 746–761.

Molden, D. J., and Gates, T. K. 1990. Performance measures for evaluation of irrigation water delivery systems. *J. Irrig. Drain. Eng.*, 116, 804–823. https://doi.org/10.1061/(ASCE)0733-9437(1990)116:6(804)

Munier, S., Belaud, G., and Litrico, X. 2010. Closed-form expression of the response time of an open channel. *J. Irrig. Drain. Eng.*, 136(10), 677–684.

Negenborn, R. R., van Overloop, P.-J., Keviczky, T., and De Schutter, B. 2009. Distributed model predictive control of irrigation canals. *Networks Heterogen. Media*, 4(2), 359–380.

Neitsch, S. L., Arnold, J. G., Kiniry, J. R., and Williams, J. R. 2004. Soil and water assessment tool—Version 2000—User's manual, Texas Water Resources Institute, College Station, TX, USA.

Pulido-Calvo, I., Roldán, J., López-Luque, R., and Gutiérrez-Estrada, J. C. 2003. Demand forecasting for irrigation water distribution systems. *J. Irrig. Drain. Eng.*, 129(6), 422–431.

Pulido-Calvo, I., Montesinos, P., Roldán, J., and Ruiz-Navarro, F. 2007. Linear regressions and neural approaches to water demand forecasting in irrigation districts with telemetry systems. *Biosyst. Eng.*, 97(2), 283–293.

Sawadogo, S., Faye, R., Malaterre, P., and Mora-Camino, F. 1998. Decentralized predictive controller for delivery canals. 1998 IEEE Int. Conf. on Systems, Man and Cybernetics, San Diego, CA, USA. Vol. 4, 3880–3884.

Sawadogo, S., Faye, R., Benhammou, A., and Akouz, K. 2000. Decentralized adaptive predictive control of multi-reach irrigation canal. 2000 IEEE Int. Conf. on Systems, Man, and Cybernetics, Vol. 5, 3437–3442.

Schuurmans, J., Bosgra, O. H., and Brouwer, R. 1995. Open-channel flow model approximation for controller design. *Appl. Math. Modell.*, 19(9), 525–530.

Schuurmans, J., Hof, A., Dijkstra, S., Bosgra, O. H., and Brouwer, R. 1999. Simple water level controller for irrigation and drainage canals. *J. Irrig. Drain. Eng.*, 4(189), 189–195. doi: 10.1061/(ASCE)0733-9437(1999)125:4(189)

Strelkoff, T. 1969. One dimensional equations of open-channel flow. *J. Hydr. Engrg. Div.*, ASCE, 95(3), 861–876.

Swamee, P. K., Mishra, G. C., and Chahar, B. R. 2000. Design of minimum seepage loss canal sections. *J. Irrig. Drain. Eng.*, 1(28), 28–32. doi: 10.1061/(ASCE)0733-9437(2000)126:1(28)

Szymkiewicz, R. 1991. Finite-element method for the solution of the Saint-Venant equations in an open channel network. *J. Hydrol.*, 122, 27–287.

Tavakoli, A., and Zarmehi, F. 2011. Adaptive finite element methods for solving Saint-Venant equations. *Scientia Iranica*, 18(6) 1321–1326

Texas Water Development Board (TWDB). 2013. Water conservation best management practices (BMP) guide for agriculture in Texas. Based on the Agricultural BMPs contained in Rep. 362, Water Conservation Implementation Task Force, Austin, TX, USA.

Torres-Rua, A. F., Ticlavilca, A. M., Walker, W. R., and McKee, M. 2012. Machine learning approaches for error correction of hydraulic simulation models for canal flow schemes. *J. Irrig. Drain Eng.*, 138(11), 999–1010. doi: 10.1061/(ASCE)IR.1943-4774.0000489

U.S. Bureau of Reclamation. 1967. Design Standards No. 3- Canals and Related Structures, Denver, Colorado, USA.

U.S. Bureau of Reclamation. 1978. The Design of Small Canal Structures, Denver, Colorado.

U.S. Bureau of Reclamation. 1988. Downstream Hazard Classification Guidelines, Denver, Colorado.

U.S. Bureau of Reclamation, Pacific Northwest Region. September 1999. Canal Lining Demonstration Project Year 7 Durability Report, 156 p, USA.

U.S. Bureau of Reclamation, Pacific Northwest Region. January 2000. Canal Lining Demonstration Project-2000 Supplemental, 46 p, USA.

U.S. Bureau of Reclamation, Pacific Northwest Region. June 2001. Construction Cost Tables – Canal Lining Demonstration Project, 5 p, USA.

USDA, Natural Resources Conservation Service. October 1980. Conservation Practice Standard, Irrigation Water Conveyance, Flexible Membrane Canal and Canal Lining, 9 p, USA.

USDA, Natural Resources Conservation Service. December 1988. Conservation Practice Standard, Irrigation Water Conveyance, Low Pressure, Underground, Plastic Pipeline, 5 p. Code 430EE, USA.

USDA, Natural Resources Conservation Service. 2001. National Engineering Handbook, Part 650, Engineering Field Handbook, Chapter 14, Water Management (Drainage), USA.

USDA, Natural Resources Conservation Service. 2007. National Engineering Handbook, Part 654, Stream Restoration Design Handbook, Chapter 8, Threshold Channel Design. USDA, USA.

USDA, Natural Resources Conservation Service. 2010. Irrigation canal or lateral code 320, USA.

Wahlin, B. T. 2004. Performance of model predictive control on ASCE test canal 1. *J. Irrig. Drain. Eng.*, 130(3), 227–238.

Wahlin, B., and Zimbelman, D., eds. 2014. *ASCE Manuals and Reports on Engineering Practice No. 131. Canal Automation for Irrigation Systems*. ASCE, Reston, VA, USA, 163–183.

Xu, M. 2017. Model predictive control of an irrigation canal using dynamic target trajectory. *J. Irrig. Drain Eng.*, 143(3), B4016004. doi: 10.1061/(ASCE)IR.1943-4774.0001084

Xu, M., van Overloop, P. J., van De Giesen, N., and Stelling, G. 2010. Real-time control of combined surface water quantity and quality: Polder flushing. *Water Sci. Technol.*, 61(4), 869–878.

Yao, L. Q., Feng, S. Y., Mao, X. M., Huo, Z. L., Kang, S. Z., and Barry, D. A. 2012. Coupled effects of canal lining and multi-layered soil structure on canal seepage and soil water dynamics. *J. Hydrol.*, 430–431(14), 91–102.

Ying, X., Khan, A. A., and Wang, S. S. 2004. Upwind conservation scheme for the Saint Venant Equations. *J. Hydraul. Eng.*, 130(10), 977–987. doi: 10.1061/(ASCE)0733-9429(2004)130:10(977)

Yussuff, S. M. H., Chauhan, H. S., Kumar, M., and Srivastava, V. K. 1994. Transient canal seepage to sloping aquifer. *J. Irrig. Drain. Eng.*, 1(97), 97–109. doi: 10.1061/(ASCE)0733-9437(1994)120:1(97)

Zhang, Q., Chai, J., Xu, Z., and Qin, Y. 2016. Investigation of irrigation canal seepage losses through use of four different methods in Hetao Irrigation District, China. *J. Hydrol. Eng.*, 22(3), 05016035. doi: 10.1061/(ASCE) HE.1943-5584.0001470

4
Ditch and Furrow Irrigation

4.1	Introduction	51
4.2	Advantages and Disadvantages of Furrow Irrigation	52
4.3	Design of Ditch and Furrow Irrigation	53
	Types of Furrow Irrigation • Width, Depth and Shape of Furrows • Slope of Furrows • Furrow Spacing and Length	
4.4	Water Management Principles and Crop Water Requirement in Ditch and Furrow	55
	Water Application and Management • Irrigation Requirements • Losses of Water in Ditch and Furrow Irrigation Method	
4.5	Controlling Water Table, Water Logging and Salinity	60
4.6	Conclusions	62
References		63

Abebech A. Beyene
Bahir Dar University

Saeid Eslamian
Isfahan University of Technology

4.1 Introduction

Irrigation is the process of applying water to soil and crops to let them grow when there is not enough rain (Bjorneberg 2013; Michael 1978; Reddy 2010). The history of irrigated farming goes back from 4,000 years to 6000 B.C. (Beyene 2018). Irrigation methods refer to the technique of water application to the soil for crop growth. Various irrigation methods have been developed over time to meet the irrigation needs of certain crops in specific areas (Bjorneberg 2013). The choice of an irrigation method is determined by certain limiting conditions which preclude one or another of the possibilities and may leave no alternative (Finkel 2018).

Water from sources (rivers, reservoirs, lakes or aquifers) is pumped or flows by gravity through pipes, canals or ditches (Bjorneberg 2013). Open channel flow occurs due to the action of gravity while flow in closed conduit or pipes are mainly controlled by the force of pressure gradient. The irrigation systems that totally rely on movement of water due to force of gravity are called gravity-fed or surface irrigation methods. Irrigation systems that rely on water pressure for the system to work with piped networks are called pressurized irrigation methods. Surface irrigation method is the application of water to the crops over the surface of the field by means of gravity flow (Albaji et al. 2020).

Furrow irrigation method is one of the surface irrigation methods, which is most widely used for irrigating row crops (Michael 1978). Furrow irrigation consists of furrows and ridges (Savva and Frenken 2002). Furrows are small channels, which carry water down the land slope between the crop rows (Reddy 2010) as shown in Figure 4.1. Furrow irrigation method is suitable for most soil types except coarse sands

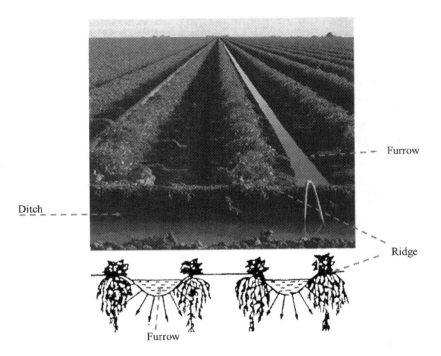

FIGURE 4.1 Example of ditch and furrow irrigation method.

due to high infiltration rates that result in excessive percolation losses. In this method, water is applied to the crops by means of that small channels or furrows which are guided by ridges.

Ditches or supply channels are laid out across the slope in order to supply water to the furrows in the field. Then, water is released from the ditches either by means of siphon tubes or by temporal ditch embankment breaching to the furrows on the lower side of the ditch. More uniform distribution of water can be achieved and the field application efficiency can be enhanced by controlling the flow into the furrows. The crops in the furrow irrigation are usually grown on ridges between the furrows (Reddy 2010), which helps to avoid direct contact/temporal ponding of water on the roots and stems of sensitive crops. As the furrow method is used to avoid ponding and the resulting root rot, this method is the most widely used surface irrigation method for most annual crops except rice.

4.2 Advantages and Disadvantages of Furrow Irrigation

In furrow irrigation, only about half to one-fifth of the field surface is being wetted, thus reducing puddling and clustering of soils and excessive evaporation of water considerably (Asawa 2008; Reddy 2010). Hence, it is the best in regard to farm water management as compared to other surface irrigation methods. Still, a high irrigation efficiency can be achieved in furrows by improving irrigation water supply techniques. The following are the advantages and disadvantages of the furrow irrigation method.
Advantages:

   i. Irrigation can be adjusted based on the available flow as the stream size can be determined based on the furrow size and number,
  ii. Possibility of achieving high irrigation efficiency,
 iii. Lower investment cost except initial labor cost,
  iv. Saves labor and time once the furrows and ditches are already prepared, and
   v. Possibility of keeping the soil structure as compared to other surface irrigation methods as furrows are suitable to soils that form surface crusts when irrigated by flood methods.

Ditch and Furrow Irrigation

Disadvantages:
  i. High possibility of tail water loss unless properly managed,
  ii. Possibility of deep percolation loss (DPL) in the upper parts of the furrow unless properly managed,
  iii. Increases soil erosion if the stream size is not properly determined,
  iv. Requires high labor for furrow and ditch preparation,
  v. Salts may concentrate in the sides of ridges when the source of irrigation water is saline and will affect the crop yields,
  vi. Difficulty of using farm equipment due to rough soil surface, and
  vii. Difficulty of irrigation adequacy in coarse-textured soils due to poor lateral spread of water.

4.3 Design of Ditch and Furrow Irrigation

4.3.1 Types of Furrow Irrigation

Depending on the alignment, furrows can be categorized into two general types as straight furrows and contour furrows. Furrows may also be classified as corrugations, deep and other several types based on their specifications and adaptability.

  i. *Level furrows:* are furrows with blocked ends and has little or no grade. This type of furrows requires huge land preparation and careful water management due to small stream velocity resulted from its low slope. Hence, irrigation water must be applied rapidly and the stream size can be as large as possible to increase stream velocity.
  ii. *Contour furrows:* are types of furrows constructed on steep lands following contours and are suitable to irrigate steep and uneven slopes. But contour furrows are not recommended in high-rainfall areas to avoid soil erosion.
  iii. *Graded straight furrows:* are small furrows constructed in a straight line along the direction of or across the slope. These furrows are built on a relatively level land.
  iv. *Graded contour furrows:* are small furrows and are mostly curved to fit the contour on sloppy and uneven land. These can be constructed on surfaces where straight furrows are not possible.
  v. *Corrugations:* are small and closely spaced furrows constructed on moderately steep and uneven land, which are made to conform to the slope of the land. Such furrows are suitable to irrigate close growing crops.

4.3.2 Width, Depth and Shape of Furrows

Width and depth of furrow depends on the soil type, crop type, stream size, and tillage equipment used to make furrows and ridges. Generally, the furrow size varies from 25 to 40 cm depth and from 15 to 30 cm ridge width (Burton 2010). Furrow cross-sectional shapes also depend on soil type and are usually V or U-shaped. Furrow stream sizes should be non-erosive and as high as possible to avoid deep percolation in the upper ends of the furrow. Larger stream sizes need larger furrow channel (wide and deep) to convey the water without overtopping. Stream sizes for individual furrows can vary from about 10 to 100 L/min depending on soil, slope, field length and management considerations (Bjorneberg 2013). It is desirable to design a furrow at a maximum velocity that is non-erosive and a minimum velocity enough to distribute the water in the channel to avoid deep percolation.

Shallow-rooted crops require shallow furrows for their roots to abstract water such as young crops with shallow rooting depth. However, the furrow can be deep with increased crop growth and this can be done by deepening (digging) the furrow when the crop grows by using hoes or any farm equipment. Shallow and parabolic furrows are required to reduce water velocity and obtain a high wetted perimeter for water to be absorbed on clay soils as water infiltrates slowly and in a wide area. On the other hand, as water infiltrates rapidly on sandy soils, water has to flow more quickly along the furrow to reduce DPL and can be achieved by providing narrow, deep, usually V-shaped furrows.

Non-erosive stream size: The maximum no-erosive stream size regardless of soil type can be determined from the empirical equation developed by U.S. Department of Agriculture (USDA) Soil Conservation Service as:

$$Q_{max} = \frac{C}{S}, \quad (4.1)$$

where:

S = Ground slope down the furrow in %
C = Empirical constant, which is 0.6 L/s.

The wetted perimeter of the furrow can be estimated by the following equation according to USDA.

$$P = 0.265 \frac{Qn^{0.425}}{S^{0.5}} + 0.227, \quad (4.2)$$

where:

P = The wetted perimeter, m
Q = Stream size (inflow rate), L/min
n = Manning's roughness coefficient
S = Furrow slope, m/m.

4.3.3 Slope of Furrows

Furrow slope should not be too large in order to avoid erosion and not too small to allow a quick water filling. Furrows may be straight laid along the land slope; if the slope of the land is small (about 5%) for lands with larger slopes, the furrows can be laid along the contours (Reddy 2010). Furrow slope and the maximum non-erosive stream size can be interrelated by the above relationship (Eq. 4.1). The bed slope of furrow is the slope (average value) along the furrow, in the direction of irrigation water flow.

The bed slope can be measured at field using simple and locally available instruments such as water level, stick, string and tape meter. Then, the horizontal distance between the two distinct points will be measured; and the differences of the vertical distance between those two points will also be measured. The ratio of the difference of the vertical distance to the horizontal is the slope between those points. Other slope measuring instruments such as clinometer can also be used. The bed slope should not be too high as it causes bed scouring and erosion. Too low slope is not also recommended in furrow irrigation since it results in deep percolation in the upper parts of the furrow and insufficient irrigation at the lower ends of the furrow due to slow advance.

4.3.4 Furrow Spacing and Length

The length of the furrow depends on the soil type, stream size, land slope and required irrigation depth, and can range from 60 to 300 m, with the shorter lengths being used on light or coarse-textured soils and longer furrows used on heavier or fine textured soils (Burton 2010). In clayey soils with poor infiltration rates, longer furrows are used, whereas in more permeable soils smaller shorter furrows can be constructed to avoid deep percolation in the upper ends of the furrows. To get improved infiltration in poorly permeable soils, furrows can be elongated by making in a zigzag manner.

Furrow spacing can depend on crop type (crop spacing), farming practice (equipment used) and water movement as it plays an important role in the distribution of water in the soil profile. The spacing of the furrows is governed by the soil type; on lighter or coarser soils, the furrows should be closer together as the lateral movement of water in the soil is significantly less than the vertical movement, but

Ditch and Furrow Irrigation

on heavier soils water moves laterally as well as horizontally and furrows can be more widely spaced (Burton 2010). Hence, the spacing can be as high as 75–150 cm on clay soils and from 30 to 60 cm for sandy soils.

For the design of furrow irrigation, all design parameters that need to be determined may require complex equations and steps. However, this can be simplified by applying deferent models such as SURDEV that have been developed by Jurriëns et al. (2001).

4.4 Water Management Principles and Crop Water Requirement in Ditch and Furrow

4.4.1 Water Application and Management

Water is applied in furrows from the ditch using siphons with enough discharge so that water can reach the tail end quickly to reduce excessive deep percolation in the upper parts of the furrow. Once water is reached the end of the furrow, the flow is cut back and hence the water is infiltrated in the entire furrow. Like other surface irrigation events, a furrow irrigation has four hydraulic phases as follows:

 i. *Advance phase*: is the time needed from the start of irrigation to arrival of water to the wetting (advancing) front at the end of furrow.
 ii. *Wetting phase*: is the time between the end of advance and flow cut-off.
 iii. *Depletion phase*: is the time interval between time of cut-off and the time when the water dries up at the inlet.
 iv. *Recession phase*: is time needed starting from the end of the depletion to the time required for water to recede from all points in the furrow.

A time versus distance graph can be plotted to show the advancing and the recession of the water in the furrow. A vertical distance between the advance and the recession curves gives the time water is present at furrow surface giving opportunity for water to infiltrate in the soil and is called the infiltration opportunity time (Figure 4.2).

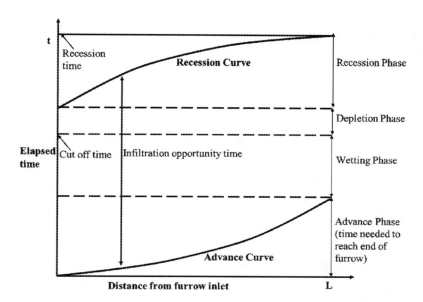

FIGURE 4.2 Phases of furrow irrigation method.

4.4.1.1 The Field Irrigation Application Efficiency (E, %)

The field irrigation application efficiency is a very important farm irrigation water management indicator. It is the ratio of depth of water added to the root zone to depth of water applied to the field (Abera et al. 2019) and can be calculated (Bos et al. 2005; Howell 2003) as follows:

$$E = \frac{D_r}{D_f}, \tag{4.3}$$

where:

D_r = The depth of water added to the root zone (mm)
D_f = The depth of water applied to the field (mm).

The depth of water applied to the field can be determined by measuring the amount of irrigation water supplied and then converting to depth (mm). The depth of water added to the root zone can be obtained by measuring the soil moisture content, and subtracting the soil moisture content just before an irrigation event from the soil moisture content after the irrigation event in volumetric basins and multiplied by depth of root zone in mm (Abera et al. 2019). The field water application efficiency is the most important parameter in any irrigation system design and management (Walker 1989).

4.4.1.2 Water Requirement Efficiency (E_s %)

This is also referred to as storage efficiency and is an indicator of how the applied irrigation water fills the crop root zone. It is related with the crop yield as it indicates soil moisture stress and can be determined as:

$$E_s = \frac{D_r}{S_m}, \tag{4.4}$$

where:

Dr = The depth of water added to the root zone (mm)
Sm = The potential soil moisture storage (mm).

The depth of water added to the root zone is described above (Eq. 4.3). The potential soil moisture storage, also called soil water storage capacity is defined as the total amount of water that is stored in the soil in the plant's root zone. The potential soil moisture storage can be determined in laboratory after saturating the soil and sucking at 300 kPa pressure using a pressure membrane.

4.4.2 Irrigation Requirements

The irrigation requirement for the furrow irrigation depends on the crop water requirement of the given climatic condition and the efficiency of the furrow irrigation method used. Among the different methods of determining the crop water retirement, the Penman method (Allen et al. 1998) is the widely used and standard method. First, the reference crop evapotranspiration is calculated using climate data to estimate the crop water requirement and then the irrigation requirement can be calculated as follows:

$$ET_o = \frac{0.408\,\Delta\,(R_n - G) + \gamma\,\dfrac{900}{T + 273}\,u_2\,(e_s - e_a)}{\Delta + \gamma(1 + 0.34 u_2)}, \tag{4.5}$$

where:

ET_o = Reference evapotranspiration (mm/day)
R_n = Net radiation at the crop surface (MJ/m²/day)

G = Soil heat flux density (MJ/m²/day)
T = Mean daily air temperature at 2 m height (°C)
U_2 = Wind speed at 2 m height (m/s)
e_s = Saturation vapor pressure (kPa)
e_a = Actual vapor pressure (kPa)
$e_s - e_a$ = Saturation vapor pressure deficit (kPa)
Δ = Slope of saturation vapor pressure curve at temperature T (kPa/°C)
γ = Psychrometric constant (kPa/°C).

The reference crop evapotranspiration (ET_o) refers to the evapotranspiration from a reference surface, not short of water and the only factor affecting it is the climatic condition. The crop evapotranspiration, also called the crop water requirement (ET_{crop}) is defined as the depth (or amount) of water needed to meet the water loss through evapotranspiration. The crop water requirement can be calculated from the reference crop evapotranspiration once the crop's total growing period and the crop factor are known using the following relationship.

$$ET_{crop} = ET_o \times K_c, \tag{4.6}$$

where:

ET_{crop} = Crop water requirement/crop evapotranspiration (mm/unit time)
ET_o = Reference crop evapotranspiration (mm/unit time) (influence of climate)
K_c = Crop factor (influence of crop type and growth stage).

4.4.2.1 Crop Growth Stages

The duration in days of the various growth stages of annual crops is divided into four major growth stages (Figure 4.3):

- *Initial stage*: this is the period from sowing or transplanting until the crop covers about 10% of the ground.

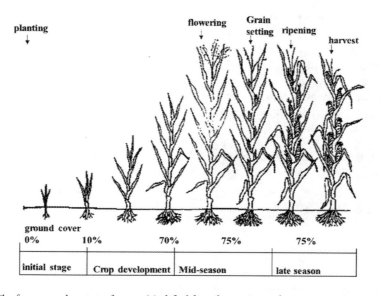

FIGURE 4.3 The four growth stages of crops. Modified from https://www.fao.org/3/S2022E/s2022e07.htm.

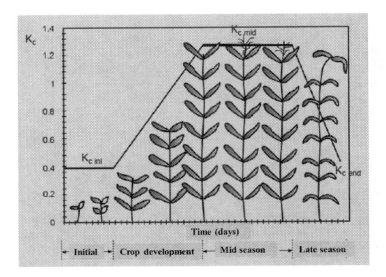

FIGURE 4.4 Crop coefficient curve over the whole crop growing season (Allen et al. 1998).

- *Crop development stage*: this stage is next to the initial stage and stays until the crop covers 70% of the ground surface. The crop does not reach its maximum height at this stage.
- *Mid-season stage*: this stage starts at the end of the crop development stage and lasts until the crop maturity. The crop mid-season stage includes flowering and grain setting. At this stage, the crop reaches its maximum height.
- *Late season stage*: this stage starts at the end of the mid-season stage and lasts until the last day of the harvest. This stage includes the crop's ripening to harvest and no crop growth is expected at this stage.

4.4.2.2 Determination of K_c

The crop factor (K_c) indicates the differences in soil evaporation and crop transpiration rate between the actual crop and the reference grass surface. Figure 4.4 shows the changes in crop coefficient over the length of the crop's growing season. The shape of the curve represents the changes in the vegetation and groundcover during plant development and maturation that affect the ratio of ET_{crop} to ET_o.

The K_c can be determined using lysimeter experiment by isolating the crop root zone from its environment and controlling the processes that are difficult to measure. Then, the soil water balance can be calculated accurately and the K_c can be determined as a difference in soil evaporation and crop transpiration of the actual crop. The K_c values for each of the growth stages of the different crops can also be obtained from Allen et al. (1998).

4.4.2.3 Calculation of Irrigation Water Requirement (IR)

4.4.2.3.1 Net Irrigation Water Requirement

The net irrigation water requirement (NIR) during the crop's growing period is the difference between ET_{crop} and the effective rainfall (P_e).

$$NIR = ET_{crop} - P_e, \qquad (4.7)$$

Determination of effective rainfall (P_e): is the amount of the rain water retained in the root zone (infiltrated and stored in the soil water zone) and can be used by the plants. The effective rainfall is, therefore, the difference between the total rainfall and the losses (runoff, evaporation and deep percolation). The

Ditch and Furrow Irrigation 59

effective rainfall can be calculated according to the formula developed by the USDA Soil Conservation Service which is as follows:

$$P_e = \frac{\text{total rainfall}}{125 \times (125 - 0.2 \times \text{total raill fall})}, \text{ if totall rainfall} < 250 \text{ mm}, \quad (4.8)$$

$$P_e = 125 + 0.1 \times \text{total rainfall, if total rainfall} > 250 \text{ mm}, \quad (4.9)$$

4.4.2.3.2 Gross Irrigation Water Requirement

The gross irrigation water requirement (GIR, IR_g) accounts the losses of water incurred during conveyance, distribution and application to the field.

$$GIR = \frac{NIR}{E}, \quad (4.10)$$

where:

GIR = The gross irrigation requirement (demand) in mm
NIR = The net irrigation requirement (demand) in mm
E = The overall irrigation efficiency of the furrow method (fraction).

4.4.3 Losses of Water in Ditch and Furrow Irrigation Method

Tail water runoff (TWR), deep percolation in the upper parts of the furrow and excessive evaporation are the common losses in this irrigation method. Large stream size can be used to spread water quickly to minimize deep percolation in the upper ends. The quarter time rule can be applied as a field water application control mechanism. TWR can be controlled by providing the end furrow by dikes.

Soil surface evaporation loss is minor in furrow and ditch as compared to other surface irrigation methods (basin and border). This is because furrow irrigation does not allow flooding of the entire field. It rather uses channeling using that small field channels called furrows along the rows of crops and the water infiltrates to the crop root zone. That is the reason why furrow irrigation has better field irrigation efficiency. The application efficiency of furrow irrigation varies from 50% to 70% (Howell 2003). For more precise irrigation management, surface evaporation in furrow irrigation can be minimized by covering the soil surface partly with different materials such as crop residues or any other which is called mulching.

4.4.3.1 Tail Water Runoff

TWR from the end of the field can be defined as:

$$TWR = \frac{\text{Volume of runoff}}{\text{Volume of water applied to the filed}}, \quad (4.11)$$

The TWR loss is one of the threats to furrow irrigation by affecting the water resources. If there is soil erosion in the upper parts of the furrow and the slope and stream size is not properly designed, it results in the TWR.

4.4.3.2 Deep Percolation Losses

The loss of water by percolation below the root zone is termed as DPLs and is determined as:

$$DPL = \frac{\text{Volume of deep percolated water}}{\text{Volume of applied water to the field}}, \quad (4.12)$$

Excessive DPLs recharge the groundwater and raise the water table and hence cause logging in the root zone. The consequences of deep percolation can lead to water logging and may result in salinity problems if the groundwater is saline. In addition to water wastage, excessive deep percolation can leach essential nutrients from the crop root zone. In some cases, the DPLs can return to gaining streams; however, this gain may be disastrous if the groundwater is laden with salts or other toxic substances.

4.5 Controlling Water Table, Water Logging and Salinity

In an intensively irrigated area for extended periods, excess water will be accumulated (ponded) on the land surface. The root zone becomes saturated when the ponded water is not percolating below the root zone. The frequent ponding and the resulting groundwater recharge result in the groundwater table to rise to root zone causing water logging. The field water balance (water balance of the root zone, mostly unsaturated zone) is essential to control waterlogging and salinity problems. The unsaturated zone, the zone from the soil surface to the water table, is suitable to most annual crops and irrigation experts are interested to manage this zone. Most often, the term soil water balance is used to refer the water balance of the unsaturated zone (Beyene et al. 2018) (Figures 4.5 and 4.6).

Water balance computations in the unsaturated zone (soil water balance) is extremely important for the given soil and climatic conditions to estimate the drainage and leaching requirements to control water logging and salinity. The major objectives of soil water balance computations in irrigated fields include:

1. To determine the soil water storage;
2. To estimate the soil water balance of the field and identify depth of the water table;
3. To estimate the evapotranspiration, surface (tail water) runoff, deep percolation and temporal groundwater movement; and
4. To determine the need for drainage and to recommend another study to quantify the drainable surplus of water below the water table.

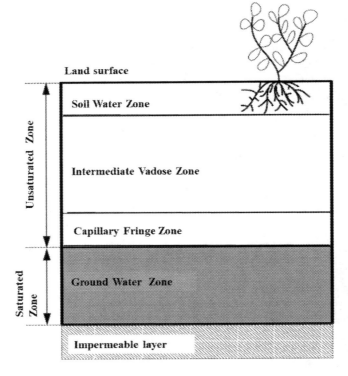

FIGURE 4.5 Sub-zones of the unsaturated zone.

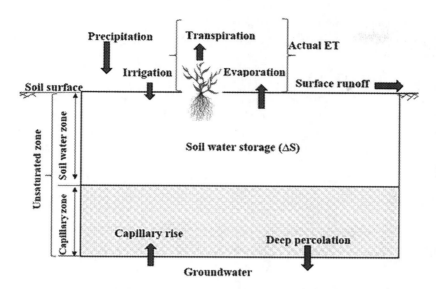

FIGURE 4.6 The soil water balance showing the important components (adapted from Beyene et al. 2018).

The rise of the groundwater table causes water logging to the root zone and the prolonged water logging seriously affects crop growth through lack of aeration and salinity development.

Salinity refers to problems with high concentration of salts. These salts must be evacuated via percolation deep down below the crop root zone by irrigation and rainfall. Generally, the water balance at the soil surface and the groundwater balance must also be computed to have a detail understanding and information about the existing field conditions of the given climate and soil to predict the drainable surplus accurately. Depth to the groundwater table should be optimum; too deep or too shallow groundwater table has significant influence on crop production by the significant response of crop yields to water table depth (Nosetto et al. 2009). To keep the groundwater table depth at optimum for crop growth, drainage is required. On the other hand, too deep groundwater table due to prolonged dry period may result drought of crops as the crop are unable to extract water far below the root zone; in this case irrigation application is required.

Removal of excess (drainage) water from the soil surface or from the sub-soil (to decline the groundwater table) is required to control water logging and to improve crop production. Surface or sub-surface drainage can be used depending on the environment to control surface ponding and water logging. Drainage is required if the depth of groundwater table is too shallow below the critical level and if the duration of surface water ponding extends from 5 to 7 days for common crops except rice. The reader may refer any drainage engineering books for further information about drainage methods, equations to estimate the drainable surplus and designs of drainage. Uncontrolled groundwater table causes salinity in addition of water logging.

Salinity is developed when the field is irrigated from groundwater of high salt content or untreated wastewater especially in arid or semi-arid regions. In areas of shallow groundwater table containing salt, the saline water rises to the root zone by capillary action. As the water is used by the crop and evaporates, it leaves the salts in the root zone. The other source of salinity is irrigating with saline water or untreated wastewater which leaves the salts on the surface of the soil due to evaporation. Effect of salinity on plant growth is expressed in the soil water plant uptake mechanisms. Plants extract water from soil by exerting an absorptive force greater than the force holding water to the soil. If the plant cannot make sufficient internal adjustment and exert enough force, it cannot be able to extract enough water and the crop faces water stress. Salts in soil water increase the force plants must exert to extract more water. Too dry or too saline soils have similar effect for the crops to exert more energy to extract water for the physiological consumption.

Generally, salinity problems due to capillary rise of the groundwater table can be controlled by drainage. Most often, salinity is controlled by leaching the salts using irrigation water in addition to irrigation to meet the crop water requirement. The additional water applied causes deep percolation and a portion of dissolved salts will leave deep below the crop root zone with the deep percolated water.

Leaching requirement is fraction of irrigation water that should be added to percolate below the root zone in order to keep the salinity levels below a certain threshold value 0 of a certain crop. The leaching requirements can be estimated by approaches from different researchers such as Ayers and Westcot (1985) or using the following equation:

$$LR = \frac{EC_w}{5EC_e - EC_w}, \quad (4.13)$$

where:

LR = The leaching requirement needed to control salts (fraction)
EC_w = The salinity of the applied irrigation water in dS/m
EC_e = The salinity of a soil saturation extract.

The total annual amount of irrigation water (depth of irrigation water) that must be applied to meet both the gross irrigation requirement and leaching requirement can be found as:

$$AW = \frac{GIR}{1 - LR}, \quad (4.14)$$

where:

AW = The depth of water to be applied to meet both the irrigation and leaching requirements (mm/year)
GIR = The depth of gross irrigation water to be applied as estimated in Eq. (4.8) above (mm/year)
LR = The leaching requirement (fraction).

Generally, higher irrigation efficiencies are required by applying optimum amount of irrigation water, by minimizing deep percolation and by controlling runoff in furrow irrigation method. On the contrary, slightly excess irrigation application and avoiding crop stress are crucial when using saline irrigation water. Care that should be taken in furrow irrigation methods in areas where salinity is a problem; in the case plants are positioned on the sides of the ridges (Burton 2010) as salts tend to accumulate on the tops of the ridges. Salinity management options are required in furrow irrigation when plants are grown in areas where saline irrigation water is supplied, and where seeds can be placed in raised beds to keep the seeds far from place of high salt concentration.

4.6 Conclusions

Furrow irrigation method is one of the most widely used surface irrigation methods, particularly used for irrigating row crops. Supply ditches are provided to supply water to the furrows so that water is released from the ditches either by means of siphon tubes or by temporal ditch embankment breaching to the furrows on the lower side of the ditch. More uniform distribution of water can be achieved and the field application efficiency can be enhanced by controlling the water flow into the furrows. The water flow into the furrows can be controlled by applying the optimum irrigation water requirement and by designing the appropriate furrows so as to minimize deep percolation and control TWR losses. Generally, proper field water management and design of furrows can control water wastage and salinity development in irrigation systems.

References

Abebech, A. Beyene, (2018). *Water balance, extent and efficiency of irrigation in the Lake Tana basin, Ethiopia*. Belgium: Ghent University.

Abebech, A. Beyene, Cornelis, W., Verhoest, N. E., Tilahun, S., Alamirew, T., Adgo, E., et al. (2018). Estimating the actual evapotranspiration and deep percolation in irrigated soils of a tropical floodplain, northwest Ethiopia. *Agricultural Water Management, 202*, 42–56.

Abebech, A. Beyene, Verhoest, N. E., Tilahun, S. A., Alamirew, T., Adgo, E., Moges, M. M., et al. (2019). Performance of small-scale irrigation schemes in Lake Tana Basin of Ethiopia: technical and socio-political attributes. *Physical Geography, 40*(3), 227–251.

Albaji, M., Eslamian, S., Naseri, A. and F. Eslamian, 2020, Handbook of Irrigation System Selection for Semi-Arid Regions, Taylor and Francis, CRC Group, USA, 317 Pages.

Allen, R. G., Pereira, L. S., Raes, D., & Smith, M. (1998). Crop evapotranspiration-guidelines for computing crop water requirements-FAO irrigation and drainage paper 56. *FAO, Rome, Italy, 300*(9), D05109.

Asawa, G. (2008). *Irrigation and water resources engineering*. New Delhi, India: New Age International.

Ayers, R., & Westcot, D. (1985). Water quality for agriculture. FAO Irrigation and drainage paper 29 Rev. 1. Food and Agricultural Organization. Rome, Italy.

Bjorneberg, D. L. (2013). IRRIGATION | Methods. *Journal of Agriculture and Ecology Research International, Reference Module in Earth Systems and Environmental Sciences*. doi:10.1016/B978-0-12-409548-9.05195-2

Bos, M. G., Burton, M. A., & Molden, D. J. (2005). *Irrigation and drainage performance assessment: practical guidelines*. London, UK: CABI Publishing.

Burton, M. (2010). *Irrigation management: principles and practices* (First ed.). Boston, MA: Cabi North American.

Finkel, H. J. (2018). *Handbook of irrigation technology* (Vol. II). Boca Raton, FL: CRC Press, Taylor & Francis Group.

Howell, T. A. (2003). *Irrigation efficiency* (Encyclopedia of water science). New York: Marcel Dekker,. pp. 467–472.

Jurriëns, M., Zerihun, D., Boonstra, J., & Feyen, J. (2001). *SURDEV: surface irrigation software; design, operation, and evaluation of basin, border, and furrow irrigation* (Vol. 59). Wageningen, Netherlands: International Institute for Land Reclamation and Improvement/ILRI.

Michael, A. M. (1978). *Irrigation: theory and practice*. New Delhi, India: Vikas Publishing House.

Nosetto, M., Jobbágy, E., Jackson, R., & Sznaider, G. (2009). Reciprocal influence of crops and shallow ground water in sandy landscapes of the Inland Pampas. *Field Crops Research, 113*(2), 138–148.

Reddy, R. (2010). *Irrigation Engineering: gene*. New Delhi, India: Gene-Tech Books.

Savva, A. P., & Frenken, K. (2002). *Irrigation manual: planning, development monitoring and evaluation of irrigated agriculture with farmer participation*. Harare, Zimbabwe: FAO.

Walker, W. R. (1989). *Guidelines for designing and evaluating surface irrigation systems*. Book Series: FAO Irrigation and Drainage Paper, 45. Italy: Rome.

5
Level-Basin Irrigation

Nasrin Azad,
Javad Behmanesh,
and Vahid
Rezaverdinejad
Urmia University

Saeid Eslamian
*Isfahan University
of Technology*

5.1	Introduction	65
5.2	Principles of Basin Irrigation Design	69
	Basin Irrigation Design Limitations • Basin Irrigation Design Methods • Basin Irrigation Design • Design Calculation Process of SCS Method	
5.3	Modern Evaluation and Design of Basin Irrigation by Modeling Tools	73
	WinSRFR Model • Application of SIRMOD Model in the Evaluation and Design of Basin Irrigation	
5.4	Application of Optimization Algorithms in the Design and Management of Basin Irrigation	106
5.5	Conclusion	107
References		107

5.1 Introduction

Basin irrigation is defined as a level field with embankment in its perimeter to prevent water overflow. Basin irrigation is the most usual form of surface irrigation, especially in regions with the small fields. Basins are typically square but they may also appear in all other forms of rectangular or irregular shapes. They may be striped with raised beds inside the basin for certain crops cultivation, but in irrigation practice as long as the inflow into these fields, they remain a unit basin. Generally, one of the main superiorities of basin irrigation is its utility in small and/or irregular fields (USDA-NRCS, 2005). Figure 5.1 illustrates a basin irrigation system.

Basin irrigation can be selected for different crops and soils of moderate to low infiltration rate with uniform and smooth slopes. This method has been well applied to fodder and grain crops in heavy soils' textures. Level basins are flat fields in which the aim is to move the inflow as fast as possible and then to let the water pond and percolate for the adequate irrigation time. Therefore, high uniformity and efficiency are achievable in this system, especially in medium to heavy textured soils. On the other hand, due to the level surface of the system, it is possible to apply irrigation water from both upstream and downstream of the field for reaching high uniformity. Because of high uniformity application in the level-basin irrigation systems, they are suitable for salinity control via leaching and subsurface drainage (Reddy, 2013). Well-designed level basins can present 90% irrigation efficiencies (Waller and Yitayam, 2015).

The low irrigation performance and application efficiency is the major problem of surface irrigation systems due to weak management and poor design (Rezaverdinejad and Norjoo, 2014; Azad et al., 2018). Different studies have emphasized the importance of infiltration uniformity and water management as the essential parameters in meeting crop water demands (Santos, 1996).

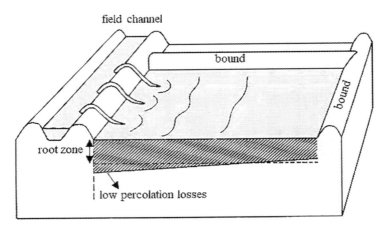

FIGURE 5.1 Basin irrigation system (Brouwer et al., 1988).

The performance of basin irrigation systems including the uniformity of the infiltrated water over the field and irrigation efficiency depends on different design and management factors. Farm management practices such as the irrigation scheduling, the inflow rate and the cutoff time can affect irrigation efficiency and performance (Santos, 1996; González et al., 2011; Miao et al., 2015). Water savings will be accessible by improving water management practices in the existing irrigation systems (Santos, 1996). However, before suitable management of irrigation in the farm, precise design of the irrigation system including dimensions of the fields (length and width) and its slope will have a great impact on high irrigation efficiency and uniformity (González et al., 2011). The importance of land surface conditions and precision leveling in advance and recession processes of basin irrigation method and its performance is well known (Playán et al., 1996a, b; Bai et al., 2011). Irregular soil surface elevation has been recognized as an important source of variability in irrigation depth (Playán and Martinez Cob, 1999) and profitability levels can reduce water use through improved infiltration uniformity (Santos, 1996). The effective results with minimal water percolation require basins without slope (Pereira et al., 2007). Zapata and Playán (2000) have shown that soil surface elevation, the spatial variability of infiltration and opportunity time are the major sources of variability in water distribution of surface irrigation systems and their performance.

In recent decades, different studies concerning water management and infiltration uniformity of applied water have emphasized the importance of designing, improving or changing irrigation technologies as important factors in supplying adequate water to meet crop demands and decreasing water deep percolation losses. Meanwhile, surface irrigation simulation models as important tools in identifying the required improvements and managements for on-farm irrigation systems, and predicting the performance of alternative strategies and new designs have attracted the attention of various researchers. A two-dimensional (2D) hydrodynamic (HD) simulation model of basin irrigation was developed by Playán et al. (1994a), and four applications of the model were presented in Playán et al. (1994b). Two hypothetical case studies were used to indicate the ability of the model to accommodate irregular field shapes and multiple inflows. Finally, the model was applied to explore the influence of field shape and spatially varied infiltration on irrigation performance.

Playán et al. (1996a) presented the extension of the 2D basin irrigation model to simulate water flow on the nonuniform slope and probe the relationship between microtopography and basin irrigation performance. In Playán et al.'s (1996b) study, experimental data were used to survey the impact of microtopography on irrigation performance. Results indicated that soil elevation was highly and significantly correlated with the times of opportunity, advance, recession and measured irrigation depth.

In the other study, Playán and Martinez Cob (1999) used the 2D irrigation model to investigate the effect of inflow discharge and quality of land leveling on basin irrigation scheduling for two crops and locations. The results showed that poor leveling and low discharge cause low application efficiency and significant deep percolation losses. In their research, decreasing the standard deviation of surface elevation from 30 to 0 mm and increasing the discharge from 0.05 to 0.20 m^3/s reduce the number of irrigations by an average 22% and 32%, respectively. Santos (1996) in on-farm management and engineering design and planning of level-basin irrigation used the SRFR-surface irrigation model to design charts which defined the optimum combination of inflow rate and cutoff time, given values of required depth and application uniformity. He reported that the optimum inflow rate and cutoff time depend on infiltration characteristics and the irrigation technology. In the research of Zapata and Playán (2000), a 2D simulation model was utilized to simulate basin irrigation with or without spatial variability of infiltration and/or elevation to evaluate the performance of the level-basin irrigation model when both sources of spatial variability are considered. The results of their study demonstrated that introduction of soil surface elevation variability improved the simulation of advance, recession and crop yield, whereas introducing infiltration variability did not.

In the other research, the Surface irrigation simulation model (SRFR) and Surface Irrigation Modeling software (SIRMOD) simulation models were used by Fabião et al. (2003) to find the best design parameters relative to current field practices. Their simulation results revealed the improvement in field sizes, land leveling, infiltration conditions and inflow rates for controlling waterlogging and water saving in the basins which water advance must be completed in a much shorter time. However, in the basins with very slow advance times due to long field lengths and poor land leveling, solutions included larger inflow discharges, improvements in land leveling and reduction of field lengths.

Khanna et al. (2003) utilized a 2D computer simulation model (COBASIM) to develop design and management guidelines for contour basin layouts. According to the obtained results, a mild slope in the basin bottom can improve irrigation performance in the situation with small depth of irrigation. Khanna and Malano (2006) reviewed different simulation models which have been developed to study the basin irrigation flow processes and improve their design and operation. Pereira et al. (2007) used Surface irrigation simulation model- Windows version (WinSRFR) and SIRMOD models in evaluation of basin irrigation scheduling strategies for irrigation water saving and concluded that improvements of basin land leveling, inflow discharges and irrigation scheduling could result in water savings of 33% and reduce percolation.

An overview of functionalities, interface, architectural elements and technical features of the WinSRFR software was introduced by Bautista et al. (2009a), and its example application was provided by Bautista et al. (2009b). González et al. (2011) analyzed the influence of the longitudinal slope of basin irrigation on the uniformity of irrigation. They provided a set of three-dimensional graphs for determining the optimal slope, length and width of a field or the inflow rate to design and manage surface irrigation systems with longitudinal slope and blocked end and achieve substantial savings of water. Their results showed a 20% savings in water by designing the field with the optimal slope. Bai et al. (2011) used WinSRFR and SIRMOD models to investigate the effect of spatial variability of basins microtopography on irrigation performance. Investigations of Chen et al. (2013) indicated that border irrigation systems in the Yucheng region along the lower reach of the Yellow River have low irrigation performances with the application efficiency <65% and applied irrigation amount of >150 mm per irrigation event, and it is often difficult to change irrigation practices in China's small-scale farmers in which rates of acceptance are usually slow. Therefore, they optimized border dimensions using agricultural irrigation survey data, field experimental data and the WinSRFR simulation model taking into consideration the existing farmers' methods of irrigation practice. In the results of their study, it was demonstrated that under optimized dimensions of borders the applied irrigation depth could be reduced by an average of 49 mm per irrigation event, and the application efficiency could be increased on average by 26.7%. They reported that as a result, the annual potential amount of water savings could be approximately $5{,}551 \times 10^4$ m^3 in the Yucheng region.

Zhang et al. (2012) established a 2D hydrodynamic numerical model of basin irrigation by using the time–space hybrid numerical method and showed that the proposed model can provide a good numerical evaluation, simulation and design of basin irrigation systems. In another study by Zhang et al. (2014a), a 2D zero-inertia (ZI) model of basin irrigation surface water flow was developed based on the standard scalar parabolic type. The performance validation of the proposed model based on experimental data showed that it has good simulation performance and can successfully simulate the basin irrigation surface water flow when the basin surface microtopography condition is relatively smooth. Also, Zhang et al. (2014b) constructed a 2D surface water flow simulation model of basin irrigation with anisotropic roughness based on the complete HD model.

The SIRMOD model was used by Reddy (2013) to design a level-basin irrigation system with the objective of reaching application efficiency of approximately 85%. The results of the mentioned research showed that the performance of level basin designed according to completion-of-advance concept was more than one that was designed based on the limiting length approach. Taghizadeh et al. (2013) used the WinSRFR model to analyze the performance of three methods of furrow irrigation system including conventional furrow irrigation, fixed alternate furrow irrigation and variable alternate furrow irrigation. Then, they optimized cutoff time and inflow rate of irrigation systems with the objective of maximizing application efficiency and distribution uniformity and minimizing runoff and deep percolation. Their results showed that by managing of the mentioned parameters, application efficiency could be enhancing from 54.5% to 74%, where the required irrigation depth is also provided. Taghizadeh et al. (2016) in the other study compared furrow irrigation design methods including the Food and Agriculture Organization (FAO), Soil Conservation Service (SCS), volume balance, Kinematic Wave (KW) and ZI using WinSRFR model. Rezaverdinejad and Norjoo (2014) in optimizing irrigation performance of firmed furrow irrigation in sugarbeet cultivation used WinSRFR model. They showed that in optimum condition of cutoff time and inflow rate, water productivity can increase about 27%. Anwar et al. (2016) used WinSRFR model and indicated that irrigation performance can be improved through changes of border strips layout in Pakistan within current irrigation services. In the other research, different approaches of infiltration parameters estimation (Elliott and Walker two-point, Valiantzas one-point, Mailapalli one-point, Rodriguez and Martos and multilevel optimization method) were compared by Rezaverdinejad et al. (2016) using the WinSRFR model under different furrow irrigation systems (traditional, variable and fixed alternate furrow irrigation) and inflow regimes (with and without cutback inflow).

Miao et al. (2015) and Miao and Shi (2017) in irrigation systems modernization, field assessment of basin irrigation performance and water saving in Hetao, Yellow River basin obtained the infiltration parameter values through the inverse mode simulation with the SIRMOD model. In another study, Miao et al. (2018) used a decision support system model for wheat basin irrigation design improvement, which considers land characteristics and soil infiltration, hydraulic simulation, crop irrigation scheduling and environmental and economic impacts. According to the obtained results of their study, the best alternatives were flat level basins with a length of 100 and 200 m and 2–4 L/s/m inflow rates. Azad et al. (2018) used the SIRMOD model to evaluate surge irrigation system performance and optimize its design and management parameters. Obtained results showed that by providing requirement efficiency of 100% in the experimented furrows with the texture of sandy loam, in inflow rate of 1.2 L/s and cutoff time of 170 min, application efficiency and distribution uniformity will be 60% and 84%, respectively. Furthermore, by considering the best length of the furrow, application efficiency could be increased up to a maximum of 65%. However, in the requirement efficiency of 90%, there is possibility to achieve application efficiency up to 90% at different inflow rates in certain length of furrow.

Due to the application of the WinSRFR simulation model in the design and management of surface irrigation, which is mentioned in various researches, in this chapter of the book, after presenting the principles and criteria of basin irrigation design, the latest version of this software will be introduced, and its application in the various stages of evaluation to the design of basin irrigation will be described in detail based on example data. At the end of the chapter, the SIRMOD simulation model and its use in the evaluation and design of basin irrigation are also mentioned.

Level-Basin Irrigation

5.2 Principles of Basin Irrigation Design

In the basin irrigation method, the soil is divided into parts that have a smooth surface without slope. The ridges are built around the basins to keep the water inside. The basins are filled to the necessary depth with water, and the water stays in the basin until it is completely infiltrated. In rice irrigation or during leaching, water may remain on the soil surface for a long time. The water flow should be able to cover the entire length of the basin in a short time. High permeability of a soil causes to choose greater flow rate or the smaller basin size. The size of the basins may vary from $1\,m^2$ (for example, in vegetables) to several hectares (for example, in rice). When the ground can be completely leveled economically, the basins are considered a rectangular square (Albaji et al. 2020).

5.2.1 Basin Irrigation Design Limitations

- Theoretically, the maximum depth of flow and also the maximum water deep percolation for a given inflow rate, both occur at the beginning of the basin. This maximum flow depth for the specified dimensions of the basin depends on the inflow rate. The depth of the water at the beginning of the basin should not exceed the height of the built ridges.
- In basin design, the average water deep percolation (the difference between net and gross irrigation requirement values) should be minimized. In other words, irrigation application efficiency (ratio of net irrigation requirement to gross irrigation requirement) should be maximized.
- Due to the blocked ends of the basins, the balance in the basin includes deep infiltration and useful storage.
- In many areas, high water deep percolation causes acute drainage problems. Therefore, it should be tried to have high irrigation efficiency.
- The depth of water flow in the basin should be such that the ridges can maintain it.
- The ridges should be made so that their upper width is at least equal to the height of the ridge. The height of the ridge after the subsidence should be at least equal to, or more than one of the following two depths: (1) gross irrigation requirement and (2) maximum depth of flow.

5.2.2 Basin Irrigation Design Methods

In general, basin design methods can be divided into the following three categories:

1. *Empirical methods*: General and empirical recommendations on the dimensions of the basin and the amount of water flow in different soils have been presented by Booher (1974). The main advantage of these methods is their simplicity. These tables are empirical and limited to specific conditions and cannot be easily generalized to other conditions. For this reason, nowadays the use of these methods is limited.
2. *Hydraulic methods*: Hydraulic design is based on solving complete or simplified of Saint Venant equations. These methods have often led to the development of mathematical models. Using the mathematical model, it is possible to study different combinations of input factors such as inflow rate, cutoff time, length and slope of the farm and even the coefficient of roughness. Therefore, the designer can improve the quality of the design.

   There are several mathematical models for simulating water flow on the soil surface. The main difference is in the form of the equations used and the method of solving them. The continuity equation is used in all models, but the main difference is how to use the momentum equation. In some models, the momentum equation is used in its complete form. However, HD models are very complex and are not easy to use under usual conditions. Therefore, in some other models, the simplified form is used, without reducing the quality of the work. Accordingly, there are complete HD, ZI, KW models, zero inertia, kinematic wave and stable flow models in resources.

Nowadays, different software have been developed for surface irrigation simulation based on the prevailing equations such as SIRMOD, NRCS-SURFACE, SURDEV and WinSRFR.

3. *Combination methods*: These methods have been created using a combination of empirical and hydraulic methods. One of the most widely used combination methods is the SCS method; its design principles are summarized in the following subsections.

5.2.3 Basin Irrigation Design

Basin irrigation is somewhat easier to design than other surface irrigation methods because water is not allowed to leave the field and the slopes are usually very low or zero. Therefore, recession and depletion are simultaneously and almost uniformly performed on the basin. Because of very low or zero longitudinal slopes, the driving force of the water is limited to the hydraulic slope of the water surface; therefore, uniformity of field surface topography is very important. The most common method of basin irrigation design is developed by SCS.

The SCS method uses the concept of infiltration curve number (IF) and the SCS infiltration equation to estimate water infiltration into the soil. The standard method of measuring infiltration in basin irrigation is the method of infiltration rings (double ring) and sometimes small basins full of water. Based on the data of the infiltration rings, the curve number (IF) is extracted and based on the SCS infiltration equation, the infiltration amount is calculated and used in the design process.

Intake Opportunity Time, T_o: This time in all parts of the basin should be more or equal to the required time for infiltration of the net irrigation value (T_n).

$$T_o \geq T_n \xrightarrow{\text{In optimal condition}} T_o = T_n \tag{5.1}$$

The hydraulic equations used in the design of level-basin parameters in the SCS method include the continuity and the Manning equations which are based on two factors including:

- First, the volume of water flowing into the basin to cover the basin width is, on average, as much as the gross irrigation requirement (I_g).
- Second, the contact time at the end of the basin is equal to the time required for the infiltration of the net irrigation requirement (I_n) (T_o be equal to T_n).

The first condition determines the gross irrigation requirement and the second condition determines the net irrigation requirement. The difference between these two values in this method is water deep percolation.

Based on the permeability characteristics (permeability group) of soil and suitable roughness coefficient (n) for the irrigated crop, the net irrigation requirement is calculated on the basis of the design and then the following items can be obtained:

1. The length of the basin is determined by knowing the flow rate and assuming the appropriate efficiency.
2. The suitable rate of flow by knowing the length of the basin and the appropriate efficiency.

Irrigation time will also be calculated in both cases.

After the design, it is necessary to calculate the maximum flow rate and the permissible flow rate as design constraints and compare it with the values of depth and flow rate of the design so that it is not more than the permissible limit.

5.2.4 Design Calculation Process of SCS Method

1. **Calculation of net infiltration time, T_n:**

   The contact time required to penetrate the net water depth (I_n) is obtained from the cumulative infiltration ratio:

TABLE 5.1 Values of SCS Infiltration Coefficients in Metric System (time: min; depth: mm) (Burt et al., 1995)

Intake Family	Soil Type	a	b	c
0.05	Clay	0.5334	0.618	7.0
0.10	Clay	0.6198	0.661	7.0
0.15	Light clay	0.7110	0.683	7.0
0.20	Clay loam	0.7772	0.699	7.0
0.25	Clay loam	0.8534	0.711	7.0
0.30	Clay loam	0.9246	0.720	7.0
0.35	Silty	0.9957	0.729	7.0
0.40	Silty	1.064	0.736	7.0
0.45	Silty loam	1.130	0.742	7.0
0.50	Silty loam	1.196	0.748	7.0
0.60	Silty loam	1.321	0.757	7.0
0.70	Silty loam	1.443	0.766	7.0
0.80	Sandy loam	1.560	0.773	7.0
0.90	Sandy loam	1.674	0.779	7.0
1.00	Sandy loam	1.786	0.785	7.0
1.50	Sandy	2.284	0.799	7.0
2.00	Sandy	2.753	0.808	7.0

$$Z = at^b + c \rightarrow I_n = aT_n^b + c \rightarrow T_n = \left[\frac{I_n - c}{a}\right]^{\frac{1}{b}} \quad (5.2)$$

where T_n is the required time or contact time of the net irrigation depth (min) and I_n (or d_n) is the net irrigation depth (mm). In the metric system, the SCS penetration equation coefficients are presented in the relevant tables (Table 5.1). The net irrigation depth (I_n) is calculated based on the water holding capacity of the soil and root depth, as well as the appropriate MAD for the plant.

2. **The relationship between the length of the basin (L) and the inflow rate per unit (Q_u):**
   It is calculated based on the volume balance equation as follows:

$$Q_u = \frac{Q}{W} \quad (5.3)$$

$$L = \frac{Q_u T_L}{Z_a + d_a} = \frac{Q_u T_L}{\frac{aT_L^b}{b+1} + c + d_a} \quad (5.4)$$

where
Z_a is the mean depth of infiltrated water (mm),
L is the basin length (m),
d_a is the mean depth of water flow (mm),
T_L is the advance time (required time to reach the flow to the end of basin) (min),
Q_u is the inflow rate per unit (m³/s/m), and
a, b and c are SCS equation coefficients.

3. **The maximum depth of flow (dmax):**
   It can be calculated in basin using the Manning equation:

$$d_{max} = 2,250 \times n^{\frac{3}{8}} Q_u^{\frac{9}{16}} T_{co}^{\frac{3}{16}} \quad (5.5)$$

where

d_{max} is the maximum depth of flow (mm),
n is the Manning roughness coefficient (–), and
T_{co} is the cutoff time of flow (min).

- In calculating the maximum flow depth, if the irrigation time (T_{co}) is less than the advance time (T_L) (the inlet flow is interrupted before the water reaches the end), in the relation of the maximum depth, T_{co} should be replaced by T_L.
- The ridge height is considered about 25.1 times the d_{max}.
- SCS experience has shown that the average depth of flow is approximately 0.8 of the maximum depth:

$$d_a = 0.8\, d_{max} \quad (5.6)$$

In this way, the length of the basin can be obtained from Eq. (5.7):

$$L = \frac{6 \times 10^4 \times Q_u T_L}{\dfrac{aT_L^b}{b+1} + c + 1798\, n^{\frac{3}{8}}\, Q_u^{\frac{9}{16}}\, T_{co}^{\frac{3}{16}}} \quad (5.7)$$

- The above equation is easily solvable for the specific flow rate and advance time. But for the specified length of the basin as well as the specific advance time, it must be solved by trial and error or numerically.

The continuity equation is used to calculate irrigation time (or cutoff time: T_{co}) for a specific application efficiency (E_a, %) as Eq. (5.8).

$$d_g \times L = Q_u \times T_{co} \rightarrow T_{co} = \frac{I_n L}{600\, Q_u E_a} \quad (5.8)$$

4. **Net required irrigation depth (In):**

$$In = (FC - PWP) * Dr * MAD \quad (5.9)$$

where

FC and PWP are the field capacity and wilting point water content, respectively (cm³/cm³);
Dr is plant root depth (mm), and MAD is management allowed depletion (%).

5. **Manning roughness coefficient (n):**
   In order to estimate the Manning roughness coefficient, the recommended values of the SCS can be used according to Table 5.2.

6. **Irrigation efficiency in basins:**
   To avoid the risk of drainwater, the design is usually conducted in such a way that irrigation efficiency is not less than 80%. This efficiency is obtained if the advance time (time required to cover the basin of water), T_L, is not more than 60% of the time required for the infiltration of the net irrigation requirement, i.e., $T_L < 0.6 T_n$ (as shown in Table 5.3). The efficiency about 70% can only be considered for soils with good drainage status. In regions where water resources are limited or expensive, it is possible to reach even 90% efficiency in the basin. To estimate the efficiency of basin irrigation, SCS has been proposed (Table 5.3). The shorter fraction of the advance time to required time for infiltration of the net irrigation requirement causes the higher the possibility of achieving the efficiency. The denominator of the fraction is fixed. Therefore, water quickly passes the length of the basin and consequently the efficiency is promoted.

TABLE 5.2 SCS-Recommended Values of Manning Roughness Coefficient (Burt et al., 1995)

Cover Type on the Soil Surface	Roughness Coefficient
Bare and smooth basins	$n=0.04$
Basins covered by cereals	$n=0.1$
Basins covered by alfalfa and similar plants	$n=0.15$
Basins covered by sugarbeet and similar plants	$n=0.25$

TABLE 5.3 Basin Irrigation Efficiencies Proposed by SCS Based on the Fraction of the Advance Time to Required Time (Burt et al., 1995)

T_L/T_n	Basin Irrigation Efficiency (%)
0.16	95
0.28	90
0.40	85
0.58	80
0.80	75
1.08	70
1.45	63
1.90	60
2.45	55
3.20	50

In order to achieve uniform water distribution within the basin, the land surface must be level. Otherwise, there will be significant variation in the applied water depth at different locations within the basin. In the sloping lands, farmers typically use small basins as small as 5×5 m, which are nearly level, as in Egypt. However, with the advent of land leveling equipment, particularly laser-leveling equipment, the size of level basins has considerably increased to fields of 5 ha or more, as in Arizona and Australia (Reddy, 2013).

Despite the principles of designing the basin irrigation system, today the use of modeling tools in the evaluation and design of surface irrigation has made significant progress and robust software have been presented in this regard. WinSRFR (Bautista et al., 2009a) and SIRMOD (Walker, 1998) are among them. In the introduction section, various research studies which have been done in this field were mentioned. Therefore, in the continuation of this chapter application of the mentioned two models in the evaluation and design of basin irrigation is described in detail based on experimental data.

5.3 Modern Evaluation and Design of Basin Irrigation by Modeling Tools

5.3.1 WinSRFR Model

Since the late 1970s, the Agricultural Research Institute (USDA) has developed hydraulic simulation models and software related to the analysis of surface irrigation systems. Results in this regard include the development of the one-dimensional model for simulation of furrow, border and basin irrigation, SRFR (Strelkoff et al., 1998), the open-end and closed-end (blocked) border system design tools, Border (Strelkoff et al., 1996) and the basin irrigation system, Basin (Clemmens et al., 1995). Since 2004, a new generation of software called WinSRFR (first version: WinSRFR1.1) has been developed to integrate previous DOS applications (SRFR, Basin and Border) into a Windows application.

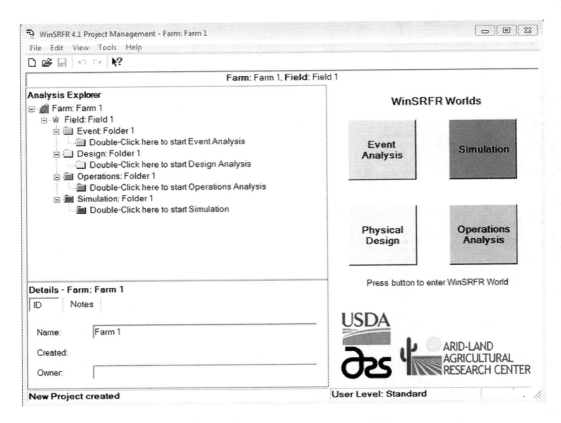

FIGURE 5.2 The main structure of the WinSRFR: Project management window.

WinSRFR software is the most comprehensive and complete surface irrigation hydraulic simulation software, and the latest version of WinSRFR5.1.1 was introduced in December 2019 (https://www.ars.usda.gov) and has features such as simulation, calibration, optimization and simulation of fertigation.

WinSRFR is a software package for analyzing surface irrigation systems based on the unsaturated flow hydraulic model. WinSRFR connects the four sub-programs including Simulation, Event Analysis, Operation Analysis and Physical Design to each other. Users can evaluate and analyze field data, estimate the infiltration properties of the field and evaluate the performance of an irrigation event with Event Analysis tools. A wide range of design and operational alternatives are easily possible with Operation Analysis and Physical Design tools. Simulation creates access to the simulation engine and can be used to test different scenarios or sensitivity analysis. This model also serves as a fund for research and development in surface irrigation hydraulics.

On the main window of this model (Figure 5.2), a project file "Farm," contains the four mentioned tools, each of which can be used depending on the objectives of the project. Farm refers to at least one particular field or land, or can represent a set of theoretical scenarios in the folders. These folders can be accessed via the Analysis Explorer on the left side of the main window. The tree control structure and related interface components create a lot of flexibility in organizing data. A specific scenario is shown in a given section by a window containing various tabs in which the user can enter and edit the data and perform the corresponding analysis and view the output data. In the following, based on the experimental data and measurements of basin irrigation, the application of the four mentioned tools is explained in detail and how to evaluate or design a basin irrigation system using this model is illustrated.

Level-Basin Irrigation

5.3.1.1 Estimation and Calibration of Infiltration Parameters and Roughness Coefficient Using the Event Analysis Tool in the WinSRFR Model

Soil water infiltration is one of the effective hydraulic parameters on surface irrigation and one of the most difficult parameters to be estimated. In fact, the rate of water infiltration into the soil determines the timing of irrigation to store a certain amount of water in the soil and is very important. In order to simulate surface irrigation with the aim of increasing irrigation efficiency and optimal use of water, it is necessary to carefully estimate soil infiltration. If the infiltration parameters are not accurately determined close to the field conditions, it may lead to unnecessary irrigation and water deep percolation and as a result, the irrigation efficiency will be low. On the other hand, imprecise estimation of infiltration may cause water scarcity than the required amount.

Importance of knowing the infiltration equation in simulating of surface irrigation, along with the problems of reliable estimation of this parameter, means spending a lot of time and cost before designing an irrigation system. There are several ways to measure infiltration, and it varies depending on the irrigation conditions. Methods such as double rings do not take into account the dynamic conditions of the field, and various methods have been proposed to better estimate the parameters of infiltration in surface irrigation. These methods include blocked basin, inflow and runoff, Elliott and Walker two-point method, Benami and Ofen advance method, Shepard one-point method, Valiantzas one-point method and multilevel optimization (Ebrahimian et al., 2010; Walker, 2005).

In the WinSRFR model, different infiltration equations were considered, including characteristic infiltration time, Natural Resources and Conservation Service (NRCS) infiltration families (USDA-NRCS, 1997), time-rated intake families, Kostiakov, modified Kostiakov and Branch function, the equations of which are described as follows.

1. *Characteristic infiltration time:* Based on this concept, infiltration can be determined by the time required to reach the required irrigation depth (Bautista et al., 2009b), which is presented as Eq. (5.10) (Strelkoff et al., 1998):

$$Z = kt_c^a \tag{5.10}$$

   where t_c is the infiltration opportunity to provide the required depth of Z.

2. *NRCS infiltration families:* In 1950, the Natural Resources Protection Organization of the U.S. Department of Agriculture proposed infiltration families as a way to classify infiltration using the concept of infiltration behavior in similar soils (Walker et al., 2006; USDA-SCS, 1974). The corresponding infiltration equation is presented as Eq. (5.11):

$$Z = kt^a + c \tag{5.11}$$

   where k and a are the empirical coefficients for different families of infiltration, but the value of c for all families is a constant number (7 mm or 0.28 inches). The NRCS infiltration families (USDA-NRCS, 1997) are widely used in surface irrigation due to their extensive performance. This equation attempts to determine each family with parameters that have a physical basis (Sayah et al., 2016). In this method, soils with the similar values of final infiltration rate, available water capacity and root zone depth are grouped together. This technique can be useful in the first phase of irrigation design where soil characteristics are estimated from similar soils (Gillies, 2008).

3. *Time-rated intake families:* This relationship is similar to the concept of characteristic infiltration time, which is used up to a depth of 100 mm. The t_{100} is the only input data required for this method. In this equation, the power a is calculated using Eq. (5.12), which is an empirical equation (Merriam and Clemmens, 1985):

$$Z = kt_{100}^a \tag{5.12}$$

$$a = 0.675 - 0.2125 \log_{10}(t100) \tag{5.13}$$

where t_{100} is the specific infiltration opportunity to a depth of 100 mm (h) and t is the infiltration time (min).

4. *Kostiakov:* The Kostiakov influence equation (Kostiakov, 1932) is one of the simplest and most common equations that is widely used in irrigation studies (Eq. 5.14):

$$Z = kt^a \tag{5.14}$$

where a (–) and k (mm/min^a) are empirical parameters that are estimated using field measurements or inverse solution. The main problem with this equation is that its derivative (the amount of infiltration) tends to zero during long duration of infiltrations (Smerdon et al., 1988). Therefore, this model is suitable for short periods of time and does not have acceptable efficiencies in long durations of time.

5. *Modified Kostiakov:* Field measurements have shown that the rate of infiltration in some soils tends to be a constant value before the end of irrigation (Smerdon et al., 1988). Since the Kostiakov infiltration equation for long periods of time with a constant amount of infiltration does not have acceptable efficiency, by adding the final infiltration value, it can correct this equation for long periods of time. The obtained equation is named modified Kostiakov or Kostiakov–Lewis (Eq. 5.15) (Walker et al., 2006). In the investigation of the effect of preferential flows on the infiltration, C is also added to the equation (Strelkoff et al., 1998).

$$Z = kt^a + bt + C \tag{5.15}$$

where t is the infiltration time (min), k (mm/min^a), a (–) are empirical parameters and f_0 (mm/min) is the final infiltration rate.

6. **Branch Function:** The modified Kostiakov equation refers to a continuous decrease in the amount of infiltration. In some soils, the rate of infiltration decreases over short periods of time and then reaches a constant value (Bautista et al., 2009b), which is corrected by branch functions (Eq. 5.16) (Clemmens, 1982).

$$Z = C + kt^a \quad t \leq t_b \tag{5.16}$$

$$Z = C_B + bt \quad t > t_b \tag{5.17}$$

where t_b is the time required to reach the final infiltration rate (Bautista et al., 2012).

This equation has two parts: the first part is the same modified Kostiakov infiltration equation that is before reaching to t_b, and the second part is a linear function valid for the infiltration opportunities more than t_b.

Event Analysis in the WinSRFR model for evaluation of irrigation and calibration of the infiltration parameters, and Manning roughness coefficient is done based on four methods including: (1) probe penetration analysis, (2) Merriam–Keller post-irrigation volume balance, (3) Elliot and Walker's two-point method and (4) EVALUE volume balance.

1. *Probe penetration analysis:* This method allows evaluation and analysis of an irrigation event based on measurements of the post-irrigation depth of the infiltration wetting front. This depth is determined by placing a Time-Domain Reflectometry (TDR) probe in the soil profile at different locations of the field. In other words, this depth determines the amount of water that must reach

the plant's root zone to meet the soil water deficit (plant's water needs) and leaching requirements. This method is effective in soils with medium to heavy texture. This method requires measurements of pre-irrigation soil water deficit, root zone's available water capacity and inflow and outflow (for open-ended systems). The outputs of this section are: (1) the depth of applied flow, runoff and infiltrated water; (2) infiltration depth profile; and (3) performance indexes, including uniformity and application efficiency. Since infiltration parameters are not determined using this method, it is not possible to provide practical recommendations for system improvement using this method. For this reason, this method does not have suitable application.

2. *Merriam–Keller post-irrigation volume balance:* The Merriam–Keller method (Merriam and Keller, 1978) is an approach for estimating the final depth of infiltration and infiltration parameters in furrow, border and basin using volumetric balance after irrigation. Infiltration parameters are determined based on farm data and the selected infiltration equation by user. In the first step, one of the infiltration equations is selected depending on the user's selection. Then in the solution section, the inputs of the equations are defined and the unknown parameter is estimated. Table 5.4 shows the parameters that must be defined by the user in each equation and the parameters that the model estimates. In this method, the volume of infiltration is calculated from the difference between the volume of inflow and outflow with the numerical integral of the infiltration depth profile after irrigation. It is necessary to explain that to determine the parameters, a trial-and-error method in the inverse solution should be used. This means that the values for the input parameters are first defined by the user and the value of the parameter is estimated by the model. The input values will be changed until the output values do not significantly differ in various iterations. WinSRFR validates the estimated infiltration function using nonuniform flow simulation. This software compares simulation results with field measurements and produces performance results for event analysis.

3. *Elliot and Walker's two-point method:* The Elliott and Walker (1982) two-point method is applied for estimation parameters k [L^2/T^a] and a [–] of the extended Kostiakov equation that uses the information of two points in the field length (the middle and end of the farm) during the advance phase of the irrigation. It also uses the differences of inflow and outflow measurements to determine the steady infiltration rate (b). This method was primarily developed for use in free-draining and sloping furrow irrigation systems. In addition, this method can be used in sloping borders. The inputs required in this method are the inflow rate, the advance time at two points on the farm (middle and end), the outflow rate, the Manning roughness coefficient and the cross-sectional area of the flow in the furrows.

4. *Analysis based on volumetric equilibrium method based on EVALUE model:* In this method, the volume balance equation is solved according to the inflow rate hydrograph, advance and recession times, flow depths as a function of time and distance and runoff rate (in the case of free-draining systems). The simplest way to estimate parameters (infiltration and roughness) is based on advance and recession data. Having more information causes to promote the estimation accuracy.

TABLE 5.4 Different Types of Infiltration Equations and Their Inputs and Outputs in the Merriam–Keller Post-*Irrigation Volume Balance*

Equation	User-Defined Inputs	Estimated Parameter by the Model
NRCS infiltration families	–	a, k
Time-rated intake families	–	a, k
Characteristic infiltration time	a	k
Kostiakov	a	k
Modified Kostiakov	a, b, c	k
Branch function	a, b, c	k

TABLE 5.5 Geometric Specifications of CRC1 Basin Along with the Advance and Recession Curves Data

Geometric and Flow Specifications		Advance and Recession Curves Data		
Parameter	Value	Distance (m)	Advance Time (hour)	Recession Time (hour)
Basin length, L (m) (Blocked end)	185.93	0	0	7.5
Basin width, W (m)	0.3	15.24	0.033	8.65
Slope, S_{ave} (m/m)	0.00003	45.72	0.2	6.7
Inflow rate, Q (L/s)	1.9	76.2	0.4	7.867
Cutoff time, T_{co} (h)	0.88	106.68	0.617	8.5
Required depth, Z_{req} (mm)	100	137.16	0.867	6.9
		167.64	1.167	8.633
		185.928	1.5	8

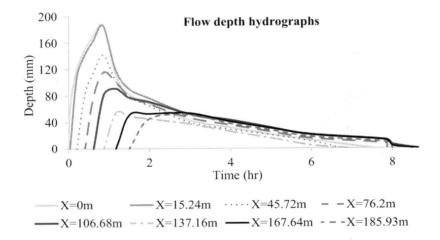

FIGURE 5.3 Flow depth hydrographs in different stations along the CRC1 basin.

Based on the farm information, the surface storage and its coefficient are calibrated. Then, other components of water balance (deep percolation) are calculated. In this method, if the flow depth information is measured at several stations, it will be the most accurate method to estimate the parameters of the infiltration and Manning coefficient.

In this section, the infiltration parameters and the Manning roughness coefficient in a basin with the following characteristics were estimated and calibrated using the Event Analysis tool in the WinSRFR model. These characteristics are related to the measured data of irrigation evaluations conducted on basin irrigation systems, labeled CRC1, presented in the WinSRFR model. In this data series, the geometric specifications of basin along with the advance and recession curves data are presented in Table 5.5. Also, measurements related to the flow depth-time in different stations along the basin are shown in Figure 5.3.

Event Analysis tool include two series of tabs that in the first series (Data Tabs), there are six tabs for entering required data including Start Event, System Geometry, Soil/Crop Properties, Inflow/Runoff, Advance/Recession and Flow Depth. Also, in the second series (Analysis Tabs), calculations related to the volume balance, parameters calibration and analyzing the performance of the irrigation event is done. To enter the required data in the Data Tabs, firstly the type of basin irrigation was selected in the Start Event tab, and the net depth of irrigation requirement was entered (Figure 5.4). Also, EVALUE

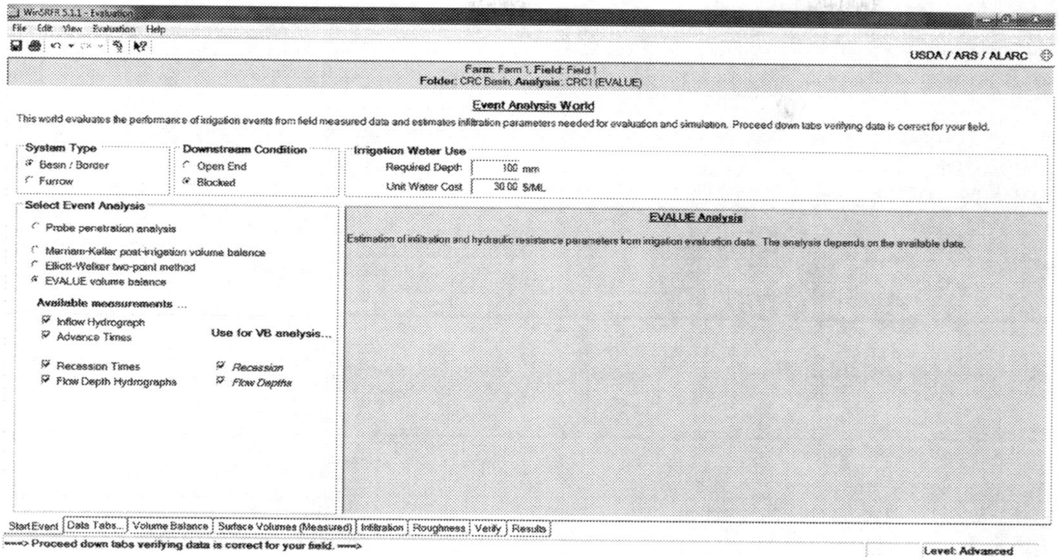

Level-Basin Irrigation

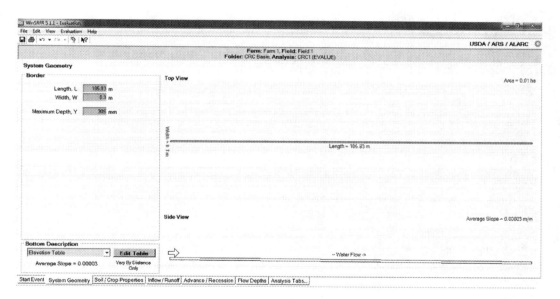

FIGURE 5.4 Start window of the Event Analysis tool in the WinSRFR.

FIGURE 5.5 System geometry tab in the Event Analysis tool of WinSRFR.

volume balance method was selected to estimate the parameters of infiltration equation and roughness coefficient. Furthermore, available data (such as the advance and recession curves, inflow hydrograph and the flow depth at the measured stations) are determined. Then, in the System Geometry tab, the specifications for length, width, maximum height and longitudinal slope of the basin were entered (Figure 5.5). WinSRFR has five options for describing the slope of the bottom:

1. *Slope*: The average longitudinal slope of basin bottom (m/m).
2. *Slope Table*: A table of longitudinal distances with the mean slope values of that distance (which is described in the Edit Table).

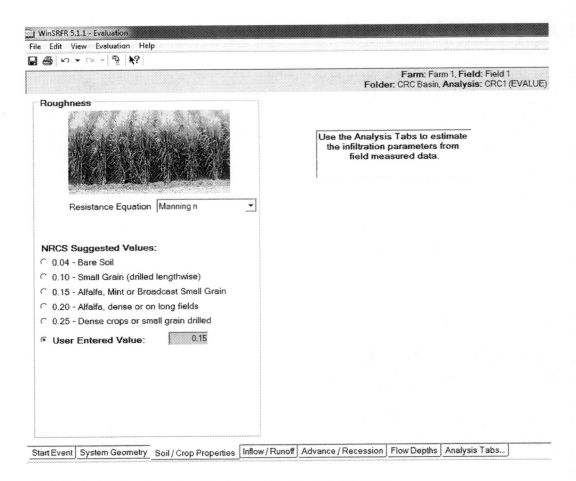

FIGURE 5.6 Soil/Crop Properties tab in the Event Analysis tool of WinSRFR.

3. *Elevation Table*: A table of distance and corresponding elevation. At least for the upstream and downstream, the elevation must be entered, and the downstream distance must be the same as the length of the basin.
4. *Average from Slop Table*: This option calculates the average slope of the tabulated slopes.
5. *Average from Elevation Table*: This option calculates the average slope from the tabulated elevations.

Here, according to the available data, the Elevation Table option is used.

In the next step, the Manning roughness coefficient (n) is defined in Soil Crop Properties tab according to Figure 5.6. In the standard mode, there are two ways to select n: by selecting User Entered Manning n, the Manning roughness coefficient is determined manually; or with selecting NRCS-Recommended n, one of the predefined options for the Manning roughness coefficient can be selected as suggested by the USDA-NRCS Surface Irrigation System Guide as follows:

0.04 for bare soils; 0.10 for small grains (drilled lengthwise);
0.15 for alfalfa, mint or broadcast small grains;
0.20 for alfalfa, dense or on long fields; and
0.25 for dense crops or small grain drilled.

Level-Basin Irrigation

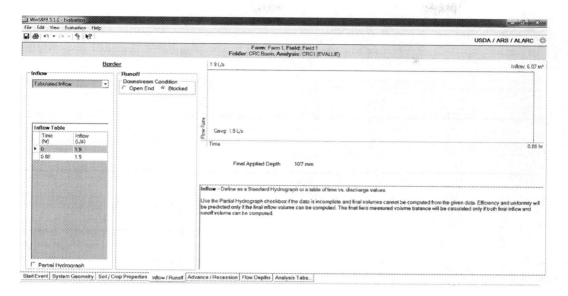

FIGURE 5.7 Inflow/Runoff tab in the Event Analysis tool of WinSRFR.

The Manning roughness coefficient will be calibrated in the next steps.

In the next step, the inflow rate and cutoff time values, along with the downstream conditions (as Blocked), were completed from the Inflow/Runoff tab (Figure 5.7). In this section, the inflow method can be selected using two approaches:

1. *Standard Hydrograph*: This option specifies a simple hydrograph of an inflow rate and cutoff time.
2. *Tabulated Inflow*: In this option, the user can enter a table of time and inflow rate data (hydrograph).

Finally, the data related to the advance and recession curves, as well as the flow depth at the measured stations were entered in the Advance/Recession and Flow Depth tabs, respectively, according to Figures 5.8 and 5.9.

In this example, complete measured data were used, including advance and recession data, flow depth at various stations and outflow hydrograph to calibrate Manning roughness coefficient and infiltration equation parameters. According to the volume balance equation, the volume of the inflow is equal to the sum of the surface storage volume and the subsurface storage volume (Eq. 5.18).

$$Qt = V_y(t) + V_Z(t) \quad t \leq t_{co} \tag{5.18}$$

In the first step of calibration, the surface storage volume is estimated using the EVALUE model and the measured flow depth values.

In the second step, an infiltration equation, such as Kostiakov or Kostiakov–Lewis, is used to estimate the volume of infiltrated water. If Z is considered the desired infiltration equation, then Eq. (5.18) is as follows:

$$Qt = v_y(t) + \int_{s=0}^{s=x} Z(t-t_s)ds \tag{5.19}$$

where

Z is the infiltrated water volume per length unit (L^3/L);
s is the advance path of flow;

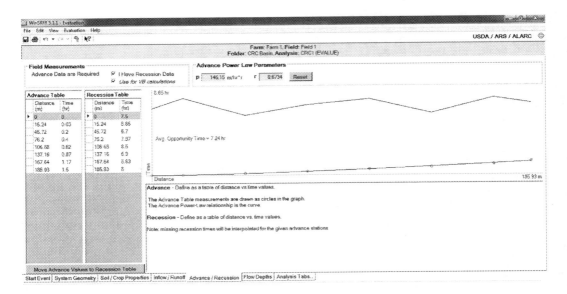

FIGURE 5.8 Advance/Recession tab in the Event Analysis tool of WinSRFR.

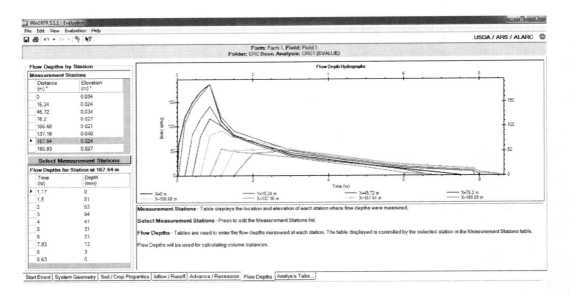

FIGURE 5.9 Flow Depth tab in the Event Analysis tool of WinSRFR.

$x(t)$ is the advance distance of flow in time t;
t_s is the time for the flow to reach the s location;
t is the advance time to the distance x; and
$t-t_s$ is the infiltration opportunity everywhere at a distance of s.

After completing the required information in different tabs, it is possible to analyze the results and estimate the infiltration parameters and roughness coefficient by trial-and-error method in the Analysis Tabs. Thus, in the Infiltration tab, by selecting the infiltration equation, infiltration equation parameters are adjusted manually to minimize the difference between the estimated infiltration volume and real

Level-Basin Irrigation

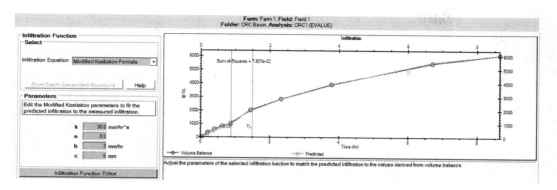

FIGURE 5.10 Infiltration tab in the Event Analysis tool of WinSRFR.

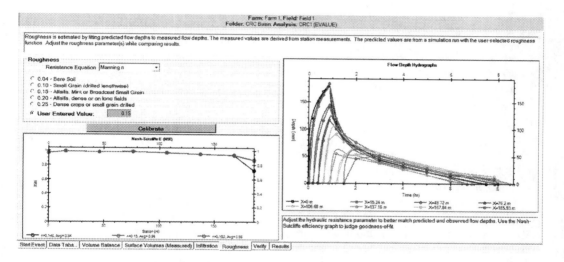

FIGURE 5.11 Roughness tab in the Event Analysis tool of WinSRFR.

measured values (the difference between the volume of inflow and surface storage) according to Eq. (5.19). In this example, considering the equation of Kostiakov–Lewis, the parameters of the equation are adjusted so that, according to Figure 5.10, the predicted values be equal to the measured values. In setting the infiltration equation parameters, each of the coefficients regulates part of the prediction function (orange line). For example, the coefficient a further adjusts the end of the prediction function. The b coefficient has the same effect on all functions and the k coefficient is effective throughout the function, but its effect is greater at the end of the infiltration function.

After calibration of the infiltration equation parameters, the Manning roughness coefficient in the Roughness tab is set to match the measured flow depth at different stations with the model estimation depth. For each value of n, the value of the Nash-Sutcliffe Efficiency (NSE) error index in different stations is estimated as shown in Figure 5.11, so that for a specific n, the NSE values be close to 1 in different stations.

Finally, after calibrating the parameters of infiltration equation and roughness coefficient, it is possible to analyze the performance of the irrigation event. For this purpose, in the Verify tab, by selecting the solution method and the Verify button, a summary of the irrigation performance is displayed in the Results tab. Regarding the solution method, the following explanations are necessary.

In the surface irrigation model, simulation is carried out on the basis of the solution of a pair of equations which is named Saint Venant equation and includes the equations of continuity (Eq. 5.20) and momentum (Eq. 5.21).

$$\frac{\partial Q}{\partial x} + \frac{\partial A}{\partial t} + I_x = 0 \tag{5.20}$$

$$\frac{1}{g}\frac{\partial v}{\partial t} + \frac{V}{g}\frac{\partial V}{\partial X} + \frac{\partial y}{\partial x} = S_o - S_f + \frac{I_x V}{gA} \tag{5.21}$$

where y is the flow depth (m), Q is the flow rate (m³/s), I_x is the infiltration rate (m³/s/m), S_0 is the bottom slope (m/m), S_f is the slope of friction (m/m), g is the gravity acceleration (m/s²), t is the time (s) and V is the velocity of flow (m/s).

To simplify and solve the Saint Venant equation, hypotheses have been used that have led to the following models.

1. *Hydrodynamic model (HD)*: This model is based on the complete solution of the Saint Venant equation. Because this model includes all the components of the Saint Venant equation, including all the sentences of inertia, its calculations are complex and long.
2. *Zero-inertia model (ZI)*: This model was first proposed by Strelkoff and Katopodes (1977). They ignored the inertia factor due to the low water velocity in surface irrigation to simplify the equation as follows:

$$\frac{\partial y}{\partial x} = S_o - S_f. \tag{5.22}$$

This simplification assumes that the forces due to the change in depth with the distance, the slope of the floor and the hydraulic drag are in equilibrium (the sum of the forces is zero and the flow depth is normal in everywhere). Such an assumption is acceptable when there is no water return effect (meaning that it can only be used for systems with open ends). Comparing this method with the Saint Venant equation shows that the accuracy of the results, despite these assumptions, is satisfactory for surface irrigation. Strelkoff and Katopodes (1977) stated that as long as the Froude number is less than 0.2, this model is valid and makes a maximum of 1% error in the calculations compared to the complete HD model. Clemmens and Strelkoff (1979) showed that in surface irrigation farms, the Froude number rarely exceeds 0.3. The accuracy of the ZI model was confirmed by field data on the basin, border and furrow irrigation by Fangmeier and Ramsey (1978) and Elliot and Walker (1982). Comparison of the results of the ZI and HD models by Clemmens and Strelkoff (1979) and Strelkoff (1977), showed that the results of these two models are very close. This model is the most widely used mathematical model of surface irrigation. Its numerical solution is simpler than a complete HD model and its accuracy is very close to the complete HD model and its maximum error is 1% (Bautista et al., 2009b).
3. *Kinematic wave model (KW)*: When the slope of the flow increases, it will be very difficult to apply the ZI formula, and it is necessary to divide the advance path of the flow into small parts at the downstream, where the slope increases fast. At a distance from the recession wave at steep slopes, the depth gradient prevents the balance of forces, and there is no longer a balance between the force at the bottom and the hydraulic drag at the top.

To prevent these problems, KW solution method is developed. Given the uniform flow in surface irrigation and ignoring changes in the flow depth and the inertia factor, the Saint Venant equation will be as follows. Due to the simplified assumptions made in this model, it responds quickly and takes less time to compute.

$$S_o = S_f \tag{5.23}$$

Level-Basin Irrigation

The assumption of uniform flow greatly simplifies analysis but limits the use of this method to sloping basins and borders and free drainage furrows.

WinSRFR performs simulation in one of two ways: (1) ZI and (2) KW.

WinSRFR automatically selects the model that is correct in the given situation, but the user can ignore the choice of KW (meaning that the user can select ZI in cases where the KW method is applicable but cannot determine the choice of the KW when that method is not feasible).

Due to the flatness of the basins, the assumption of uniform flow (KW model) in the hydraulic simulation of the basin is associated with an error, and therefore the ZI model is used in the simulation process.

In this section, the Event Analysis tool is used to estimate the infiltration parameters and roughness coefficients of a basin irrigation, which are the main inputs in the surface irrigation models. The estimated parameters in this section can be used as input in other parts of the software including simulation, operational analysis and physical design sections. It is also possible to compare the performance of different methods of Merriam–Keller, Elliot and Walker and EVALUE volume balance and also the performance of different infiltration equations in this section.

5.3.1.2 Evaluation and Modern Design of Basin Irrigation Using WinSRFR Model

Researches showed that many variables have an important role in increasing the performance and efficacy of basin irrigation. The most important of which are flow variables including inflow rate (Q) and cutoff time (T_{co}) and geometric variables (basin dimensions and longitudinal slope of the bottom) (Bautista et al., 2009b). In cases where it is not possible to change the geometry of the basins (longitudinal slope, width and length of the basin) to increase irrigation efficiency, increasing the irrigation performance is carried out only based on the optimal management of inflow rate and cutoff time. To achieve this goal, using the diagrams of the model performance evaluation indicators [such as irrigation adequacy (AD), application efficiency (AE), uniformity distribution (DU) and deep penetration percentage (DP)] against inflow rate and cutoff time, can help to achieve the best combination of Q and T_{co} with the maximum performance of the irrigation system.

In addition to improving the performance of existing irrigation systems (by changing the management or changing the geometry of the basin), this model can also be used in the design of new irrigation systems. In this section, in order to simulate operation analysis and physical design of basin irrigation system, WinSRFR model is used based on a series of measured basin irrigation data (labeled Basin1) summarized in Table 5.6.

5.3.1.2.1 Evaluation of the Performance of the Basin Irrigation System Using the Simulation Tool in the WinSRFR Model

The first step in the surface irrigation analysis process is to evaluate the performance of the irrigation system, which is based on actual field-measured data. Simulation creates a tool for summarizing, graphically displaying and analyzing field data. The primary purpose of this section is to evaluate

TABLE 5.6 Summary of the Experimented Basin Irrigation Characteristics (Basin 1)

Geometric and Flow Parameters	Value	Infiltration Parameters (Modified Kostiakov)	Value
Basin length, L (m) (Blocked end)	200	K (mm/h$^{a(\cdot)}$)	16.14
Basin width, W (m)	10	a (-)	0.57
Slope, S_{ave} (m/m)	0.001	b (mm/h)	27.5
Manning, n	0.05	c	0
Inflow rate, Q (L/s)	22	Required depth, Z_{req} (mm)	50
Cutoff time, T_{co} (h)	1.4		

the performance of the irrigation system (application efficiency, distribution uniformity and irrigation adequacy) based on actual measured data.

In this section, Simulation tool was used in WinSRFR model to simulate and evaluate the state of basin irrigation in terms of water infiltration along the basin and irrigation efficiency, uniformity and adequacy. The tool includes seven tabs for entering the required data (in the tabs of Start Simulation, System Geometry, Soil/Crop Properties and Inflow/Runoff), a summary of the input data (in the Data Summary tab), model run (in the Execution tab) and displaying the model outputs (in the Results tab). First, in the Start Simulation tab, which is the first tab of this section (Figure 5.12), the type of basin irrigation is selected and the required depth of irrigation (Z_{req}) is entered. Then in the System Geometry tab (Figure 5.13), the specifications for length, width, maximum height and slope of the basin are entered

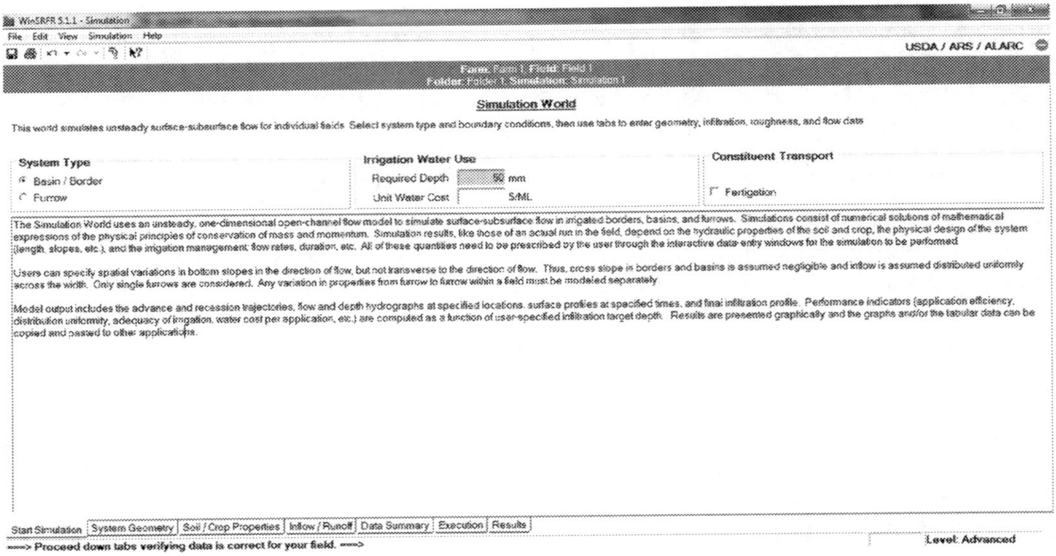

FIGURE 5.12 Start window of the Simulation tool in the WinSRFR.

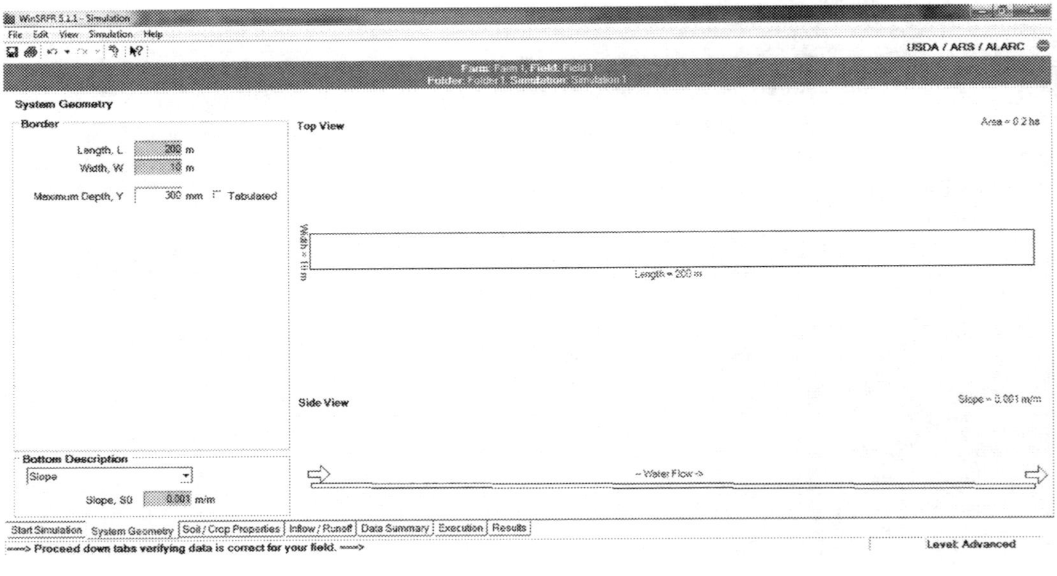

FIGURE 5.13 System Geometry window of the Simulation tool in the WinSRFR.

Level-Basin Irrigation

(this window is similar to the corresponding window in the Event Analysis tool). In order to define the slope of basin, it is possible to introduce several slopes in different distance of basin by selecting various options in Bottom Description (such as Slope Table and Elevation Table).

In the next step, the Manning roughness coefficient and infiltration parameters are completed in the Soil/Crop Properties tab according to Figure 5.14. Various options available in this section to describe infiltration and roughness coefficient are discussed in detail in the Event Analysis section (Section 5.3.1.1). Then, the inflow rate and cutoff time values, along with the downstream conditions, are completed in the Inflow/Runoff tab (Figure 5.15). The data related to the flow cutback (if there is a decreasing

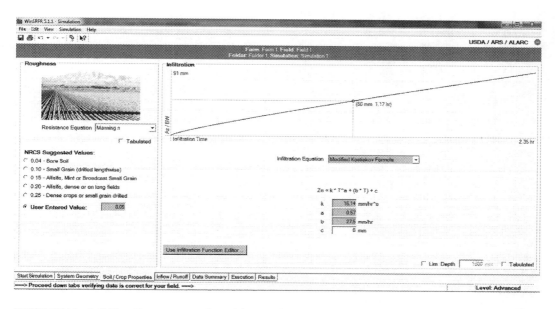

FIGURE 5.14 Soil/Crop Properties window of the Simulation tool in the WinSRFR.

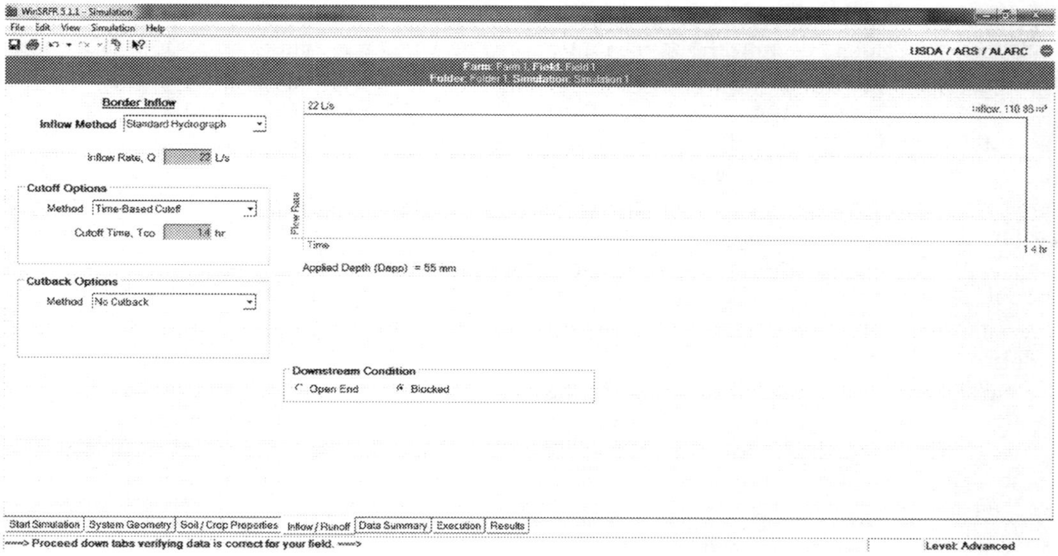

FIGURE 5.15 Inflow/Runoff window of the Simulation tool in the WinSRFR.

flow) must be entered in this section. Finally, by observing the summary of the previous steps input data in the Data Summary tab, the model is run in the Execution tab by selecting the Run Simulation option based on the ZI solution model.

In the Results tab, a summary of irrigation efficiency, advance and recession curves along the basin, inflow and outflow hydrograph, infiltrated water depth during the basin and profile of water depth at ground surface can be observed and analyzed. In Figure 5.16, the advance and recession curves and the depth of the infiltrated water along the basin are shown along with different irrigation efficiencies. According to the simulation results, based on the characteristics of the basin and the irrigation scheduling, the depth of infiltrated water at the end of the basin (about a quarter of the basin length) is less than the required irrigation depth and irrigation deficit prevails. Indicators of irrigation efficiency and uniformity in Figure 5.16c are as follows:

Application efficiency, $AE = 84\%$
Deep percolation fraction, $DP = 16\%$
Distribution uniformity of the minimum, $DU_{min} = 31\%$
Adequacy of the low quarter, $Ad_{lq} = 71\%$.

In order to observe the progress of the flow at the soil surface as well as the infiltrated water depth as an animation over time (Figure 5.17), users can select View Simulation Animation Window from the Simulation tab in the main menu bar of the model.

The summary of the simulation results showed that despite the relatively high irrigation efficiency (84%), the irrigation adequacy is not complete (71% at one quarter of the end of the basin) and the distribution uniformity at the quarter of the end of the basin is also low. Therefore, it is necessary to change the irrigation program into the optimum scheduling. Users can change the design parameters (such as length, width and slope of the basin) and the parameters of irrigation operation (such as inflow rate and cutoff time) to investigate their impact on performance and uniformity indicators, as well as the depth of infiltrated water and irrigation adequacy, and choose the best results. However, the WinSRFR model in the Operation Analysis tool provides the possibility of simultaneous optimization of inflow rate and cutoff time to maximize irrigation efficiency and uniformity along with 100% irrigation adequacy. In the following, this section is discussed.

5.3.1.2.2 Improving Operational Techniques in Basin Irrigation by Operation Analysis Tool in the WinSRFR Model

Operation Analysis World defines the potential irrigation performance under the range of inflow rates and cutoff times for a system with specific dimensions, slope and soil characteristics to identify solutions that satisfy the irrigation requirement. This analysis is performed using performance contours that provide an overview of performance changes as a function of the operational variables (inflow rate, Q and cutoff time, T_{co}). This tool allows the user to search for combinations of decision variables that lead to high levels of uniformity and efficiency, while justifying practical and hydraulic limitations (such as maximum allowed velocity of the flow and proportion of shift hours). The performance contours created by the software consist of performance indicators include application efficiency (AE), minimum distribution uniformity (DU_{min}), runoff (RO) and deep penetration (DP).

In this section, the performance contours analysis of a basin irrigation system with the features provided in Table 5.6 is performed. To obtain the outputs of this section, it is necessary to enter the required inputs, the process of which is very similar to the Simulation section 5.3.1.2.1. In this way, selecting the type of basin irrigation and adjusting the required water depth are done in the first tab, i.e., Start Operation (Figure 5.18), length, width, maximum depth and slope of the basin are determined in System Geometry tab, Manning roughness coefficient and infiltration equation parameters are defined in Soil/Crop Properties tab and downstream conditions (Blocked or Open End) are determined in Inflow/Runoff tab. Finally, the model was run in the Execution tab by setting the Q and T_{co} range (taking into account practical and hydraulic constraints) to draw the performance contours.

Level-Basin Irrigation

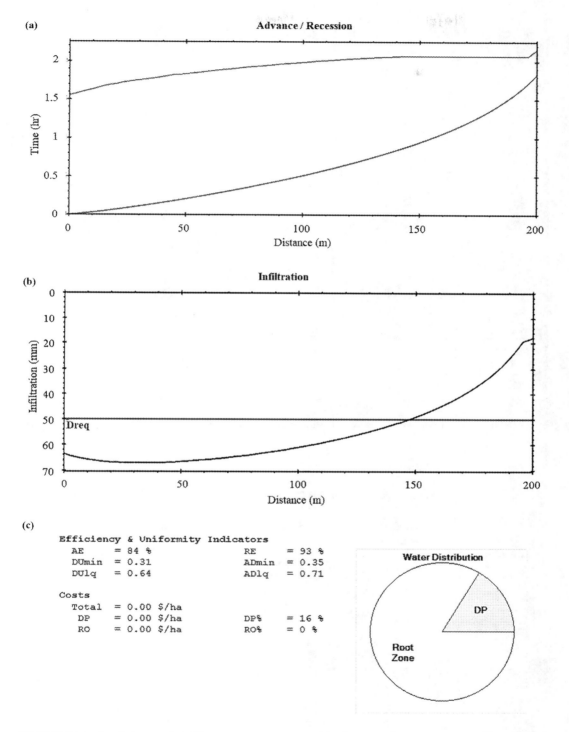

FIGURE 5.16 Simulation results of Basin1 irrigation system: (a) advance and recession curves, (b) infiltrated water depth along the basin and (c) different irrigation efficiencies.

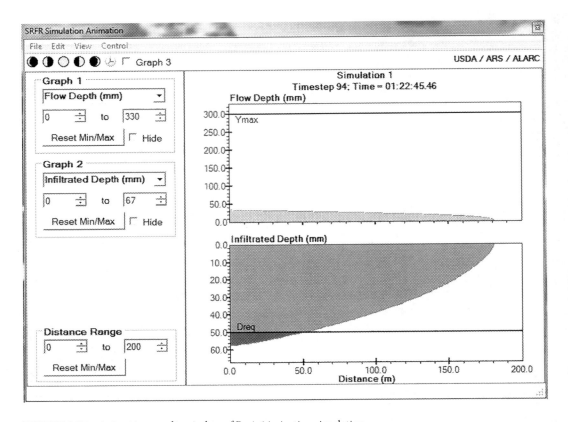

FIGURE 5.17 Animation results window of Basin1 irrigation simulation.

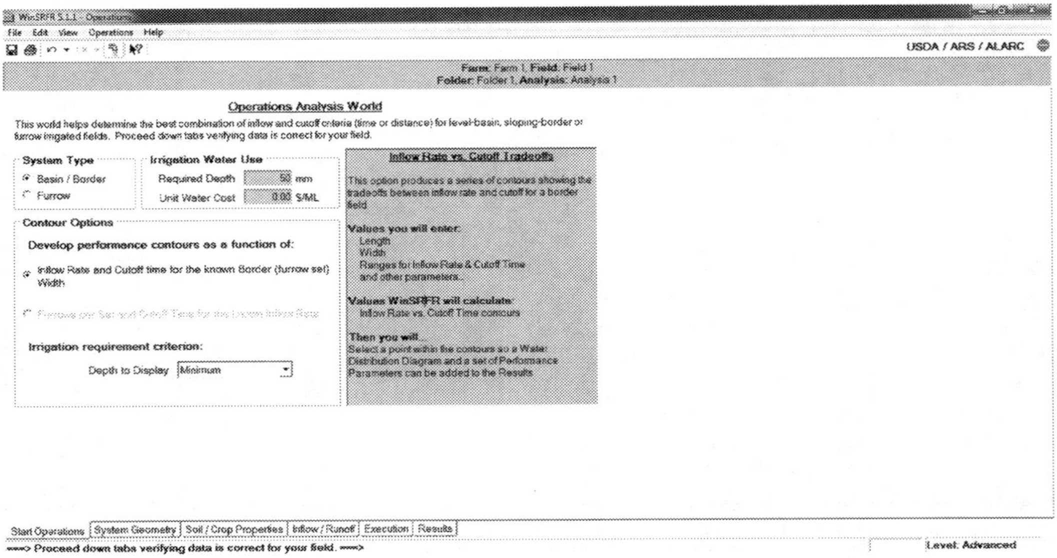

FIGURE 5.18 Start window of the Operations Analysis tool in the WinSRFR.

Level-Basin Irrigation

The Operations Analysis output consists of the performance indicators displayed in several Results tab pages. Figure 5.19a shows application efficiency (AE) contours in which the dotted line ($D_{req} = D_{min}$) crossing through is the locus of points of Q and T_{co} for which minimum applied depth (D_{min}) is equal to the irrigation requirement (D_{req}) and the required irrigation depth is provided during the basin. As expected, the AE is maximum at the bottom left corner of the diagram where basin is faced with deficit irrigation. AE is decreasing with increasing Q and T_{co}. In fact, $T_{co}-Q$ combinations on the left side of the dotted line indicate $D_{min} < D_{req}$, while on the right, the opposite is true. Additional details can be obtained by navigating over the contours with the cursor in which the Q and T_{co} are displayed along with the AE value next to the cursor. Considering any point on this diagram, the infiltration diagram can also be extracted and interpreted in those conditions using the Simulation tool. Figure 5.20 shows the status of different points in this AE contours along with the infiltration diagram. Figure 5.5 shows that in the left side of the dotted line, although it is possible to achieve high application efficiency with some combinations of Q and T_{co}, the required irrigation depth will not be provided (Figure 5.20a and b). As Figure 5.5 shows, water scarcity may occur at the beginning of the basin (Figure 5.20a) or the end of the basin (Figure 5.20b), depending on the different combinations of Q and T_{co}. In the AE contour diagram, there are also combinations of Q and T_{co} that in addition to low irrigation adequacy, deep percolation losses are high and their application efficiency is low (Figure 5.20c and d). Therefore, irrigation should be strictly avoided under such conditions. On the right side of the dotted line, the irrigation efficiency is low and applied water throughout the basin is greater than the required depth (Figure 5.20e and f). Even in some places along the dotted line, there are combinations of Q and T_{co} where application efficiency is very low (Figure 5.20g and h).

The black dot in the left side of the dotted line in the contour region identifies the current condition ($Q = 22$ L/s, $T_{co} = 1.4$ h) with $AE = 84\%$. As discussed in the previous section (Figure 5.14), in the current situation, $D_{min} < D_{req}$ and the entire length of the basin did not receive enough water. Therefore, despite the high application efficiency of this project, irrigation adequacy is low, and it is necessary to optimally manage the inflow rate and cutoff time. In this regard, by selecting $Q = 33$ L/s and $T_{co} = 1$ hour on the dotted line in Figure 5.19a, the irrigation adequacy is increased to 100%, while maintaining the high application efficiency of 84%. This point is an optimal choice for irrigation scheduling of this basin. The infiltration graph is also shown in Figure 5.19c, which is extracted by performing the Simulation tool in this optimal condition. The performance contour of the minimum distribution uniformity (DU_{min}) is also presented in Figure 5.19b. As Figure 5.5 shows, and based on the results of the Simulation tool (Figure 5.16c), in the current situation, the DU_{min} is 31%. However, by selecting the best combination of inflow rate and cutoff time ($Q=33$ L/s and $T_{co}=1$ hour), DU_{min} increased to 83% (Figure 5.19b).

The Operation Analysis tool made possibility of selection of the optimal combination of inflow rate and cutoff time as the operation parameters of basin irrigation to achieve the maximum application efficiency, distribution uniformity and irrigation adequacy. In the design phase of the irrigation systems, many variables affect the performance of the system to achieve a suitable design. However, in operating irrigation systems, the effective variables will be limited to the inflow rate and cutoff time. In this case, the performance contours are considered as a suitable tool to improve the performance of the irrigation system. However, as shown in Figure 5.19a, the application efficiency cannot be increased to more than 84%, in the case of irrigation adequacy of 100%, because basin's physical condition, such as slope, length, width and permeability does not allow this. However, the WinSRFR model in the Physical Design tool has made possibility of changing the design parameters of the basin by simultaneously optimizing the length and width to maximize irrigation efficiency and uniformity with 100% irrigation adequacy. In the following, this section is discussed.

5.3.1.2.3 *Appropriate Design of Basin Irrigation by Physical Design Tool in the WinSRFR Model*

In the final step, the analysis may test irrigation performance under an alternative design, which includes a change in the length and width of the basin (if field conditions allow). For this purpose, the Physical Design World is used to optimize the length and width of the basin. This section is similar to that of

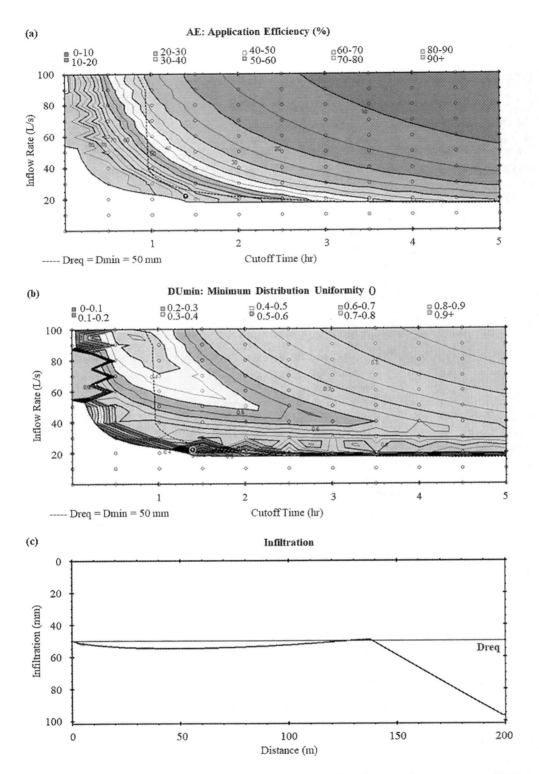

FIGURE 5.19 Operation Analysis results of Basin1 irrigation system: (a) application efficiency contours, (b) minimum distribution uniformity contours and (c) infiltrated water depth profile.

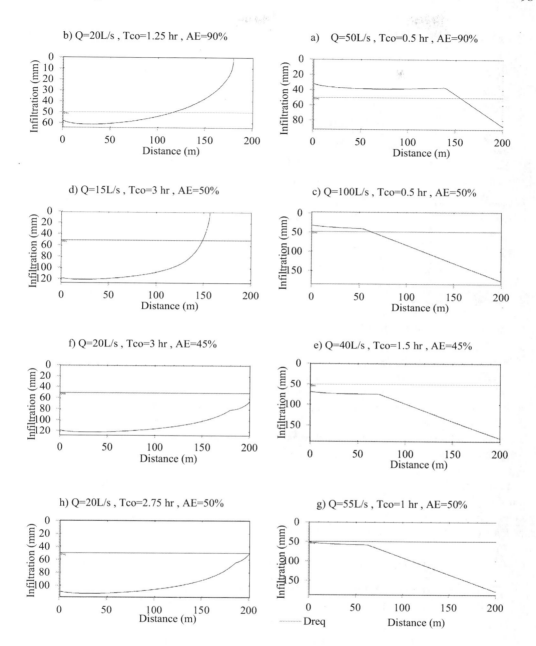

FIGURE 5.20 Infiltration diagram of different points of the AE contours.

operation analysis based on performance contours as a function of decision variables (basin length and width). The process of entering the input data in this section is the same as in the Operation Analysis section. This process starts from selecting the type of field irrigation and adjusting the required water depth in the first tab, i.e., Start Design (Figure 5.21) and after entering the maximum depth and the basin slope in the System Geometry tab, the Manning roughness coefficient and the infiltration equation parameters are entered in the Soil/Crop Properties tab. Then, the inflow rate was determined in the Inflow/Runoff tab, and the downstream flow conditions are selected as Blocked. Finally, the model was run in the Execution tab by setting the range of the length and width of the basin (taking into account the practical and hydraulic limits) to draw the contour lines.

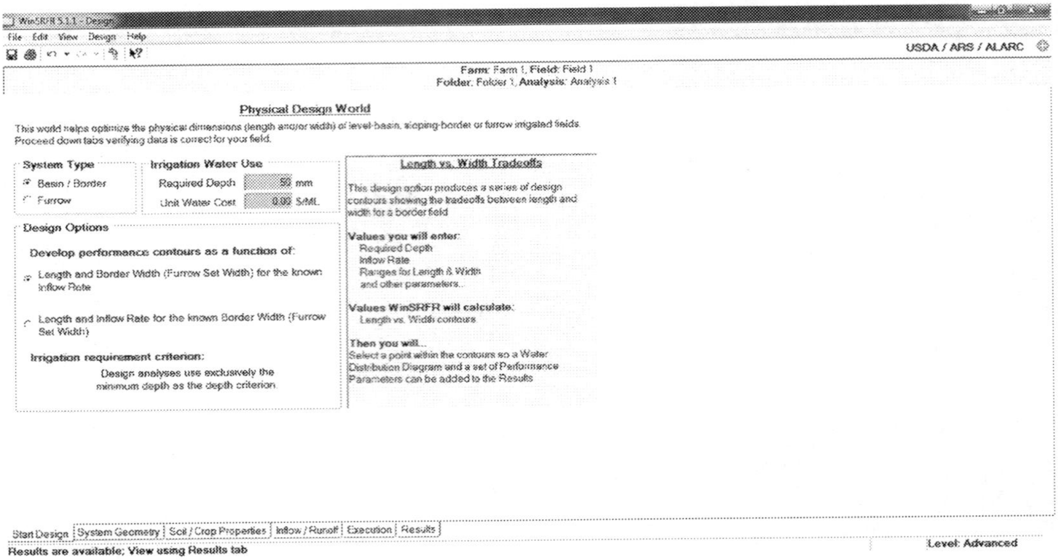

FIGURE 5.21 Start window of the Physical Design tool in the WinSRFR.

The results of the optimization of the basin length and width with the aim of maximizing the application efficiency and distribution uniformity and adequacy of 100% have been shown in Figure 5.22. According to the obtained results from this figure, by selecting $L=50$ m for the basin length and $W=50$ m for the basin width and in the optimized condition of the inflow rate and cutoff time ($Q=33$ L/s and $T_{co}=1$ hour, which were optimized in the previous section), application efficiency increased from 84% to 95% and distribution uniformity increased from 83% to 94%, while irrigation adequacy is 100%. Furthermore, lesser water deep percolation losses in this optimization (Figure 5.22c) are comparable with the results of previous section (Figure 5.19c).

In this section, Physical Design tool is used to optimize the length and width of the basin for a given land slope, hydraulic roughness properties and the desired infiltration depth. If optimized operations do not lead to acceptable performance, changes must be made to the existing design. In cases where potential changes in the slope of the field should be considered as part of the design, separate analyses should be performed by the user. Basin irrigation leveling is a very effective option in increasing irrigation efficiency. To illustrate the importance of this issue, a practical description is provided in the next section.

5.3.1.3 Investigating the Effect of Leveling of Basin Irrigation on Increasing Irrigation Efficiency Using WinSRFR Model

The basis of the level-basin irrigation system is to quickly exert the required amount of water using a maximum (nonerosive) flow rate, and allow the water to reach the end of the basin and infiltrate into the soil. Level-basin irrigation systems are usually used in heavy soils with low permeability and high water holding capacity. In this method, the field is divided into blocks without slope. Basins are filled with the required water depth, and due to the high ridges, the water stays in the basin until it penetrates. Thus, if the basin is flat, the advance and recession curves in the basin will be as shown in Figure 5.23.

If the basin is not flat and has a longitudinal slope or variable elevation, after cutting off the water flow, the water depth at the end of the basin or at the points of the pit will be higher and as a result infiltration in such places will be considerable.

To investigate the effect of leveling on increasing the performance of basin irrigation, a basin is considered with the Table 5.7 specifications which are related to the measured data in WinSRFR model labeled CRC4.

Level-Basin Irrigation

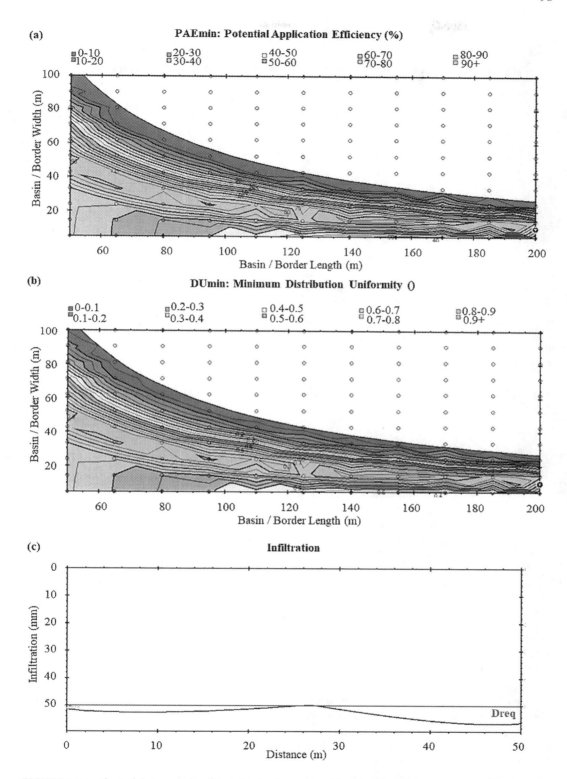

FIGURE 5.22 Physical design results of Basin1 irrigation system: (a) application efficiency contours, (b) minimum distribution uniformity contours and (c) infiltrated water depth profile.

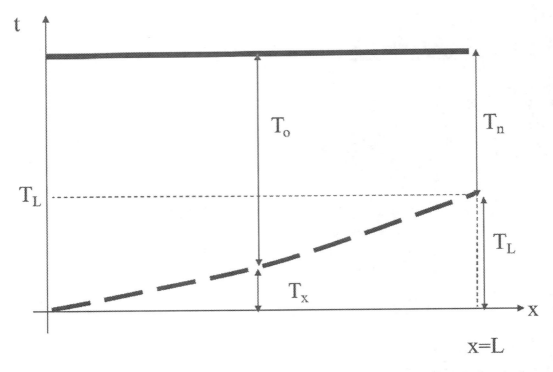

FIGURE 5.23 Schematic shape of the advance and recession curves in level-basin irrigation with a length of L.

TABLE 5.7 Summary of the Examined Basin Irrigation Characteristics (CRC4)

Geometric and Flow Parameters	Value	Infiltration parameters (Modified Kostiakov)	Value
Basin length, L (m) (Blocked end)	188.98	K (mm/h$^{a\,(-)}$)	30
Basin width, W (m)	1	a (-)	0.5
Slope, S_{ave} (m/m)	0.00084	b (mm/h)	7
Manning n	0.17	c	0.5
Inflow rate, Q (L/s)	6.3	Required depth, Z_{req} (mm)	100
Cutoff time, T_{co} (h)	0.62		

Investigation of this condition of basin irrigation in the Simulation section of WinSRFR model shows that the amount of deep penetration losses is 5% and the application efficiency is 95% (Figure 5.24a). But as Figure 5.24a shows, DU_{min} is about 59% and the irrigation adequacy is low. So that the depth of infiltrated water in about 75% of basin length is less than the required depth of irrigation (Figure 5.24b). Therefore, the Operations Analysis tool was used to investigate the possibility of changing the inflow rate and cutoff time to increase irrigation adequacy. The results of this survey are presented in Figure 5.25. To provide the required depth of irrigation along the basin, the pair values of Q and T_{co} should be selected on the dotted line in AE contour plot (Figure 5.25). As shown in this figure, by moving on the dotted line, it is not possible to increase the application efficiency to more than 57% (in different combinations of Q and T_{co}). Meanwhile, the minimum uniformity distribution is 57% in such cases (the results are not shown). Therefore, in this irrigation, due to the low rate of infiltration in the soil, water faster reaches the end of the basin and due to the slope in the basin, it accumulates

Level-Basin Irrigation

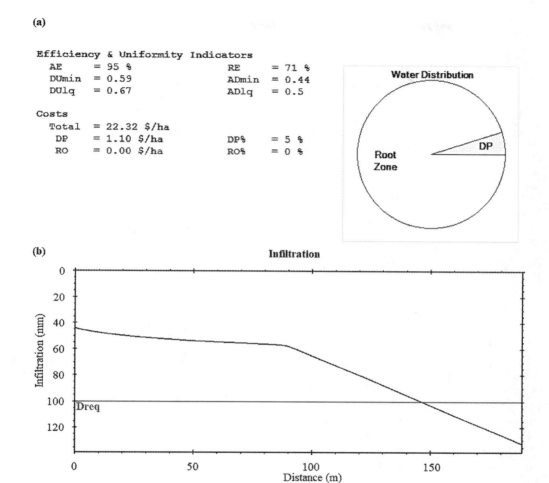

FIGURE 5.24 Simulation results of CRC4 irrigation system with the sloping bottom: (a) irrigation performance and (b) infiltrated water depth profile.

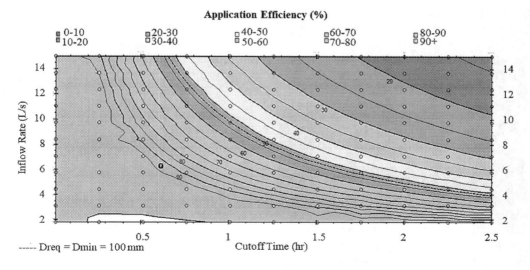

FIGURE 5.25 Operation Analysis results of CRC4 irrigation system with the sloping bottom.

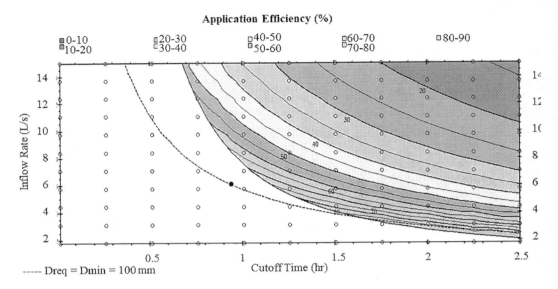

FIGURE 5.26 Operation Analysis results of CRC4 irrigation system with the zero-slope bottom.

at the end of the basin. In this situation, with the continuation of irrigation with the aim of providing the required depth of irrigation, the infiltration of water at the end of the basin increases and causes a lot of deep losses at the end of the basin and reduces the irrigation efficiency. Under these conditions, changing the dimensions of the basin will not improve irrigation efficiency. In such cases, leveling the basin is a good option to increase irrigation efficiency.

The results of the Operation Analysis section on the slope of zero for the conditions of this project are shown in Figure 5.26. With the simultaneous optimization of inflow rate and the cutoff time, it is possible to increase the irrigation efficiency to 91% by selecting $Q=3.5$ L/s and $T_{co}=1.65$ h on a zero slope. By implementing these conditions in the Simulation section, user can also see the result of leveling on increasing irrigation efficiency, distribution uniformity and adequacy of irrigation in Figure 5.27.

5.3.1.4 New Options in the Last Version WinSRFR5.1.1

Recently, a new option has been added to the WinSRFR model to simulate the non-reactive solute transport in surface water flow using the Fertigation option. Fertigation is the practice of distributing fertilizer with the irrigation flow. This modeling tool couples the advection–dispersion equation to the equations of water flow. The simulation predicts the solute transport in the water flow and into the soil as a function of time and distance and its final mass density distribution along the field. To simulate the solute distribution in the soil profile in the surface irrigation, the fertigation simulation is linked to HYDRUS (Šimůnek et al., 2011) model in which water movement in the soil is computed using the Richards equation.

5.3.2 Application of SIRMOD Model in the Evaluation and Design of Basin Irrigation

SIRMOD software has been developed under Windows for designing, evaluating and simulating surface irrigation systems. The first version of the SIRMOD model was developed in 1987 by Walker at Utah State University. This model uses three approaches, namely the full HD, ZI and KW to solve the Saint Venant equations and simulate the hydraulics of surface irrigation (border, furrow and basin) at the field scale. SIRMOD helps in management practices like water application rates and cutoff times and evaluation of alternative field layouts, i.e. field length and slope (Walker, 1998). This model allows the user to specify continuous or surged flow regimes and cutback options.

(a)

```
Efficiency & Uniformity Indicators
   AE    = 91 %         RE    = 100 %
   DUmin = 0.91         ADmin = 1
   DUlq  = 0.94         ADlq  = 1.03
Costs
   Total = 33.00 $/ha
   DP    = 3.09 $/ha    DP%   = 9 %
   RO    = 0.00 $/ha    RO%   = 0 %
```

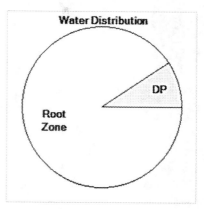

(b)

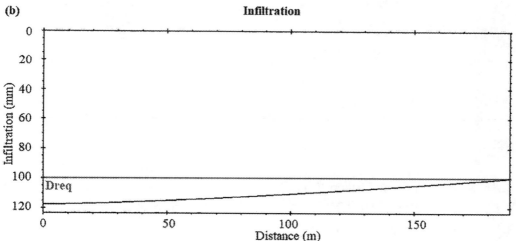

FIGURE 5.27 Irrigation performance (a) and infiltrated water depth profile (b) of leveled CRC4 basin irrigation system.

SIRMOD has a main window or panel (Figure 5.28), which is divided into five sub-windows (Tab), including:

- Field Topography/Geometry
- Inflow Controls
- Infiltration Characteristics
- Hydrograph Inputs
- Design Panel.

For all model applications, entering some of the data in the three sub-windows of Inflow Controls, Field Topography/Geometry and Infiltration Characteristics is necessary. The Hydrograph Inputs tab is optional for entering field data into the model and comparing with the simulation results with goals such as estimation of infiltration parameters. The Design Panel window is also used for design applications.

The following methods can be used to access discussed five main tabs:

- Selecting the Input menu in the model main menu.
- Using the quick selection button in the toolbar menu that will display the main panel.

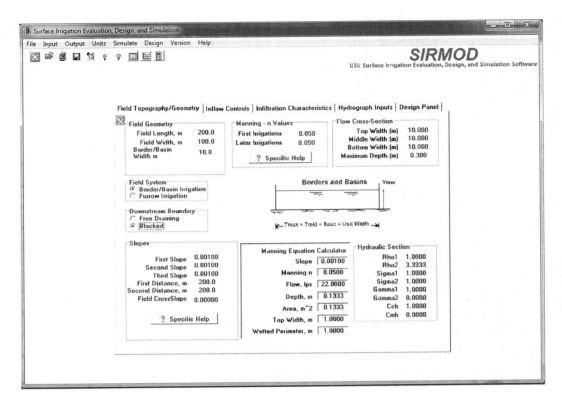

FIGURE 5.28 Main window of SIRMOD model.

In the following, how to enter information and work with the five sub-windows is described. For this purpose, the basin irrigation data presented in Table 5.6 have been used.

5.3.2.1 Field Topography/Geometry

The data entry page for topography and farm geometry is shown in Figure 5.29. In this section, it is necessary to enter the following information:

- *Field system*: includes two options of Border/Basin Irrigation and Furrow Irrigation.
- *Field geometry*: includes farm length and width and basin width.
- *Downstream boundary*: In this section, due to the closed end of the basin and the absence of downstream runoff, the Blocked option was selected.
- *Manning n values*: Freshly plowed soil is usually hydraulically rougher than the surface of soils that have been irrigated and smoothed. On the other hand, the surface of the basin and the strip becomes more hydraulically rough with increasing crops size and density. This model can consider the Manning roughness coefficient for the first irrigation conditions and later irrigations. In this section, by simulating an irrigation event, 0.05 value is considered for this parameter.
- *Field slope*: SIRMOD is able to simulate farms with combined slopes, which can be determined by specifying three slopes and their distance.
- *Flow cross section, hydraulic section and Manning equation calculator*: Two methods are provided in this software to define the geometric characteristics of the flow cross section. In the first method, user can define the geometric characteristics of the flow cross section with four parameters of top width, middle width, bottom width and maximum depth. By entering the values of these parameters, the eight empirical coefficients of the Hydraulic Section (which connect the

Level-Basin Irrigation

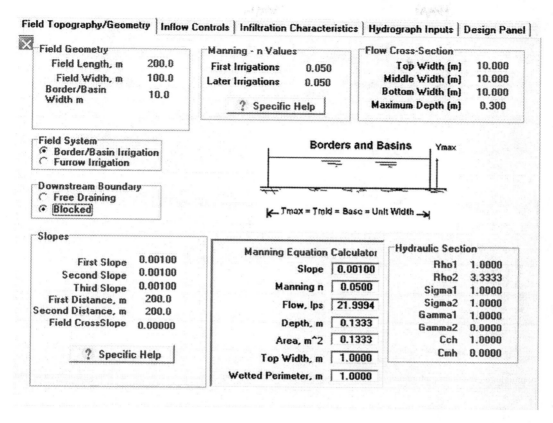

FIGURE 5.29 Field Topography/Geometry window of SIRMOD.

flow depth, flow cross section area, surface width and wetted perimeter) are automatically calculated. Here, where the selected system is a basin, the values of top width, middle width and bottom width are the same and equal to the width of the basin. The user can also enter the slope, Manning roughness coefficient and flow rate values in the Manning equation calculator for calculating flow depth, flow cross section area, surface width and wetted perimeter values.

5.3.2.2 Inflow Controls

In this section, different options should be determined including Simulation Shutoff Control (By Elapsed Time or No. of Surges or By Target Application, Z_{req}), Inflow Regime Control (including Continuous Inflow, Continuous Inflow Hydrograph, Cutback or different status of surge irrigation), Type of Simulation Model (including HD, ZI and KW Kinematic Wave, Zero Inertia and Hydrodynamic), Run Parameters (including Inflow rate, Time of cutoff and Cutback and surge irrigation parameters). Also, in this window (Figure 5.30), it is possible to control the Simulation Speed and Graphic Slope and also Special Numerical Coefficients.

5.3.2.3 Infiltration Characteristics

The infiltration equation used in this model is the Kostiakov–Lewis equation, which requires the coefficients of this equation to be entered in this section (Figure 5.31). For this purpose, if the user intends to do the simulation only for the initial continuous flow, the infiltration parameters should be entered only in the corresponding first column. If the surge irrigation is assessed, the entry of infiltration coefficients in the first and third columns is required. The user can also simulate the next irrigation conditions by changing the options.

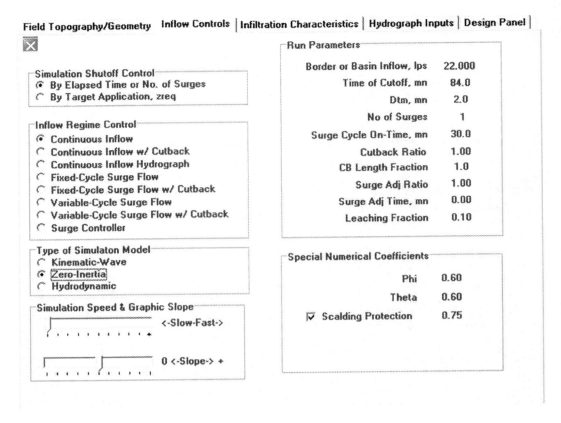

FIGURE 5.30 Inflow Controls window of SIRMOD.

Below the infiltration parameters, the Qinfilt (Reference Discharge) parameter shows the amount of flow on which the infiltration parameters are determined. If the user does not know its value, the flow rate used in the simulation must be entered in this box.

In this window, the software calculates the amount of required intake opportunity time (t_{req}) according to the infiltration equation by entering the amount of root zone moisture depletion (Z_{req}).

If the infiltration equation coefficients are not predetermined, the model considers the following three methods to determine them:

I. It is possible to use four buttons called Tables below the infiltration parameters to select data from the default infiltration tables based on the USDA-SCS adjusted infiltration family curves.
II. Using the Two-Point method [explained in Section 5.3.1.1 (Elliott and Walker, 1982)].
III. Using of multilevel method (Walker, 2005).

5.3.2.4 Surface Irrigation Simulation

When the data from the previous three steps are entered, the simulation is performed by clicking on the button from the toolbar menu or selecting the Simulation option from the main menu bar. With the start of the simulation, the simulation screen appears and various plots of advance and recession profiles, runoff hydrograph (which is not present in basin irrigation), infiltrated water depth and required depth (Z_{req}) are drawn as animation (Figure 5.32). If the infiltrated depth exceeds the required depth, the user can see the loss of water deep percolation. Also, at the bottom right of the screen, a summary of the performance indices of the simulated irrigation event can be observed such as application

Level-Basin Irrigation

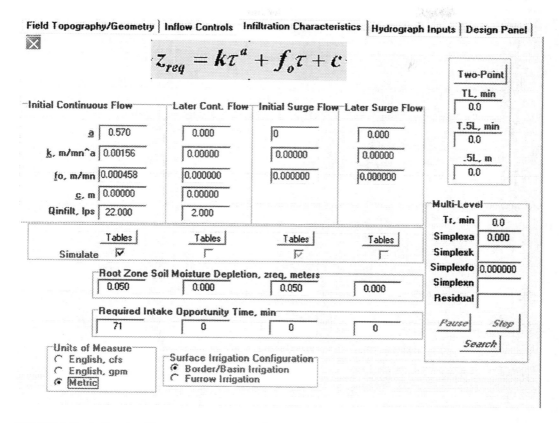

FIGURE 5.31 Infiltration Characteristics window of SIRMOD.

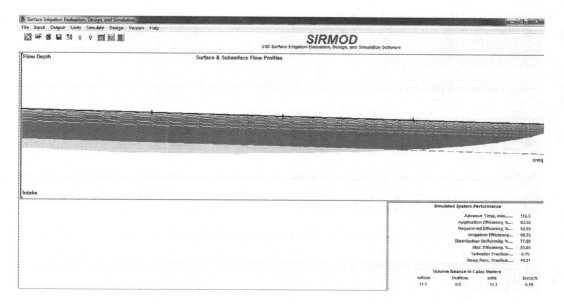

FIGURE 5.32 Simulation results of Basin1 irrigation system in SIRMOD.

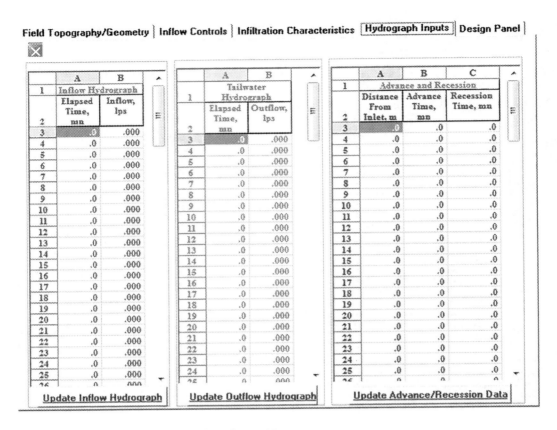

FIGURE 5.33 Hydrograph Inputs window of SIRMOD.

efficiency, distribution uniformity, requirement efficiency, deep percolation fraction and tailwater fraction (Figure 5.32). In addition, the four numbers at the bottom of the mentioned area show the results of the volumetric balance of the simulated flow, so that the error percentage indicates the difference between the input flow, infiltration and runoff (if the farm end is not closed). Errors less than 5% are acceptable.

5.3.2.5 Hydrograph Inputs

In this window, user can enter the data related to inflow and tailwater hydrographs and advance and recession curves (Figure 5.33). If these data are available, the measured and simulated values can be compared after the simulation process, and the infiltration parameters can be calibrated by trial and error. In setting the infiltration equation parameters, each of the coefficients regulates some of the prediction variables. In this regard, at first, the coefficients a and k are estimated by comparing measured and simulated advance and recession curves, and then the coefficient f_0 is estimated by adapting the tailwater hydrographs.

5.3.2.6 Design Panel

In this section, it is possible to observe the Field Layout at the right side of this window and total available flow, total time flow, maximum water velocity, design flow and cutoff time in the section of Input Data for Design (Figure 5.34). By setting these operation and design parameters, and then, by pressing the Simulate Design button, the results of the irrigation system's performance are displayed in the

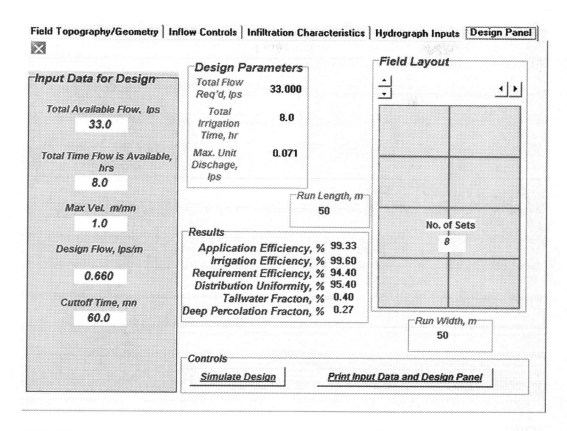

FIGURE 5.34 Optimum design results of Basin1 irrigation system using SIRMOD.

TABLE 5.8 Comparison of the Current Status of the Basin 1 Irrigation System with Optimum Designed Results of the SIRMOD Model

	Q (L/s)	T_{co} (h)	L (m)	T (m)	AE (%)	RE (%)	DU (%)
Current condition	22	1.4	200	10	83.50	92.59	77.68
Optimum design	33	1	50	50	99.33	94	95

Results section in this window (such as application efficiency, distribution uniformity, requirement efficiency, deep percolation fraction, tailwater fraction). In this regard, whenever the numbers in the Design Parameters are red, it means that the division of the field in the Field Layout is not correct, and if both are black, it means that the division is correct. Changing design and management parameters in the trial-and-error method continues until a design with the desired efficiency is obtained.

The results of the optimal physical design of the specified basin irrigation are shown in Figure 5.34. Also, the results of comparing the current implementation and optimal design of the irrigation system are indicated in Table 5.8. As the results of this table show, by optimal designing and management of this irrigation (including changing the two operation parameters of Q and T_{co} and two design parameters of basin length and width), irrigation efficiency is increased from 88.5% to about 99%, and the distribution uniformity is enhanced from 77.68% to 95% and irrigation adequacy is 94%. In this section, the effect of land slope change on increasing irrigation efficiency can be also examined by changing the maximum water velocity that is affected by the basin slope.

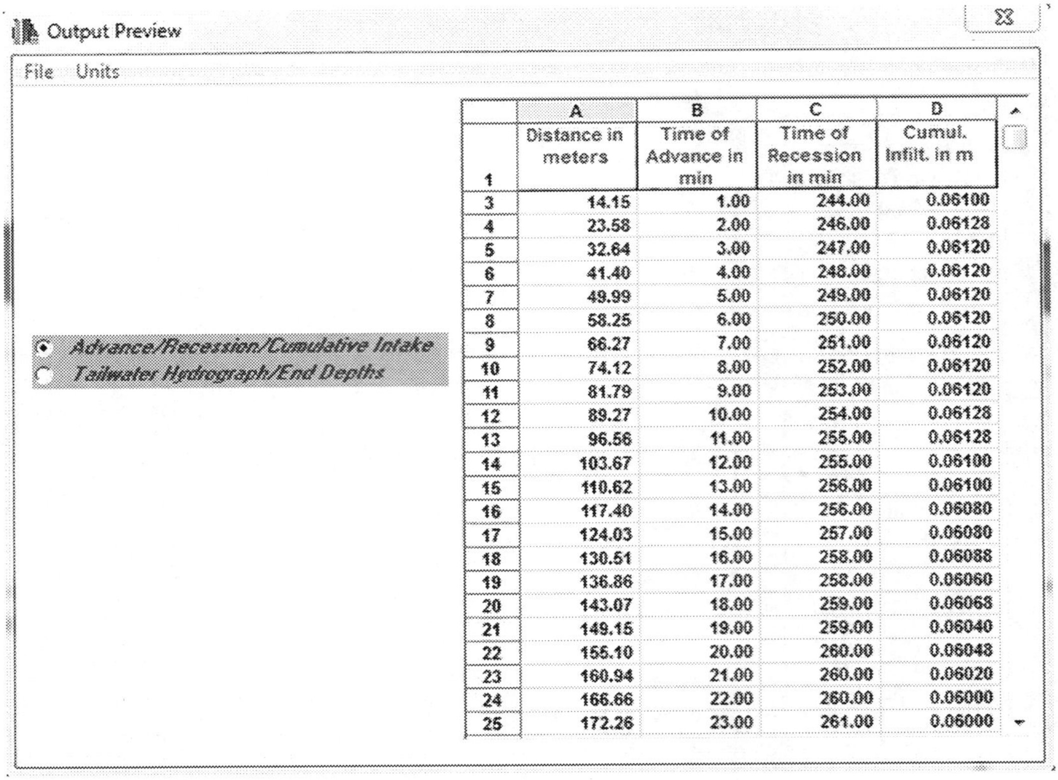

1	A Distance in meters	B Time of Advance in min	C Time of Recession in min	D Cumul. Infilt. in m
3	14.15	1.00	244.00	0.06100
4	23.58	2.00	246.00	0.06128
5	32.64	3.00	247.00	0.06120
6	41.40	4.00	248.00	0.06120
7	49.99	5.00	249.00	0.06120
8	58.25	6.00	250.00	0.06120
9	66.27	7.00	251.00	0.06120
10	74.12	8.00	252.00	0.06120
11	81.79	9.00	253.00	0.06120
12	89.27	10.00	254.00	0.06128
13	96.56	11.00	255.00	0.06128
14	103.67	12.00	255.00	0.06100
15	110.62	13.00	256.00	0.06100
16	117.40	14.00	256.00	0.06080
17	124.03	15.00	257.00	0.06080
18	130.51	16.00	258.00	0.06088
19	136.86	17.00	258.00	0.06060
20	143.07	18.00	259.00	0.06068
21	149.15	19.00	259.00	0.06040
22	155.10	20.00	260.00	0.06048
23	160.94	21.00	260.00	0.06020
24	166.66	22.00	260.00	0.06000
25	172.26	23.00	261.00	0.06000

FIGURE 5.35 Output Preview window in SIRMOD model.

5.3.2.7 SIRMOD Outputs

In the SIRMOD, there are tools for graphical display as well as output printing. To display the outputs, user can select Output in the main menu, which includes three sections: Display Output Results, Plotted Results and Print Input Data. Selecting the Display Output Results option opens the Output Preview window in the form shown in Figure 5.35. It is also possible to access this window via the quick button. In this window, two tables including Advance/Recession/Cumulative Intake and Tailwater Hydrogaph/End Depths are displayed.

In the Plotted Results in the Output tab of the main menu (or the quick selection button) there is two tabs including Files and Current Data Plot Options. In the Current Data Plot Options tab, it is possible to display graphical data including Advance Data, Tailwater Data, End Depth Data and Applied Depth Data. Open Existing Output Files and Save Outputs are also possible in the Files tab.

In the Print Input Data tab of the Output menu, user can print input data using a computer printer.

5.4 Application of Optimization Algorithms in the Design and Management of Basin Irrigation

In recent years, optimization tools have been used by researchers to optimize various issues. In addition to the direct use of surface irrigation models in the management and design of basin irrigation, which was reviewed in the previous sections, by linking surface irrigation models to optimization programs, they can be used in the management and optimal design of surface irrigation as single and multi-objective programs. Garemohamadlou et al. (2020) used WinSRFR link to the genetic and Particle Swarm

Optimization (PSO) algorithm in the estimation of infiltration parameters and multi-objective optimization of closed-end border irrigation performance in the Ramshir irrigation and drainage network. Such a procedure can also be used in basin irrigation and is recommended.

5.5 Conclusion

In the world, most of the agricultural lands are under surface irrigation. Basin irrigation, a level field with ridge in its perimeter, is the most usual form of surface irrigation in heavy soils' textures with moderate to low infiltration rate, especially in regions with the small and/or irregular fields. Water scarcity problems, especially in arid and semi-arid regions, necessitate agricultural water management. Due to existence of many effective parameters on surface irrigation, it is necessary to use modeling tools which can simultaneously manage the effective parameters and provide a proper management and design framework of irrigation system with maximum water efficiency. In this chapter, surface irrigation simulation models were utilized in modern evaluation, management and design of basin irrigation system based on experimental data. The results showed that the WinSRFR model can be used to select the optimal combination of inflow rate (Q) and cutoff time (T_{co}) as the operation parameters and basin length (L) and width (W) as design parameters of basin irrigation for obtaining the maximum application efficiency, distribution uniformity and irrigation adequacy. According to the obtained results, by simultaneously selecting the optimum values of $Q=33$ L/s, $T_{co}=1$ hour, $L=50$ m and $W=50$ m, the irrigation adequacy, minimum distribution uniformity and application efficiency are increased from 71% to 100%, 31% to 94% and 84% to 95%, respectively. Furthermore, in the sloping basin with high deep percolation losses at the end of the basin, leveling of basin bottom increased irrigation efficiency from 57% to 91% with 100% irrigation adequacy. Therefore, simulation models are a useful tool for basin irrigation design and management to achieve the best performance of irrigation.

References

Albaji, M., Eslamian, S., Naseri, A. and F. Eslamian, 2020, Handbook of Irrigation System Selection for Semi-Arid Regions, Taylor and Francis, CRC Group, USA, 317 Pages.

Anwar, A. A., Ahmad, W., Bhatti, M. T., and Haq, Z. U. (2016). The potential of precision surface irrigation in the Indus Basin Irrigation System. *Irrigation Science*, 34(5), 379–396.

Azad, N., Rezaerdinejad, V., Besharat, S., Behmanesh, J., and Sadraddini, A. A. (2018). Optimization of surge irrigation system based on irrigation and furrow geometric variables using SIRMOD model. *Journal of Water and Irrigation Management*, 7(1), 151–166. (In Persian)

Bai, M. J., Xu, D., Li, Y. N., and Pereira, L. S. (2011). Impacts of spatial variability of basins microtopography on irrigation performance. *Irrigation Science*, 29(5), 359–368.

Bautista, E., Clemmens, A. J., Strelkoff, T. S., and Schlegel, J. (2009a). Modern analysis of surface irrigation systems with WinSRFR. *Agricultural Water Management*, 96(7), 1146–1154.

Bautista, E., Clemmens, A. J., Strelkoff, T. S., and Niblack, M. (2009b). Analysis of surface irrigation systems with WinSRFR—Example application. *Agricultural Water Management*, 96(7), 1162–1169.

Bautista, E., Strelkoff, T. S., and Clemmens, A. J. (2012). Errors in infiltration calculations in volume-balance models. *Journal of Irrigation and Drainage Engineering*, 138(8), 727–735.

Booher, L. J. (1974). *Surface Irrigation*. FAO, Rome, Italy.

Brouwer, C., Prins, K., Kay, M., and Heibloem, M. (1988). *Irrigation water management: irrigation methods*. FAO Land and Water Development Division, Rome, Italy.

Burt, C., Appling, D. H., and Norton, L. (1995). *The surface irrigation manual: A comprehensive guide to design and operation of surface irrigation systems*. CA: California, Waterman Industries, Inc. United States of America.

Chen, B., Ouyang, Z., Sun, Z., Wu, L., and Li, F. (2013). Evaluation on the potential of improving border irrigation performance through border dimensions optimization: A case study on the irrigation districts along the lower Yellow River. *Irrigation Science*, 31(4), 715–728.

Clemmens, A. J. (1982). Evaluating infiltration for border irrigation models. *Agricultural Water Management*, 5(2), 159–170.

Clemmens, A. J., Dedrick, A. R., and Strand, R. J. (1995). BASIN—a computer program for the design of level-basin irrigation systems, version 2.0, WCL Report 19. USDA-ARS U.S. Water Conservation Laboratory, Phoenix, AZ, USA.

Clemmens, A. J., and Strelkoff, T. (1979). Discussion of Strelkoff and Katapodes (1977). *Journal of Irrigation and Drainage Division, ASCE*, 104(IR3), 337–339.

Ebrahimian, H., Liaghat, A., Ghanbarian-Alavijeh, B., and Abbasi, F. (2010). Evaluation of various quick methods for estimating furrow and border infiltration parameters. *Irrigation Science*, 28(6), 479–488.

Elliott, R. L., and Walker, W. R. (1982). Field evaluation of furrow infiltration and advance functions. *Transactions of the ASAE*, 25(2), 396–400.

Fabião, M. S., Gonçalves, J. M., Pereira, L. S., Campos, A. A., Liu, Y., Li, Y. N., and Dong, B. (2003). Water saving in the Yellow River Basin, China. 2. Assessing the potential for improving basin irrigation. *Agricultural Engineering International: CIGR Journal of Scientific Research and Development*. Manuscript LW 02 008.

Fangmeier, D. D., and Ramsey, M. K. (1978). Intake characteristics of irrigation furrows. *Transactions of the ASAE*, 21(4), 696–0700.

Garemohamadlou, H., Rezaverdinejad, V., Lalezari, R., and Azad, N. (2020). Muli-objective optimization performance of closed-end border irrigation using WinSRFR and genetic algorithm (Case Study: Ramshir Irrigation and Drainage Network). *Journal of Soil and Water Research*, 51(2), 427–440. (In press)

Gillies, M. H. (2008). Managing the effect of infiltration variability on the performance of surface irrigation, Doctoral dissertation, University of Southern Queensland. Australia.

González, C., Cervera, L., and Moret-Fernández, D. (2011). Basin irrigation design with longitudinal slope. *Agricultural Water Management*, 98(10), 1516–1522.

Khanna, M., and Malano, H. M. (2006). Modelling of basin irrigation systems: A review. *Agricultural Water Management*, 83(1–2), 87–99.

Khanna, M., Malano, H. M., Fenton, J. D., and Turral, H. (2003). Design and management guidelines for contour basin irrigation layouts in southeast Australia. *Agricultural Water Management*, 62(1), 19–35.

Kostiakov, A. N. (1932). On the dynamics of the coefficient of water percolation in soils and the necessity of studying it from the dynamic point of view for the purposes of amelioration. *Transactions on 6th congress international soil science, Russian Part A*, 1, 7–21.

Merriam, J. L., and Clemmens, A. J. (1985). Time rated infiltrated depth families. In Proceedings of the Development and Management Aspects of Irrigation and Drainage Systems, pp. 67–74, ASCE.

Merriam, J. L., and Keller, J. (1978). *Farm irrigation system evaluation: A guide for management*. Agricultural and Irrigation Engineering Department, Utah State University, Logan, UT, USA, 271 p.

Miao, Q., and Shi, H. (2017). Field assessment of basin irrigation performance in Hetao, Inner Mongolia. *Irrigation & Drainage Systems Engineering*, 6(3), 193.

Miao, Q., Shi, H., Gonçalves, J. M., and Pereira, L. S. (2015). Field assessment of basin irrigation performance and water saving in Hetao, Yellow River basin: issues to support irrigation systems modernisation. *Biosystems Engineering*, 136, 102–116.

Miao, Q., Shi, H., Gonçalves, J. M., and Pereira, L. S. (2018). Basin irrigation design with multi-criteria analysis focusing on water saving and economic returns: application to wheat in Hetao, Yellow River basin. *Water*, 10(1), 67.

Pereira, L. S., Gonçalves, J. M., Dong, B., Mao, Z., and Fang, S. X. (2007). Assessing basin irrigation and scheduling strategies for saving irrigation water and controlling salinity in the upper Yellow River Basin, China. *Agricultural Water Management*, 93(3), 109–122.

Playán, E., and Martínez-Cob, A. (1999). Simulation of basin irrigation scheduling as a function of discharge and leveling. *Investigación Agraria: Producción y Protección Vegetal*, 14(3), 545–554.

Playán, E., Walker, W. R., and Merkley, G. P. (1994a). Two-dimensional simulation of basin irrigation. I: Theory. *Journal of Irrigation and Drainage Engineering*, 120(5), 837–856.

Playán, E., Walker, W. R., and Merkley, G. P. (1994b). Two-dimensional simulation of basin irrigation. II: Applications. *Journal of Irrigation and Drainage Engineering*, 120(5), 857–870.

Playán, E., Faci, J. M., and Serreta, A. (1996a). Modeling microtopography in basin irrigation. *Journal of Irrigation and Drainage Engineering*, 122(6), 339–347.

Playán, E., Faci, J. M., and Serreta, A. (1996b). Characterizing microtopographical effects on level-basin irrigation performance. *Agricultural Water Management*, 29(2), 129–145.

Reddy, J. M. (2013). Design of level basin irrigation systems for robust performance. *Journal of Irrigation and Drainage Engineering*, 139(3), 254–260.

Rezaverdinejad, V., Ahmadi, H., Hemmati, M., and Ebrahimian, H. (2016). Evaluation and comparison of different approaches of infiltration parameters estimation under different furrow irrigation systems and inflow regimes. *Journal of Water and Soil Science (Science & Technology of Agriculture and Natural Resources)*, 20(76), 161–176. (In Persian)

Rezaverdinejad, V., and Norjoo, A. (2014). Optimization of furrow irrigation performance using WinSRFR under furrow firming conditions of sugarbeet cultivation. *Journal of Water and Soil*, 27(6), 1281–1293. (In Persian)

Santos, F. L. (1996). Evaluation and adoption of irrigation technologies: management-design curves for furrow and level basin systems. *Agricultural Systems*, 52(2–3), 317–329.

Sayah, B., Gil-Rodríguez, M., and Juana, L. (2016). Development of one-dimensional solutions for water infiltration. Analysis and parameters estimation. *Journal of Hydrology*, 535, 226–234.

Šimůnek, J., van Genuchten, M. T., and Šejna, M. (2011). The HYDRUS software package for simulating two- and three-dimensional movement of water, heat, and multiple solutes in variably-saturated media. In: Technical Manual, Version 2. PC Progress, Prague, Czech Republic.

Smerdon, E. T., Blair, A. W., and Reddell, D. L. (1988). Infiltration from irrigation advance data. I: theory. *Journal of Irrigation and Drainage Engineering*, 114(1), 4–17.

Strelkoff, T. S. (1977). Algebraic computation of flow in border irrigation. *Journal of the Irrigation and Drainage Division*, 103(3), 357–377.

Strelkoff, T., and Katapodes, N. D. (1977). Border irrigation hydraulics with zero-inertia. *Journal of the Irrigation and Drainage Division*, 103(3), 325–342.

Strelkoff, T. S., Clemmens, A. J., and Schmidt, B. V. (1998). SRFR, Version 3.31—A model for simulating surface irrigation in borders, basins and furrows. US Department of Agriculture Agricultural Research Service, U.S. Water Conservation Laboratory, Phoenix, AZ, USA.

Strelkoff, T. S., Clemmens, A. J., Schmidt, B. V. and Slosky, E. J. (1996). BORDER—A Design and Management Aid for Sloping Border Irrigation Systems. WCL Report 21. US Department of Agriculture Agricultural, USA.

Taghizadeh, Z., Rezaei, H., and Verdinejad, V. R. (2016). Comparison and improvement of surface irrigation design methods (Case study furrow irrigation). *Water and Soil Science*, 26(1/4), 53–66. (In Persian)

Taghizadeh, Z., Verdinejad, V. R., Ebrahimian, H., and Khanmohammadi, N. (2013). Field evaluation and analysis of surface irrigation system with WinSRFR (Case study furrow irrigation). *Journal of Water and Soil*, 26(6), 1450–1459. (In Persian)

USDA-SCS (US Department of Agriculture, Soil Conservation Service), 1974. National Engineering Handbook. Section 15. Border Irrigation. National Technical Information Service, Washington, DC, USA, Chapter 4.

USDA-NRCS (US Department of Agriculture, Natural Resources and Conservation Service), 1997. National Engineering Handbook. Part 652. Irrigation Guide. National Technical Information Service, Washington, DC, USA.

USDA-NRCS (US Department of Agriculture, Natural Resources and Conservation Service), 2005. National Engineering Handbook, Part 623, Surface Irrigation. National Technical Information Service, Washington, DC, USA, Chapter 4

Walker, W. R. (1998). SIRMOD – surface irrigation modeling software. Utah State University, USA.

Walker, W. R. (2005). Multilevel calibration of furrow infiltration and roughness. *Journal of Irrigation and Drainage Engineering*, 131(2), 129–136.

Walker, W. R., Prestwich, C., and Spofford, T. (2006). Development of the revised USDA-NRCS intake families for surface irrigation. *Agricultural Water Management*, 85(1–2), 157–164.

Waller, P., and Yitayew, M. (2015). *Irrigation and drainage engineering*. Springer, Switzerland.

Zapata, N., and Playán, E. (2000). Simulating elevation and infiltration in level-basin irrigation. *Journal of Irrigation and Drainage Engineering*, 126(2), 78–84.

Zhang, S., Xu, D., Bai, M., Li, Y., and Xia, Q. (2014a). Two-dimensional zero-inertia model of surface water flow for basin irrigation based on the standard scalar parabolic type. *Irrigation science*, 32(4), 267–281.

Zhang, S., Xu, D., Bai, M., and Li, Y. (2014b). Two-dimensional surface water flow simulation of basin irrigation with anisotropic roughness. *Irrigation Science*, 32(1), 41–52.

Zhang, S., Xu, D., and Li, Y. (2012). Two-dimensional numerical model of basin irrigation based on a hybrid numerical method. *Journal of Irrigation and Drainage Engineering*, 138(9), 799–808.

III

Pressure Irrigation

6
Sprinkler Irrigation Systems

Neil A. Coles
University of Leeds
The University of Western Australia
Verdant Earth

Mark R. Rivers
ClearWater Research and Management Pty. Ltd.

Saeid Eslamian
Isfahan University of Technology

6.1	Introduction	113
6.2	Broadacre Sprinkler Delivery Systems	114
	Solid-Set (or Fixed) Sprinklers • Lateral or Periodically Moved Irrigation • Centre Pivot and Automatic Linear Move Irrigation • Floppy (Flexible) Sprinklers • Advantages and Disadvantages • Other System Components and Controllers	
6.3	Irrigation Performance Measures	122
	Efficiency • Uniformity • Other Performance Measures	
6.4	Comparison of Irrigation Systems	129
6.5	Domestic Sprinkler Systems	131
6.6	Broadacre Irrigation System Design	133
	Factors Affecting the Irrigation Interval • Monitoring Irrigation Performance	
6.7	Sprinkler Irrigation Research	136
6.8	Conclusions	137
Notes		137
References		137

6.1 Introduction

Irrigation systems are used extensively around the world to alleviate soil water deficiency and are a key component in aiding food production (FAO & WWC 2015). This makes irrigated agriculture inherently the largest user of developed water resources in agricultural production systems at global and many national scales.[1] Water availability and quality for irrigation is critical to agricultural productive capacity and, therefore, food security, and can be severely affected by variations in climate including those caused by global heating (Steffen et al. 2018). In an increasingly developed and populated world, water use efficiency and optimal distribution will become progressively significant factors to ensure maintenance of, and, indeed, increases to, agricultural production from existing irrigated land to feed a growing global population (Brennan 2008; FAO and WWC 2015; Rivers et al. 2015).

Irrigation is simply the artificial application of water to soil for agricultural or horticultural production and is generally, on a global scale, delivered via three main systems: sprinkler, micro-sprinklers or drippers, and flood (or surface) irrigation. The first two methods involve large, often complex, systems of pipes, pumps, valves, controllers, sensors and sprinklers, with surface irrigation being a much simpler and less technically demanding method (Rivers et al. 2015; Martin, Kincaid, and Lyle 2007). Owing to its lower investment cost, surface or flood irrigation, is deployed on approximately 85% of irrigated lands worldwide with the remainder using either sprinkler or micro-irrigation systems, with a smaller

DOI: 10.1201/9780429290152-9

113

percentage using subsurface irrigation systems (Bjorneberg 2013). This aside, sprinkler systems are one of the most commonly used agricultural irrigation methods that require specialist knowledge in understanding the production requirements of the system, its design, operational components, water pressure and quality to ensure optimal performance (Zhang, Merkley, and Pinthong 2013; Keller and Bliesner 2000).

The aim of this chapter is to provide a brief review of sprinkler systems, focusing only on the use of mounted sprinkler systems which are generally considered to be those where water is applied above the ground surface, under pressure through various types of openings or nozzles to distribute "rain-like" droplets over the soil surface to mimic rainfall (Burger and ARC-Institute for Agricultural Engineering 2003; James 1988). Such systems are basically lengths of pipe or hose, a standpipe, sprinkler outlet, pump and water source (Hoffman and Martin 1993). From these basic, initial components continuous improvement has occurred over many years in the various sprinkler application methodologies that promote uniformity of water intake, distribution and water use efficiency, minimise runoff losses and optimise spatial variability while allowing the monitoring of system performance (ANSI and ASAE 1989; Hoffman and Martin 1993; Ascough and Kiker 2004; Merwe and Johannes 2008).

Significant improvements in the operation of sprinkler systems have been made particularly over the last two decades with a focus on improving water use efficiency and reducing water losses through evaporation, runoff, leakage or deep percolation (beyond the root zone). Water losses through runoff are associated with the increased redistribution of water-borne pollutants such as sediment or nutrients (from fertilisers and manures) with the associated environmental issues driving improvements to irrigation systems (Zia et al. 2013). This continued development has been accompanied by technological and analytical innovations, adoption of operational best practice, improved farmer education and training, and the use of enhance sensor, automation and efficiency measurement techniques that have transformed the design of irrigated agriculture to better match soil, crop and resource conditions (Martin, Kincaid, and Lyle 2007; Elwadie, Mao, and Bralts 2010; Ferrarezi, Dove, and Van Iersel 2015; Koech and Langat 2018; Adeyemi et al. 2017). Innovative sprinkler systems are now designed, installed, and operated very specifically to deliver more efficient irrigation in a wide array of environmental conditions and agricultural systems (IA 2017). This has facilitated the expansion of irrigated agriculture in regions and topologies that were previously designated as unsuitable for irrigation (Martin, Kincaid, and Lyle 2007).

As more efficient and complex – but static – systems were developed, their distribution (and uptake) was initially impeded by the labour and expense necessary to design, install and move these systems. Gradually these labour requirements and expenses have significantly decreased through the introduction of automated and travelling (or moving) sprinkler systems. Automation and moving systems, coupled with improved efficiency monitoring technologies has enabled agricultural producers to irrigate more frequently, with lower amounts of water that are better matched to soil moisture levels and plant water requirements (Rivers et al. 2015; Zia et al. 2013). This has allowed for the improved management of unintentional surface runoff and subsurface leaching to groundwater while also increasing the potential to better store water in the crop root zone while continuing to satisfy crop growth requirements (Martin, Kincaid, and Lyle 2007).

In the following sections, we describe the most common sprinkler irrigation systems used in agriculture and provide a brief review of sprinkler irrigation in a domestic setting.

6.2 Broadacre Sprinkler Delivery Systems

Sprinkler or spray irrigation applies water to an area (usually for crops or pasture) in a controlled manner, to simulate rainfall and remove limitations to plant growth because of water scarcity. Water, which is usually extracted from secure surface or groundwater sources, is distributed via a system or network of channels, pumps, pipes and sprinklers for coordinated residential, horticultural, industrial and agricultural applications (Figure 6.1).[2] Sprinkler systems normally have a higher investment cost and level

Sprinkler Irrigation Systems

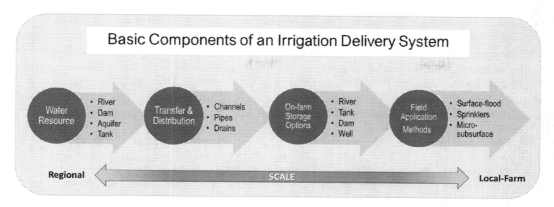

FIGURE 6.1 Basic components of an irrigation delivery system: water supply, transfer mechanism, localised storage and field delivery options.

of complexity than simpler systems such as surface or flood irrigation, but they offer a greater level of water use efficiency and automation through monitoring technology, efficient application rates and reduced losses beyond the field (Bjorneberg 2013). Important considerations where water costs are high and availability is limited.

The water resource or supply options, often referred to as the "head works" (Figure 6.1), are dependent on a range of factors including:

a. regional and local climate, topography, and hydrology;
b. existing water storage and transfer infrastructure;
c. available energy options and requirements;
d. financial costs, capacity and investment options; and
e. water quality, security, reliability and requirements.

Water sources and on-farm water supplies also vary across a scale from regional (e.g. lake, rivers, large dam, deep aquifer) to local options (e.g. small dams, storage tanks, stream, near-surface aquifer) (Figure 6.1).

Large-scale sprinkler irrigation systems are commonly used for agricultural production and can be considered to be permanent, semi-permanent or portable according to design, components and application (Burger and ARC-Institute for Agricultural Engineering 2003). The choice of irrigation system selected will depend on the soil type, local topography, crop water requirements, installation costs, maintenance needs and above all, water supply factors (including water quality, availability and achievable flow rate).

Irrigation water can be supplied to end users ("Local Farm" in Figure 6.1) in a variety of ways. Irrigators may be part of a regional irrigation "cooperative" or may be "self-suppliers". They may access water from engineered surface-water distribution systems, from natural water courses or from groundwater bores or wells. Each of these options presents management issues for suppliers, regulators and irrigators. For the various forms of sprinkler systems, irrigation laterals (or more local distribution networks running from larger, more regional networks) are generally classed as fixed set, periodic move, or continuous, self-move systems (USDA and NRCS 1997) that water is supplied to under pressure via:

a. overhead or on-ground sprinkler packages or sets;
b. in-ground, pop-up systems (generally domestic/horticultural);
c. high-pressure sprinklers or guns from a fixed, central location in the field; or
d. from sprinklers on a moving platform (including manual move lateral systems, wheeled lateral systems, centre pivot systems, linear move or boom sprays and travelling gun types).

If the various systems are optimally designed, operated and managed then application efficiencies of between 50% and 95% can be attained (USDA and NRCS 1997). These on-farm systems are generally supplied by a mainline pipe or channel which delivers water from the regional water source (e.g. feeder pipes, small dam, river, lake, holding tank, groundwater aquifer) (Figure 6.1).

Most notably of all the variables within irrigation systems, poor on-farm management is often recognised as one of, if the not most, significant contributors to reduced water application efficiencies and crop productivity losses (Zhu et al. 2018). Other inefficiencies and losses occur through:

a. evaporative losses from spray, soil surface and the surface of plant leaves at contact points;
b. leaks in the delivery system or at the sprinklers themselves;
c. subsurface percolation – i.e. drainage below the root zone;
d. off-site surface-water loss from over-irrigation or non-uniform application; and
e. wind drift beyond the area of intended application (these losses can be up to 10% depending on wind speed temperature and droplet size) (Hoffman and Martin 1993; Martin, Kincaid, and Lyle 2007).

While there are many types and configurations of irrigation equipment, a few systems predominate within irrigated agricultural production and these are described below.

6.2.1 Solid-Set (or Fixed) Sprinklers

A solid-set system can simply be a collection of spray or sprinkler distribution hardware consisting of long hoses or pipes with sprinklers and nozzles along their length, linked to pumps and pressure or flow-control devices that are designed for a specific set of operating parameters (ANSI and ASAE 1989). Solid-set sprinkler irrigation (Figure 6.2a and b) is widely used as the first level of improvement from furrow or surface (flood) irrigation (other than increasing efficiencies within the surface irrigation system). While still being relatively simple and cheap to establish and operate, these systems deliver greater water application efficiency, increase crop productivity (per litre) and reduce the environmental impact of irrigation for relatively low installation costs (Elwadie, Mao, and Bralts 2010). Solid-set systems are usually set out along lateral pipes that are connected to a main supply line using a hydrant valve coupling (Figure 6.2a). The pipes can be aluminium or light steel (generally laid on the soil surface and portable) or made of polyethylene or uPVC if they are to be more permanent (and installed either on the soil surface or buried). Standpipes connect individual sprinklers to the lateral lines (Figure 6.2b).

Pipes are usually of a consistent style, size and connection to enable simple interconnection and ease of transfer between different irrigation operations and fields, providing uniform pressure responses and, therefore, irrigation application efficiency (James 1988; Hoffman and Martin 1993; Martin, Kincaid, and Lyle 2007). In most cases, the system is placed in the field at the start of the irrigation season and left until the crop is harvested; however, using subsurface installation will allow the system to remain in the fields more permanently and improve vehicle and livestock access (Figure 6.2b). During irrigation events, sprinklers in one or more fields or crop zones can be turned on or off using a control valve on the mainline (Figure 6.2a). Valve operation can be achieved manually, automatically scheduled or timer-controlled relatively easily. Application efficiency for solid-set systems are thought to be around 60%–85% depending on design and system management (*see Section 6.3 Irrigation performance measures*), with triangle or diamond zone sprinkler patterns thought to improve uniformity of water delivery (USDA and NRCS 1997).

6.2.2 Lateral or Periodically Moved Irrigation

Periodically moved systems are similar to solid-set systems but generally use a series of moveable pipes and impact sprinklers that are regularly shifted a set distance – usually on a planned watering schedule – until the whole field is irrigated (Albaji, 2020,). They require less system hardware than solid-set systems but require more labour. Water is pumped through lateral pipes or booms each with a wheel and a set

Sprinkler Irrigation Systems

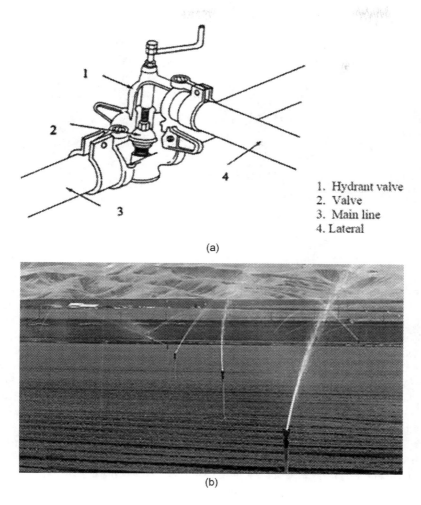

FIGURE 6.2 (a) Main supply hydrant that is connected to the laterals of a fixed set system and transfers water from the main reservoir to the sprinkler set (ARC-ILI 2004). (b) Photo of fixed or solid-set sprinkler irrigation layout in operation with each sprinkler set in a specific pattern to ensure maximum coverage and minimum overlap and water losses (Rivers 2014).

of sprinklers, which are rotated either by hand or with a purpose-built mechanism such as a travelling sprinkler system (Figure 6.3a and b). In most cases, the sprinklers are either programmed to move a set distance or they are hand-rolled into position across the field. The water supply is then reattached to lateral lines. While these types of systems can be less expensive to implement than some other systems, they are more labour-intensive.

6.2.3 Centre Pivot and Automatic Linear Move Irrigation

Centre pivot systems are essentially a fixed-point pump with water distributed through a lateral pipe to a system of sprinklers that, together, rotate on wheeled towers from the fixed, centred end point outward to create a circular pattern (Figure 6.4a). Linear move systems use, essentially, the same hardware as centre pivot systems but instead of rotating around a central point, they move in a linear fashion, up and down a field, drawing water either from a channel running parallel to the field or via a flexible hose fixed to a stationary pump (Figure 6.4b).

FIGURE 6.3 (a) Periodically hand-rolled sprinkler system (Rivers 2014). (b) A computer-controlled travelling irrigator with mounted gun spray (N. Coles).

Sprinkler "packages" are individually designed for each centre pivot irrigator with each sprinkler optimised for the water pressure applied to it at its individual location (Martin et al. 2012). At each individual location, water pressure varies along the span of the pivot system due to pressure losses caused by in-pipe friction and pressure drops at each individual sprinkler. The efficient installation of a centre pivot irrigator requires the correct calculation of flow restrictions for each sprinkler to account for this lateral variability and to ensure uniform water application.

As this system is circular in nature, the correct sprinkler "packages" must also account for the speed of movement at any given point over the soil surface at various distances from the centre pivot (Martin et al. 2012). That is, sprinklers close to the pivot will be moving over the soil surface at a much lower ground speed than those at the end of the pivot for any given irrigator speed simply due to variations in the distance from the pivot and water application rates must be calculated to account for this variation (see Figure 6.5).

Another consideration for centre pivot irrigator design is that they are generally retrofitted into traditional square or rectangular fields. This often means that some fields have "dry corners" where the end wheels and last sprinklers of the irrigator do not reach. In some instances, these corners are irrigated by the use of an "end gun", but it is generally accepted that the loss of productivity in these corners is more than compensated for by improved agricultural production under the main irrigator (Martin, Kincaid, and Lyle 2007; Martin et al. 2012).

6.2.4 Floppy (Flexible) Sprinklers

The floppy sprinkler system is a relatively new application system and is constructed using a lightweight plastic nipple on which a flexible silicon tube is mounted. A series of sprinklers are suspended on cables above the area to be irrigated allowing easy passage of traffic and livestock. The distribution pattern of the sprinkler is established as water passes through the silicon tube, with the essentially random flexing of the tube and its free, 360-degree rotation apportioning the water flow into uniform, medium-sized water droplets with little or no mist formation (Ascough and Kiker 2004). The droplets produced from this type of system are thought to best represent natural rainfall. Each floppy sprinkler head is fitted with a flow controller that can modify the flow rate to achieve up to 730 L/h with variable water pressures from 2 to 6 bar (El-Sayed et al. 2009).

These are adjusted and optimised for each sprinkler to balance pump head pressures and frictional losses along feeder lines to ensure uniform water application throughout the system. A series of sprinklers are attached to the main feeder lines which are permanently installed onto overhead support

Sprinkler Irrigation Systems

FIGURE 6.4 (a) Crop irrigation using centre pivot systems consisting of a long metal boom (or frame) that revolves around the central point with water sprinklers paced along the boom that delivers water at defined application rate (left) (M. Rivers). An aerial view (top) showing centre pivot system irrigating crops in Arizona using groundwater (N. Coles). Rain gun mounted on the end of the boom to increase the reach of the boom spray (right) (M. Rivers). (b) Crop irrigation using linear move irrigation systems consisting of the same long metal boom used in centre pivot systems but moving in a lateral manner rather than rotating (M. Rivers).

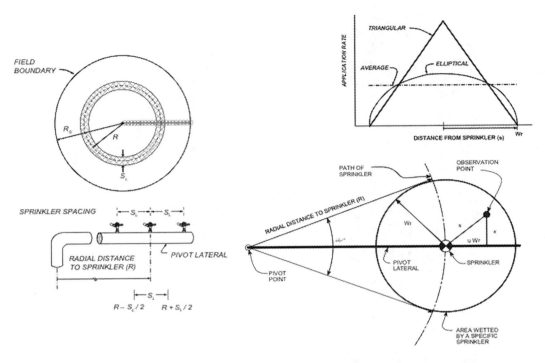

FIGURE 6.5 Calculation of the area for determining the discharge and sprinkler spacing for lateral arm of centre pivot (L) and representative diagram of the parameters used compute the uniformity of water application rates (Martin et al. 2007).

frames and poles. This setup can achieve an average soil surface-water application rate of 5 mm/h and generally requires little or no maintenance.

The design parameters of the overhead floppy sprinkler system may affect the uniformity of water distribution and water application. For example, if the spacing is incorrect or if the flow hydraulics have been calculated incorrectly providing either insufficient water volumes or too much pressure. Nozzle wear, riser height and wind drift are some aspects of floppy system design that may affect their efficiency and uniformity (Griffiths and Lecler 2001; Ascough and Kiker 2004; El-Sayed et al. 2009).

6.2.5 Advantages and Disadvantages

Sprinkler irrigation systems allow better management of irrigation water distribution in terms of both timing and spatial uniformity of application when compared with simpler surface or flood systems (Elwadie, Mao, and Bralts 2010). There are many different sprinkler system designs available and the potential advantages and disadvantages of these systems in a general sense are shown in Table 6.1. Furthermore, this type of irrigation methodology delivers advantageous water distribution uniformity, high water use efficiency, precise monitoring of application depth and the prospect of utilising variable soil landscapes for agriculture (Bernardo, Soares, and Mantovani 2006; Prado 2016). These are important aspects to consider with increasing emphasis now placed on water source security, environmental requirements and the investments associated with sprinkler installations, maintenance and pumping water– particular rising energy costs – which has provided impetus for the adoption of improved technologies and computer simulations by sprinkler manufacturers, designers and irrigation users (Faria et al. 2013; Prado 2016; Zia et al. 2019).

One of the most common influences on application uniformity and therefore, distribution efficiency, for sprinkler systems is wind direction and strength (Brennan 2008; Moazed et al. 2010). Wind can distort spray patterns parallel to the wind direction which can in turn result in over- and under-watering of the

Sprinkler Irrigation Systems

TABLE 6.1 Advantages and Disadvantages of Sprinkler Systems Relative to Flood or Surface Irrigation Methods

System	Advantages	Disadvantages
Sprinklers	• Significant reduction in the system water losses • Small frequent water applications can be achieved • High uniformity of water application • Leakage or deep percolation below the root zone is avoided • Reduced water requirements • No land levelling required • Fertigation is possible • Reduced land loss for system layout • High level of automation and monitoring is possible	• Application uniformity can be affected by wind • High capital and operating costs can limit uptake in areas with low water costs (i.e. surface irrigation preferred) • Operational knowledge and maintenance scheduling required • Water quality requirements are high (i.e. no sediments or carbonates) to reduce clogging • On-ground pipe system may reduce farm vehicle or livestock access • Increased understanding of technology and system simulations

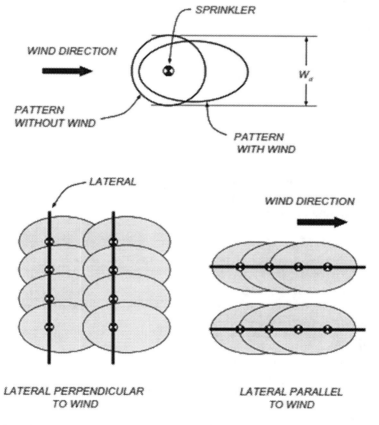

FIGURE 6.6 Effect of wind speed and direction on sprinkler spray uniformity and soil moisture distribution. Note that setting laterals perpendicular to the prevailing wind direction and spacing sprinkler appropriately can deliver the most consistent, productive and cost-effective water distribution (Martin et al. 2007).

crop depending on the direction the prevailing winds as shown in Figure 6.6. If the seasonal direction of prevailing winds is known, then the sprinkler (or distribution laterals) can be spaced and positioned on the ground to take advantage of this situation (Brennan 2008). Computer simulations for sprinkler systems are becoming increasingly important in designing irrigation sets, assigning irrigation equipment (i.e. sprinkler heads, nozzle diameters, pressures and spacing) and the development of new sprinkler types, and evaluating the accuracy of operational or scheduling procedures (Faria et al. 2013; Prado 2016). For all systems, monitoring of soil moisture, prevailing wind conditions, sprinkler and pressure combination performance, will assist in determining best scheduling times, sprinkler water distribution radial curves and application rates relative to soil moisture and crop requirements and nutrient loss management (Martin, Kincaid, and Lyle 2007; Nega 2011; Zia et al. 2013; Rivers et al. 2015). In this way, the optimal working conditions that result in acceptable Christiansen Uniformity Coefficients[3] can be determined.

6.2.6 Other System Components and Controllers

For any standard irrigation system, a major design objective is to apply a known volume of water uniformly over a defined area for a set time period, thus delivering a uniform depth of water per unit area (Prado 2016). System components are then specified to meet the irrigation design requirements so as to promote the efficient operation of the system, conserve water and optimise crop production (Brennan 2008). These components include: valves, pressure regulators, flow meters, backflow prevention systems, filters, pressure gauges, chemical injection equipment and irrigation controllers.[4] All aspects of each component should be appraised against system design specifications and existing standards for performance (e.g. ASAE Standards, ISO 11738:2000).[5]

6.2.6.1 Flow Controllers

Pressure or flow controllers are of particular interest as they impact the effectiveness of sprinkler systems. Uneven or uncontrolled pressure distributions and the subsequent flow and application deviations can result in spray distribution pattern distortions that disrupt water application uniformity (Zhang, Merkley, and Pinthong 2013). Pressure fluctuations can be caused by uneven field topography, excessive pipe friction losses or pumping pressure variations, resulting in uneven water delivery via any sprinkler type, but more specifically impact-drive rotating sprinklers (ARC-ILI 2004). Flow controllers are usually made from shaped brass or moulded plastic and will only reduce flow rates (Figure 6.7) not increase them if the flow is deficient. Examples of some flow-control devices are given in Table 6.2 and are used to meet specific flow and pressure requirements with the expectation that well-designed irrigation system should not exceed pressure variations of more than 20%.

6.2.6.2 Sprinklers

There are a wide variety of sprinkler systems, sprinkler heads and delivery devices (i.e. floppy tubes, rotating) that can be attached to the water delivery/conveyance systems described above. However, there are four general sprinkler types on the market – rotating, floppy, fixed nozzles, and perforated pipes – with impact-drive rotating sprinklers most in use (Figure 6.8). The design of the sprinkler system and the sprinkler heads chosen depends on:

   a. expected uniformity of application;
   b. application rate – a function of discharge, wetted diameter and spacing;
   c. drop size distribution – a function of nozzle diameter, pressure and pressure variations; and
   d. cost of system relative to the expected returns (ARC-ILI 2004).

As the soil infiltration capacity limits the application rate, this is a major determinant of the appropriate sprinkler, nozzle sizes, flow rates, pressure and sprinkler spacing (USDA and NRCS 1997) for the design of the irrigation system. In this sense the discharge rate (or requirement) of the sprinkler will determine the sprinkler size which should be in line with the recommended lateral spacing and dispersion pattern.

6.3 Irrigation Performance Measures

An optimal irrigation system delivers water to the soil surface uniformly at a rate equal to or less than the soil infiltration rate. The water is then able to percolate into the soil profile and down to the root zone in sufficient quantities to optimise plant uptake and growth but not at such a high rate that water is lost to the subsoil drainage, runs off or evaporates or results in water logging.

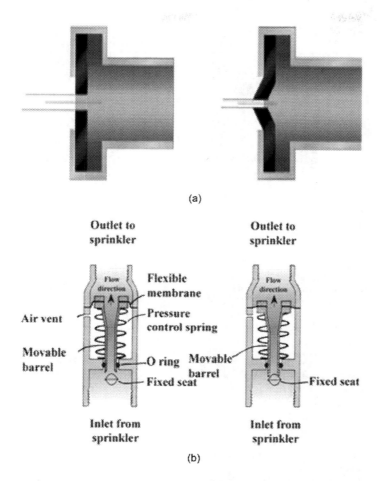

FIGURE 6.7 Examples of (a) unregulated (*L*) and regulated (*R*) flows through a piped system to ensure correct pressure delivery and constant flow rate to sprinkler heads. (b) Pressure regulators showing unregulated (*L*) and the preferred regulated (*R*) flows to ensure constant flow rates through sprinkler (Kranz 1988).

Irrigation efficiency is notoriously difficult to define consistently with efficiency defined at either the field, farm or system scale – or a combination of all three (Figure 6.9). "Irrigation efficiency" needs to be defined appropriately before it can be assessed in a meaningful way. Many definitions of irrigation efficiency at a system or farm scale already exist and may include for example:

- tonnes of agricultural product per ML of water applied,
- tonnes of agricultural product per ML of water lost to drainage,
- financial returns per ML of water applied, and
- benchmarking against respected land managers or against well-researched values.

Efficiency is also measured in terms of the operational efficiencies or water management given that poor irrigation efficiency and scheduling can result in excess water losses though over-application. There are, similarly, various measures of the application efficiencies of the various types of irrigation systems. These include:

- application efficiency,
- soil water storage efficiency,
- irrigation efficiency, and
- effective irrigation efficiency.

TABLE 6.2 Advantages and Disadvantages of Different Flow Controllers to Manage Operation Efficiencies of Sprinkler Irrigation Systems by Managing Pressure Fluctuations within the Water Delivery Pipeline

Advantages	Disadvantages
<td colspan="2" align="center">Flexible Flow Control</td>	

Flow-Control Nozzle

• Designed for low-pressure operations • Single opening and self-flushing action reduce potential blockages • Maintain flow at +/−10% of expected performance curves • Limited or no pressure drop caused by device • Teardrop shape of orifice designed to deliver desired distribution pattern and droplet size at low pressures • Stream diffusion improves water distribution uniformity and droplet sizes at low pressures	• When operated under high wind speeds the wetting patterns can become distorted • Not suitable for excessively wide spaces (optimally 12 m × 15 m) as wider spacing diminishes application uniformity • Orifice is easily blocked altering distribution patterns • Devices are prone to wear and need regular maintenance • Distribution patterns are distorted with increased pressure

Flexible Orifice

• Device is self-cleaning • Device is not prone to scaling • Responses to pressure variations is instantaneous • Flow rate delivered within +/−10% of normal flow rate • In considering normal flow controls the rates stay within 5%–8% of the mean flow rate though the full range	• Orifice is easily blocked altering distribution patterns • Devices are prone to wear and need regular maintenance

Spring Loaded Flow Controls

• At a pressure below the set regulation pressure, the throttling action is negligible • Can operate at a maximum continuous pressure of 70 M but can also withstand pressure surges of extremely short durations • If placed at the beginning of each lateral, the reducer enables the designer to set the same pressure at the beginning of each line of the system • Allowable pressure variation can be restricted to the lateral alone and allow the use of longer, thinner pipes	• Devices are prone to wear and need regular maintenance • Expensive relative to flexible flow controllers

Source: Adapted from ARC-ILI 2004.

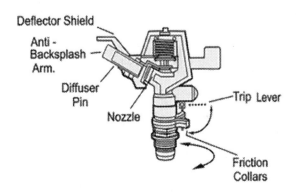

FIGURE 6.8 Diagram of basic impact sprinkler (garden use) with limited moving parts for ease of installation, operation and maintenance (Butts 2019). These can be made from metal or plastic and vary according to purpose, application rates, uniformity requirements, pressure and cost (Photo: Paul Moody_unsplash, 2022).

Sprinkler Irrigation Systems

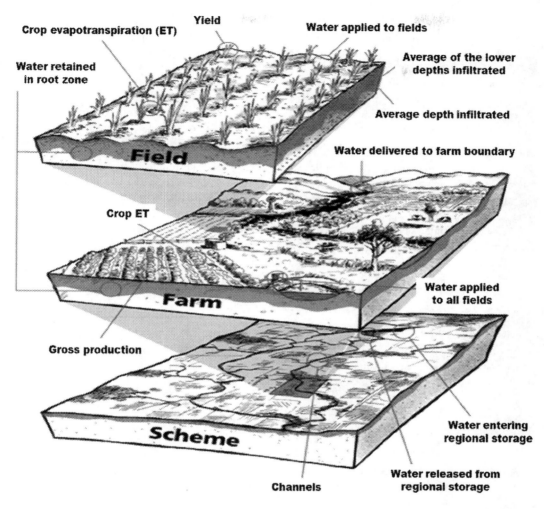

FIGURE 6.9 Different measures of irrigation efficiencies can be applied at different scales (Field, Farm, Scheme) using alternate metrics and are often difficult to determine (Rivers 2014).

Various definitions and the "best practices" which impact irrigation efficiency including improved scheduling are discussed in the following sections.

6.3.1 Efficiency

In the planning phase it is necessary to determine the performance of a proposed irrigation system that quantifies the design and management criterion, as well as the field operating procedures (Burt et al. 1997). In this way we can effectively utilise water to enhance plant growth using an application system that accounts for the variability in both physical (water quality, plant and soil types) and technological factors (pipe diameter, pressure, nozzle size). These factors combined with irrigation system management influence the water application efficiencies that can be obtained. As early as the mid-1940s, Israelsen et al. (1944) considered that there were internal and external factors that influence efficiency, some of which can be managed by irrigators or farm managers. These included:

1. preparation of land for irrigation to mitigate runoff and erosion,
2. methods of water application system design, spacing and radial cover,

3. time rate of water application – amount applied per sprinkler and duration of application,
4. surface runoff losses – matching rate of application to soil infiltration capacity,
5. soil moisture content before irrigation – determination of soil water deficits prior to irrigation,
6. volume of water applied at each irrigation – using infiltration and storage capacity to avoid deep percolation beyond the root zone,
7. available water supply – ensures pressure continuity and water quality, and
8. personal attention or monitoring of water distribution – ensures system is correctly designed and functioning as expected to avoid over- or under-watering of crops (Israelsen et al. 1944).

Therefore, in understanding irrigation and efficiency, it is important to determine what constitutes efficiency – relative to managed water application, plant use and accumulated losses – i.e. water applied but not available for plant use. There are many inter-related parameters that can be combined into the various measures (or metrics) of efficiency or use (Figure 6.10); these include equations defining irrigation efficiency (*IE*), distribution efficiency (*DE*), distribution uniformity (*DU*) and application efficiency (*EA*). Some are based on volumetric efficiency measures which describe the losses of water occurring en-route from the supply point to the point of consumptive use, and generally take the generic form.

$$\text{Efficiency} = \frac{\text{Volume delivered to the next stage of system} \times 100}{\text{Volume entering any given stage}}$$

The distribution of water from the initial water source to the irrigated field can be achieved using natural drains, constructed earthen or lined channels and/or closed conduits. For example, the efficiency with which water is distributed within an irrigation district or delivered in a field can be determined using the following equation:

$$\text{Distribution Efficiency} = \frac{\text{Volume delivered to the farm} \times 100}{\text{Volume delivered to the district}}.$$

Alternatively, further refinements can yield the water conveyance efficiency (E_c), defined as the percentage ratio of the volume of water delivered to the field boundary to the volume of diverted from the source and can be expressed as:

$$E_c = 100 \left(\frac{W_f}{W_c} \right), \tag{6.1}$$

where W_f is the water delivered to the field and W_c is the water diverted from source (Rogers et al. 1997).

Observed sources of water loss from irrigation are surface runoff and deep percolation (soil moisture infiltrated below the root zone). If evaporative losses that occur during irrigation are ignored, then Israelsen et al. (1944) suggested that the volume of water applied should equal the sum of (1) the volume stored as soil moisture in the root zone, plus (2) the volume lost by deep percolation below the root zone, plus (3) the volume lost as surface runoff. This enabled Israelsen et al. (1944) to define application efficiency E_a as "the ratio of the volume of water that is stored in the soil root zone and ultimately consumed (transpired or evaporates or both) to the amount water delivered to the farm", given as:

$$E_a = 100 \left(\frac{W_c}{W_f} \right), \tag{6.2}$$

Sprinkler Irrigation Systems

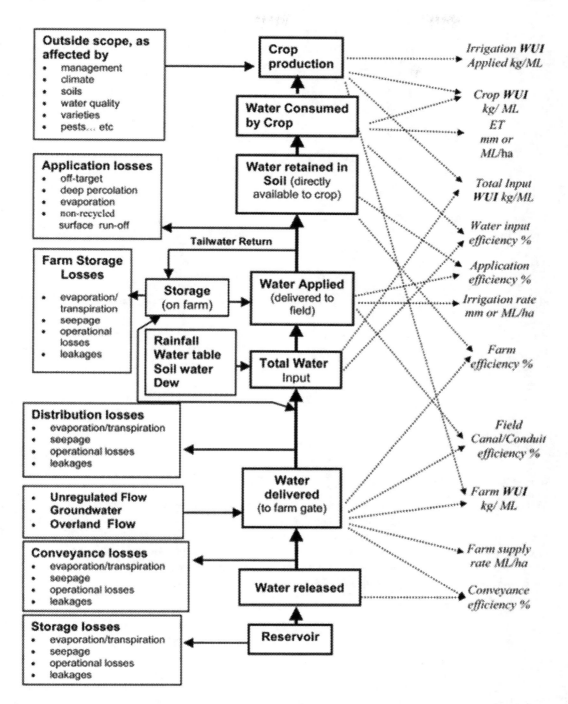

FIGURE 6.10 Measures of efficiency used for irrigation systems at various scales, points of water distribution or application using alternative metrics as a measure of performance (Rivers 2014).

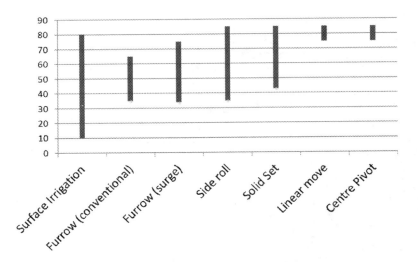

FIGURE 6.11 Typical efficiency percentage ranges for common irrigation methods. Note efficiencies improve with increasing system complexity, investment, technical requirement and operational procedures.

where W_c is water available for use by the crop and W_f is water delivered to the field. Therefore, in broad terms, the water application efficiency is the percentage of water delivered to the field that is used by the crop derived from.

$$\text{Application Efficiency} = \frac{\text{Volume added to the soil moisture store} \times 100}{\text{Volume delivered to the field}}$$

Typical application efficiencies for different irrigation systems are given in Figure 6.11.

6.3.2 Uniformity

Variable rate irrigation systems can improve overall water productivity (cubic metre water applied per kg of plant produced) though improvements in technology for application and monitoring to ensure precision in the amounts delivered, thus avoiding under- or over-irrigation (O'Shaughnessy et al. 2013). The uniformity of water application by a particular system is best measured by an assessment of the uniformity of applied depths (Burt et al. 1997). For sprinkler irrigation, there are a number of measures of uniformity, one being the Uniformity Coefficient (U_c) calculated as a coefficient percentage by using Christiansen's formula:

$$U_c = 100 \left[1 - \frac{\sum_{j=1}^{n} |V_i - \bar{V}|}{\sum_{j=1}^{n} V_i} \right], \tag{6.3}$$

where V_i is the depth of water in an individual catch can and $\bar{V}$ is the average depth of water in all catch cans (Christiansen 1942). Further measures of uniformity are obtained by determining the distribution uniformity (U_d). It is defined here as a measure of the uniformity with which irrigation water is distributed to different areas in a field (Burt et al. 1997). This can be calculated as the percentage of average application amount received in the least-watered quarter of the field, expressed as:

$$U_d = 100 \left(\frac{D_{lq}}{D_{av}} \right), \tag{6.4}$$

TABLE 6.3 Different Economic and Water Use Indices That Can Be Utilised to Assess Irrigation System Performance

Index	Output / Input	Units
Crop Water Use Indices (WUI)		
Crop economic WUI	Gross return / Evapotranspiration	\$ / mm
Crop WUI	Yield / Evapotranspiration	kg / mm
Irrigation Water Use Indices (WUI)		
Irrigation WUI	Yield / Irrigation water applied	kg / ML
Gross production economic WUI	Gross return / Total water applied	\$ / ML
Irrigation economic WUI	Gross return / Irrigation water delivered to the field	\$ / ML
Yield per drainage volume WUI	Crop production / Drainage volume	kg / ML

Source: Rivers (2014).

where D_{lq} is the average low-quarter depth of water infiltrated and D_{av} is the average depth of water infiltrated (Rogers et al. 1997). This can only be achieved through effective design, maintenance and management of irrigation systems. The uniformity with which an irrigation system applies water has an effect on the efficiency of the system, but measures are often required to be accounted for if on-farm management is to be adequately evaluated (Ascough and Kiker 2004). These include variations in applied depth due to sprinkler spacings, wind effects and variations in nozzle outflow or pressure.

6.3.3 Other Performance Measures

There are many other measures of efficiency particularly for irrigation business performance (Table 6.3) which can define the best use and application of water, the most appropriate crops and the best returns on investment relative to the type of irrigation systems installed. As has been demonstrated, soil type, topography, source and cost of water, the delivery system and the value of the crop will all influence the type of investment made on-ground. Overarching these components are elements of land water and system management, particularly irrigation scheduling that affect returns.

6.4 Comparison of Irrigation Systems

The system efficiencies for the various forms of sprinkler irrigation (for both scheduling and distribution) each allow differing levels of water management or systemic application to be utilised. However, increasing management potential may be associated with increasing cost (Figure 6.12). The selection of an appropriate irrigation system in any location relies on a balanced approach to this relationship. Extensive comparative research has been undertaken globally to assess the relative "efficiencies" of the various systems using the definitions outlined in Section 6.3. Although there are large variations at the international scale on what can be defined as cheap or costly, there are some generally applicable studies.

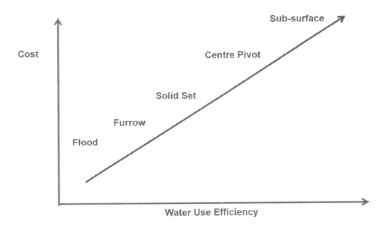

FIGURE 6.12 Relative costs and water use efficiencies for different irrigation systems. Note costs rise as system complexity increase but this is offset by improvements in water use efficiency.

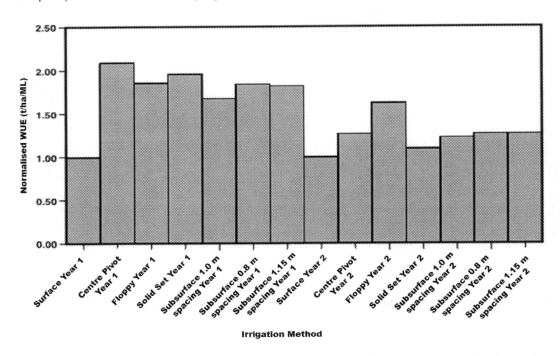

FIGURE 6.13 Relative water use efficiencies for different irrigation systems, measured as tonnes of productive dry matter produced per litre of water and normalised to flood (surface) irrigation.

One such study undertaken in Australia has been used at a whole-of-system scale to drive the change from surface irrigation to more expensive (but more productive) irrigation technologies (Rivers 2011; Rivers et al. 2011). Several irrigation systems were installed and monitored over a period of a number of years with various parameters of Water Use and Water Use Efficiency measured at a high frequency. This research showed clearly that all "alternative irrigation systems" (including the various sprinkler systems) used water more efficiently than surface irrigation to produce plant growth over a number of growing seasons – and, therefore, more food for human consumption, milk and stock feed per litre of water (Figure 6.13). Associated research showed, effectively, that optimally managed sprinkler irrigation systems can produce twice as much plant matter than surface irrigation systems while using half as much water (Rivers 2008; Rivers et al. 2010).

6.5 Domestic Sprinkler Systems

Domestic application of sprinkler system necessarily follows the same basic traits as larger scale operations; that is, the need to ensure optimal performance, water use efficiency and plant yield. A garden system is defined as a network of permanent (mostly below ground) piping connected to sprinklers or drips that are designed and installed to water a specific area, and can be utilised to provide evaporative cooling, nutrient transport and dispersion of any plant expelled soluble wastes (IA and RUWSP 2019). As with large-scale systems, a basic knowledge of site conditions is required including but not limited to:

- soil texture, structure and estimated infiltration rates;
- predisposition of the soil to developing water repellence;
- water sources, quality, cost and pressure;
- estimated areas to be irrigated, elevations or slope conditions;
- plant types and distributions; and
- rainfall and expected evapotranspiration rates.

The next step in the process is the design of the irrigation system based on the chosen site data with basic elements an irrigation installation plan given in Figure 6.14. A site plan is drawn up based on the data obtained for the site and based on the required irrigation delivery requirements. Types of sprinkler system deployed in a domestic setting include driplines, micro-sprays, pop-up sprinklers and rotary systems and are dependent on the soils, area to be watered and plant type. Different sprinkler types can be installed within the same system but may be divided into different hydrozones. Rotary

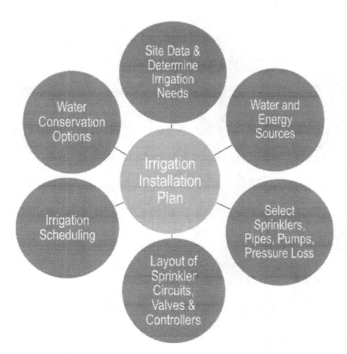

FIGURE 6.14 The basic elements of a site irrigation plan for gardens, using an installation guide for which an irrigation schedule should be developed. A testing and maintenance procedure should also be developed.

Simple Calculation for determining irrigation application rates

$$Rate\ (mm/h) = Total\ Flow\ (L/h) / Total\ Area\ Irrigated\ (m2)$$

For Example:

If we use 10 sprinklers operating together then

$$Total\ Flow\ Rate = (10*5L/min\ emitters) \times 60\ min = 3000\ L/h$$

Area irrigated is 100 m²

The watering rate of the system is calculated as:

$$3000\ L/h\ /\ 100\ m^2 = 30\ mm/hr$$

$$Run\ time\ (min) = \frac{Depth\ (mm) \times 60\ min}{Application\ Rate\ mm/h} = \frac{10mm \times 60\ min}{30\ mm/h} = 20min$$

FIGURE 6.15 An example calculation of the water application rate and running time, to determine the optimal watering rate for a given depth of water and known area for 10-head watering system.

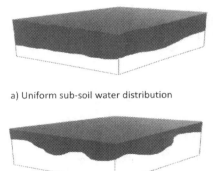

a) Uniform sub-soil water distribution

b) Non-Uniform sub-soil water distribution

FIGURE 6.16 (a) Expected subsoil wetting pattern for uniform water application across the targeted area. (b) Indicative subsoil wetting pattern resulting from non-uniform water applications.

sprinklers such as those used for lawns use impact- or gear-driven sprinklers to achieve a uniform distribution system with the calculation of water application rates for different systems and conditions given in Figure 6.15.

A uniform water distribution within the subsoil (Figure 6.16a) can be achieved by creating an intersection of non-uniform wetting patterns that can be defined for different sprinklers, spray heads (no moving parts) or pop-ups. As with broadacre systems, much depends on the soil texture, evaporation losses and the infiltration capacity, coupled with expected evapotranspiration and volume of moisture that can be stored within the root zone. The intake or infiltration rate and storage (or retention) capacity of the soil will determine the optimal rate at which water can be applied. Non-uniform applications will be reflected in the subsoil distribution patterns (Figure 6.16b) that haves implications for plant growth, yields or cover. These factors, among others, will determine the irrigation (watering) schedule required to provide a uniform subsurface distribution of water within the soil profile.

A basic tenet of irrigation efficiency is to follow these simple principles: (1) know the amount of water to apply, (2) when to apply it, (3) apply it uniformly and (4) limit waste (or loss) through wind drift, runoff or deep drainage (or percolation) owing to excessive water applications (IA 2017). Losses can also accrue from design problems, for example inadequate sprinkler placement or spacing, inappropriate spray or distribution patterns, incorrect nozzle selection and poor system pressure management (too high or too low) which can cause over- or under-application of water to the soil surface (IA and RUWSP 2019).

To improve efficacy of garden or urban systems, sprinkler controllers are becoming increasingly common and more sophisticated. Controllers can be used to manage time and length of irrigation, synchronise water applications with local weather data, or cross-referenced with rain sensors that can adjust timing of irrigation or switch systems off. Sprinkler controllers are manufactured for a range of functions from quite basic to more advanced models, and can be battery-powered, web-based, solar-powered to decoder controllers (IA and RUWSP 2019). They can also be used to manage different hydrozones or stations,[6] within the garden or urban setting, and may be used to manage up to 12 zones, depending on the complexity of the system, the area covered and the cost of installation.

6.6 Broadacre Irrigation System Design

Sprinkler irrigation provides a greater versatility than surface or flood irrigation, as it may be used on various landscapes, topographies and a broader range of soil types. However, this versatility presents its own problems in terms of designing and selecting the right irrigation components (i.e. piping, pumps, sprinklers, valves and nozzles) to match the conditions and the prospective crops or pastures to be grown. Recently published guidelines for Australia have proposed that optimal sprinkler systems are often associated with the following:

- Course or sandy soils with high infiltration rates;
- Shallow soils that may restrict land levelling required for surface irrigation methods;
- Areas with steep slopes that may be prone to erosion hazards;
- Growing high priced crops and where water is scarce and costly, where water supplied may be for main irrigations, supplemental irrigations or for protective irrigations;
- For inland regions where sprinklers may be used to supply the full crop requirement as irrigation water is scarce and this type of irrigation system ensures a high efficacy of water use, or where the system is able to be modified to meet demand deficits;
- In humid areas irrigation can provide supplemental water during the periods of drought or peak demand; and
- Sprinkler irrigation can be used for protecting crops from freezing temperature or frost damage (IA 2017).

Additional basic design and management parameters for irrigation developments, their design and maintenance are given in Table 6.4. Performance criteria recognised as industry standards fall into one of six categories of measurable performance, as follows:

1. water use efficiency,
2. energy use,
3. labour,
4. capital,
5. system effectiveness, and
6. environment.

6.6.1 Factors Affecting the Irrigation Interval

Often water use and soil type present challenges in terms of determining irrigation application rates for crops or plant needs (Figure 6.17). These issues can be managed through irrigation scheduling which

TABLE 6.4 Parameters and Metrics That Should Be Considered When Evaluating and Design a Broadacre Irrigation System, from Basic Site Information to Operational Capacity and Management Options

Steps	Design Parameters	Unit (s)	Other Attributes and Information
1	Irrigated area	Ha	• Area-size and shape • Size glasshouse or polytunnel • Size and quantity of pots (internal) • Land restrictions • Topography • Subsurface obstructions (pipes, cables, etc.)
2	Water supply	L/s	• Water allocation • Water sources (weir, dam, bores) • Supply rate or restrictions • Re-use potential • Energy hub/source for pumping options
3	Plant options		• Type of plant (crop, turf, pasture) • Growth seasonality/variation on water demand • Cultivation/harvest practice • Rooting depth • Crop water use coefficient
4	Soil water (Readily Available Water (RAW))	mm/min	• Soil type • Soil structure • Field capacity and wilting point • Infiltration rates
5	Local climate		• Evapotranspiration rates • Rainfall (seasonal totals, rates) • Wind • Managed environments (polytunnel, glasshouse)
6	Management system		• System integration • Type of application system (sprinkler, pivot, drip) • System compatibility and process controls • Budgeting and cost control • Risk assessment (operational, financial)
7	Irrigation system design elements	mm/day	• Irrigated area • Peak water use • Gross application rate (depth vs time) • Application efficiency (%) • Operational cycle (duration/frequency) • Pump utilisation ratios • Flow rate of system (delivery capacity)

Source: Adapted from Irrigation Australia notes (IA 2017).

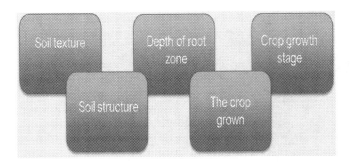

FIGURE 6.17 Different soil and crop parameters that affect the determination of water application rates and scheduling during crop/pasture growth. These should be considered in regard to irrigation design (see Tables 6.4 and 6.5) and irrigation method (see Figure 6.18) to determine the optimal application efficiency.

Sprinkler Irrigation Systems

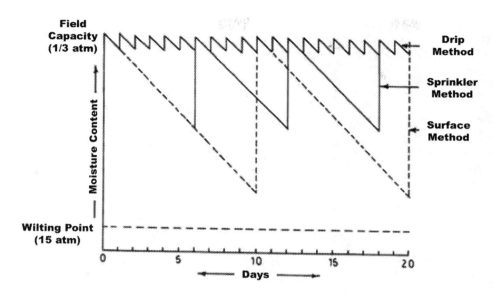

FIGURE 6.18 Demonstration of change in soil moisture content and successive irrigations using different application methods and irrigation intervals for each method, highlighting the level of moisture retention, depth of wetting front and potential loss from deep drainage (Rivers 2014).

TABLE 6.5 The Length of Interval and Amount of Water Applied to a Crop Via Irrigation Scheduling Can Be Determined Using Different Methods and Soil Condition Parameters

Method	Parameter	Equipment	Advantages	Disadvantages
Hand feel and appearance of soil	Soil moisture from feel	Hand probe	Easy can be accurate with experience	Poor accuracy Labour required
Gravimetric soil moisture	Soil moisture from samples	Auger, oven	High accuracy	Labour Time
Tensiometers	Soil moisture tension	Tensiometers, incl. vacuum gauge	Accurate Timely	Labour Maintenance
Electrical resistance blocks	Soil moisture from electrical resistance	Resistance blocks, metre	Instant Large range Can be remote	Sensitive to salinity Labour Maintenance
Water budget	Climate: temp, radiation, wind humidity, etc.	Weather station or data	No feild work Felxible Can forecast Used for mutliple sites	Calibration Adjustment Difficult calculations

determines when and how much water to apply to a field so as to maximise irrigation efficiency by applying the exact amount of water needed to replenish the soil moisture to the desired level. Irrigation scheduling saves water and energy, thus reducing production costs.

The interval between irrigations and the amount of water to apply at each irrigation depend on how much water is held in the root zone and how fast it is used by the crop (Figure 6.18). All irrigation scheduling procedures consist of monitoring indicators that determine the need for irrigation using various methods and parameters (Table 6.5).

6.6.2 Monitoring Irrigation Performance

While it is recognised that the initial performance of any irrigation system design is dependent on determining the appropriate sprinkler and water application rate to meet the expected soil properties and crop characteristics during the season, an effective subsequent performance monitoring system is essential. All

TABLE 6.6 Some of the Benefits of Monitoring and Reporting on Irrigation System Performance

Monitoring	Reporting
• Improved intra-system connectivity and distribution networks • Better calibration of sprinkler systems • Improved fertiliser management • Improved farm management and profitability • Refined application and distribution efficiencies • Understanding flow rates and differential water pressures • Water quality and sediment loads	• More appropriate water allocations • Catchment-scale assessments • Compliance with regulations • Integrated water resource management • Provides for comparative assessment • Assessment of potential for crop or soil contamination

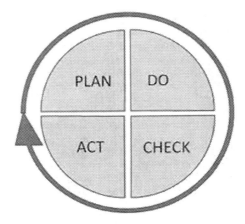

FIGURE 6.19 This shows an illustration of suggested action sequence that is undertaken to design, evaluate and monitor an irrigation system to ensure that is operates at optimal efficiency, productivity and cost-effectiveness.

irrigation scheduling procedures consist of monitoring indicators that determine the need for irrigation. So, any improvements in irrigation efficiency and water management and therefore productivity and costs, rely on accurate, representative and timely data. There are many parameters that can be monitored (and reported on if necessary) which will improve irrigation system efficiency (see Table 6.6). In all cases, it should be remembered that irrigation water can be supplied to end users in a variety of ways, with irrigators or irrigation managers being part of a regional irrigation "cooperative" (e.g. Harvey Water Irrigation Area in Western Australia) or they may be "self-suppliers" (i.e. from bores or river/stream systems).

Water could be accessed from engineered surface-water distribution systems, from natural water courses or from groundwater bores or wells. Each of these options presents management issues for suppliers, regulators and irrigators. In any case, once the water is supplied there is a need to design and monitor the efficiency of the water application system to deliver the best outcome for crops and the environment. In all cases monitoring, evaluation and reporting are key elements of this and should follow a sequence of actions to deliver the best outcome such as the steps shown in Figure 6.19.

6.7 Sprinkler Irrigation Research

Technological developments in irrigation technology continue and are especially important with the growing recognition of water resource scarcity. While improvements continue to be made into irrigation hardware and field equipment, research is also now being undertaken into smart optimisation of irrigation systems at the farm, district and regional scales. As an example, ongoing multi-national research is now examining the use of smart, networked, self-powered soil-moisture micro-sensors to be linked to wireless networks and to autonomously measure, monitor and analyse soil moisture and nutrient data within irrigated fields and to use this data to trigger irrigation events at the local scale and water resource management activities at the regional scale (Zia et al. 2013; Coles et al. 2013; Rivers et al. 2015; Zia et al. 2019).

6.8 Conclusions

This chapter has provided a brief overview of the various sprinkler irrigation systems with surface and subsurface irrigation systems discussed elsewhere in the handbook. In this chapter we have introduced an outline of the major sprinkler irrigation systems, together with some of the basic measures of irrigation efficiency, as well as the general levels of "efficiency" measured in these irrigation systems. The basic design criteria – using soils, crops, spacing and scheduling – have been introduced with pointers to the use of technology and scheduling as a means to improving water use efficiency.

In terms of uniformity and water application, close attention to the performance measures is recommended, because if these are not measured accurately, then the system cannot be effectively managed. Irrigation systems should be properly designed, installed and maintained to achieve high efficiencies, with an action plan put in place to ensure that the systems are performing according to their design parameters, to promote the efficient application, availability and use of water to enhance crop growth, productivity and reduce environmental impacts.

Notes

1. http://www.fao.org/water/en/.
2. https://www.usgs.gov/special-topic/water-science-school/science/.
3. Christiansen (1942) defined a numerical index that computed the system uniformity for overlapping sprinklers. This was designated the uniformity coefficient (UC) and "*a percentage on a scale of 0 to 100 (absolute uniformity)*". Using this scale to evaluate a system performance values of 80 and above are nominally considered to be the minimum value.
4. Reference BUL294, one of a series of the Agricultural and Biological Engineering Department, University of Florida. Original publication date March 1994. Reviewed July 2002. Visit the EDIS Web Site at http://edis.ifas.ufl.edu (Smajstrla et al. 2002).
5. American Society of Agricultural Engineering (ASAE) Standards – International Standard Organisation (ISO).
6. A hydrozone is an irrigation area that is covered by a single sprinkler valve.

References

Adeyemi, Olutobi, Ivan Grove, Sven Peets, and Tomas Norton. 2017. "Advanced monitoring and management systems for improving sustainability in precision irrigation." *Sustainability* 9 (3). doi: 10.3390/su9030353

Albaji, M., Eslamian, S., Naseri, A. and F. Eslamian. 2020. *Handbook of Irrigation System Selection for Semi-Arid Regions*, Boca Raton, Taylor and Francis, CRC Group, USA, 317 Pages.

ANSI, American National Standards Institute, and American Society of Agricultural Engineers ASAE. 1989. Test Procedure for Determining the Uniformity of Water Distribution of Center Pivot and Lateral Move Irrigation Machines Equipped with Spray or Sprinkler Nozzles American Society of Agricultural Engineers.

ARC-ILI Irrigation Design Manual. 2004. *Agricultural Research Council-Institute for Agricultural Engineering*. Silverton, Pretoria, RSA.

Ascough, G. W., and G. Kiker. 2004. "The effect of irrigation uniformity on irrigation water requirements." *Water SA* 28. doi: 10.4314/wsa.v28i2.4890

Bernardo, S., A. A. Soares, and E. C. Mantovani. 2006. *Manual de Irrigação*. 8th edition ed: Editora UFV, Viçosa.

Bjorneberg, D. L. 2013. "IRRIGATION | Methods☆." In *Reference Module in Earth Systems and Environmental Sciences*. Elsevier.

Brennan, Donna. 2008. "Factors affecting the economic benefits of sprinkler uniformity and their implications for irrigation water use." *Irrigation Science* 26 (2): 109–119. doi: 10.1007/s00271-007-0077-9

Burger, J. H., and ARC-Institute for Agricultural Engineering. 2003. *Irrigation Design Manual: A Manual for the Planning and Design of Irrigation Systems*. Silverton: ARC-Institute for Agricultural Engineering.

Burt, C. M., A. J. Clemmens, T. S. Strelkoff, K. H. Solomon, R. D. Bliesner, L. A. Hardy, T. A. Howell, and D. E. Eisenhauer. 1997. "Irrigation performance measures: efficiency and uniformity." *Biological Systems Engineering: Papers and Publications* 38. https://digitalcommons.unl.edu/biosysengfacpub/38

Christiansen, J. E. 1942. "Irrigation by sprinkling." *California Agriculture Experiment Station Bulletin* (No. 670): 124.

Coles, N. A., J. Camkin, N. Harris, A. Cranny, P. Hall, and H. Zia. 2013. "Water- boundaries and borders- the great intangibles in water quality management: Can new technologies enable more effective compliance?" TWAM 2013 International Conference: Transboundary water management across borders and interfaces: present and future challenges, Aveiro, Portugal, 16–20 March 2013.

El-Sayed, A. S., M. M. Hegazi, I. H. El-Sheikh, and A. F. Khader. 2009. "Performance evaluation of floppy sprinklers." *Misr Journal of Agricultural Engineering : Irrigation and Drainage* 26(2): 766–782.

Elwadie, M., L. Mao, and V. Bralts. 2010. "A simplified method for field evaluation of solid set sprinkler irrigation systems." *Applied Engineering in Agriculture* 26: 589–597 doi: 10.13031/2013.32064

FAO, Food and Agricultural Organisation, and World Water Council. WWC. 2015. Towards a Water and Food Secure Future: Critical Perspectives for Policy-Makers. Seventh World Water Forum in Daegu, South Korea.

Faria, Lessandro C., Giuliani do Prado, Alberto Colombo, Henrique F. E. de Oliveira, and Samuel Beskow. 2013. "Simulação da distribuição de água em diferentes condições de vento e espaçamentos entre aspersores." *Revista Brasileira de Engenharia Agrícola e Ambiental* 17: 918–925.

Ferrarezi, Rhuanito, sue Dove, and Marc Van Iersel. 2015. "An automated system for monitoring soil moisture and controlling irrigation using low-cost open-source microcontrollers." *Hort Technology* 25: 110–118. doi: 10.21273/HORTTECH.25.1.110

Griffiths, B., and N. Lecler. 2001. "Irrigation system evaluation." *Proceedings of the South African Sugar Technology Association* 75: 58–67.

Hoffman, G. J., and D. L. Martin. 1993. "Engineering systems to enhance irrigation performance." *Irrigation Science* 14(2): 53–63. doi: 10.1007/BF00208398

IA, Irrigation Australia. 2017. *Irrigation System Design Guidelines*. Brisbane: Queensland Department of Natural Resources and Mines and Irrigation Australia Limited.

IA, Irrigation Australia, and Regional Urban Water Supply Planning. Department of Natural Resources RUWSP, Mines and Energy. 2019. Efficient irrigation for water conservation: guideline for water efficient urban gardens and landscapes. Brisbane, Queensland: Queensland Government.

Israelsen, Orson W., Wayne D. Criddle, Dean K. Fuhriman, and Vaughn E. Hansen. 1944. "Bulletin No. 311- Water-Application Efficiencies in Irrigation." *UAES Bulletins* (Paper 273). doi: https://digitalcommons.usu.edu/uaes_bulletins/273

James, L. G. 1988. *Principles of Farm Irrigation System Design*. New York, NY: Wiley.

Keller, J., and R. D. Bliesner. 2000. *Sprinkler and Trickle Irrigation*. Caldwell: The Blackburn Press.

Koech, Richard, and Philip Langat. 2018. "Improving irrigation water use efficiency: a review of advances, challenges and opportunities in the Australian context." *Water* 10: 1771. doi: 10.3390/w10121771

Martin, D. L., Dennis Kincaid, and William Lyle. 2007. "Design and operation of sprinkler systems." *Design and Operation of Farm Irrigation Systems*. doi: 10.13031/2013.23699

Martin, D. L., William Kranz, A. Thompson, and H. Liang. 2012. "Selecting sprinkler packages for center pivots." *Transactions of the ASABE* 55: 513–523. doi: 10.13031/2013.41397

Van der Merwe, FPJ. 2008. "A method of evaluating an irrigation water use in terms of "efficient, sustainable and beneficial use of water in the public interest". MEng. Dissertation. Faculty of Engineering. University of Pretoria.

Moazed, H., A. Bavi, S. Boroomand-Nasab, A. Naseri, and M. Albaji. 2010. "Effects of climatic and hydraulic parameters on water uniformity coefficient in solid set systems." *Journal of Applied Sciences* 10: 1792–1796.

Nega, A. 2011. "Efficiency of different irrigation systems for sustainable management of water and nutrient flows in the Harvey irrigation area." M. Eng. Dissertation, Environmental Engineering, University of Western Australia.

O'Shaughnessy, Susan, Yenny Urrego Pereira, Steve Evett, Paul Colaizzi, and Terry Howell. 2013. "Assessing application uniformity of a variable rate irrigation system in a windy location." *Applied Engineering in Agriculture* 29: 497–510. doi: 10.13031/aea.29.9931

Prado, Giuliani do. 2016. "Water distribution from medium-size sprinkler in solid set sprinkler systems." *Revista Brasileira de Engenharia Agrícola e Ambiental* 20: 195–201.

Rivers, M. 2014. *APAS irrigation & WRM short course: irrigation scheduling and efficiency*. Perth, WA: University of Western Australia.

Rivers, M., N. A. Coles, H. Zia, N. R. Harris, and R. Yates. 2015. "How could sensor networks help with agricultural water management issues? Optimizing irrigation scheduling through networked soil-moisture sensors." *2015 IEEE Sensors Applications Symposium (SAS)*, 13–15 April 2015.

Rivers, M. R., K. P. Smettem, P. A. Davies, A. Allen, S. Vivian, and D. Weaver. 2010. *A comparison of water and nutrient utilisation in surface and centre pivot-irrigated beef properties in Western Australia – Final Report*. Perth, WA: Centre of Excellence for Ecohydrology, School of Environmental Systems Engineering, University of Western Australia.

Rivers, M. R. 2008. *Determining in-farm water and nutrient use, incorporation and loss mechanisms on beef farms. Final Report*. Perth, WA: Alcoa Foundation.

Rivers, M. R. 2011. *Alternative Irrigation Trials show promising results for a drying climate. Dairy-Trough*. Bunbury, Western Australia: Western Dairy.

Rivers, M. R., D. M. Weaver, K. R. J. Smettem, and P. M. Davies. 2011. "Estimating future scenarios for farm-watershed nutrient fluxes using dynamic simulation modelling." *Physics and Chemistry of the Earth* 36: 420–423.

Rogers, D. H., F. R. Lamm, M. Alam, T. P. Trooien, G. A. Clark, L. P. Barnes, and K. Markin. 1997. *Irrigation management series: efficiencies and water losses of irrigation systems*. Manhattan: Cooperative Extension Service, Kansas State University.

Smajstrla, A. G., B. J. Boman, G. A. Clark, D. Z. Haman, D. S. Harrison, F. T. Izuno, D. J. Pitts, and F. S. Zazueta. 2002. Efficiencies of Florida Agricultural Irrigation Systems. University of Florida, Gainesville, Florida: Institution of Food and Agricultural Science, Cooperative Extension Service.

Steffen, Will, Johan Rockström, Katherine Richardson, Timothy M. Lenton, Carl Folke, Diana Liverman, Colin P. Summerhayes, Anthony D. Barnosky, Sarah E. Cornell, Michel Crucifix, Jonathan F. Donges, Ingo Fetzer, Steven J. Lade, Marten Scheffer, Ricarda Winkelmann, and Hans Joachim Schellnhuber. 2018. "Trajectories of the earth system in the anthropocene." *Proceedings of the National Academy of Sciences* 115(33): 8252. doi: 10.1073/pnas.1810141115

USDA, United States Department of Agriculture, and Natural Resource Conservation Service NRCS. 1997. *Irrigation Guide*. Vol. Part 652, *National Engineering Handbook*.

Zhang, Lin, Gary P. Merkley, and Kasem Pinthong. 2013. "Assessing whole-field sprinkler irrigation application uniformity." *Irrigation Science* 31(2): 87–105. doi: 10.1007/s00271-011-0294-0

Zhu, X. Y., P. Chikangaise, W. D. Shi, W. H. Chen, and S. Q. Yuan. 2018. "Review of intelligent sprinkler irrigation technologies for remote autonomous system." *International Journal of Agricultural & Biological Engineering* 11(1): 23–30.

Zia, H., N. R. Harris, G. V. Merrett, A. Cranny, M. Rivers, and N. A. Coles. 2013. "A review on the impact of catchment-scale activities on water quality: a case for collaborative wireless sensor networks." *Journal of Computers and Electronics in Agriculture* 96(96): 126–138.

Zia, H., N. R. Harris, G. V. Merrett, and M. R. Rivers. 2019. "A low-complexity machine learning nitrate loss predictive model–towards proactive farm management in a networked catchment." *IEEE Access* 7: 26707–26720. doi: 10.1109/ACCESS.2019.2901218

7
Mini-Bubbler Irrigation

K. Y. Raneesh
Sri Shakthi Institute of Engineering and Technology

Saeid Eslamian
Isfahan University of Technology

7.1 Introduction 141
7.2 Types of Irrigation 142
7.3 Mini-Bubbler Irrigation 144
    System Layout and Components • Mini-Bubbler Emitters • Irrigation Scheduling • Design Criteria and Considerations • Advantages • Disadvantages • Basic Design Considerations
7.4 Conclusions 147
Bibliography 147

7.1 Introduction

Pressurized irrigation systems, in the form of sprinklers and micro-irrigation, have played an important role in improving irrigation efficiency and water application uniformity during the past several decades (Sadeghi et al. 2012). Micro-irrigation is the most recent technique that uses closed-conduit pipes to apply irrigation water to the soil near the plant root zone. Micro-irrigation achieves higher irrigation efficiencies and higher yields than traditional surface irrigation systems, but at the expense of increased energy consumption, higher capital costs, and higher maintenance requirements to keep mechanized pumps and filtration systems operational. Micro-irrigation systems can be broadly categorized into four types: drip, spray, mini-bubbler, and subsurface systems, based on their difference in hydraulic design or the method used to apply water to the soil. In addition, mini-bubbler systems can be further subdivided into high- and low-pressurized systems. Low-head mini-bubbler systems are based on gravity flow, do not require mechanical pumps or filtration systems, and can operate at pressure heads as low as 1 m (3.3 ft). They are particularly well suited for the irrigation of orchard crops, and traditional irrigation systems, such as furrows, can be easily converted into mini-bubbler systems.. Sadeghi et al. (2012) presented a new analytical procedure taking into account the non-uniform outflow profile for hydraulic analysis and design of multiple outlets pipelines. The method was developed based on presenting a new friction head loss distribution along the lateral and simulates pressure and outflow profiles along the trickle or sprinkler irrigation laterals and manifolds, as well as gated pipes. Mini-bubbler irrigation systems consist of a mainline connected to a water source, a constant head device, manifolds, laterals, and small-diameter delivery hoses. The laterals are laid midway between two rows of trees, and small-diameter hoses (called delivery hoses or tubes) are inserted in the laterals to deliver water to the trees. Hoses are anchored to a tree or stake, and hose heights are adjusted so that water flows out from all hoses at equal rates. The name of the system, mini-bubbler, is derived from the fountain of water streaming out from the hoses and from the bubbling noise made as air escapes from the pipelines when the system is turned on. The distinguishing feature of mini-bubbler systems is the flexible delivery hoses (Figure 7.1),

FIGURE 7.1 Mini-bubbler irrigation.

in contrast to small manufactured emitters commonly used with other micro-irrigation systems. These hoses allow greater rates of water to discharge into the small basins, and they do not require a filtration system because of their large orifice openings. The basins are usually circular or rectangular in shape, and are bordered by low embankments, or levees, so that water is uniformly distributed over the root zone. Despite their simplicity and advantages, mini-bubbler systems are not extensively used in developing countries. Researchers speculate that mini-bubblers are not used in this country because no manufactured components are needed and micro-irrigation companies would rather sell and install systems requiring component parts such as emitters, pressure regulators, and filters. In addition, the low use of mini-bubbler systems may be related to the demonstrated suitability of these systems for orchard crops, but not for other crops, and to the possibility of air locks occurring in the delivery hoses and laterals that block or restrict water flow through the system. For developing countries, the advantages of mini-bubbler systems are probably more important because many of these countries have energy shortage and plastic emitters are not readily available. It appears that mini-bubbler systems are not used in both developed and developing countries because engineers and farmers are not aware of this technology and there is no well-defined design procedure available to facilitate design and installation.

The aim of this chapter is to showcase the importance of mini-bubbler system for application in irrigating crops. In mini-bubbler irrigation, the application of irrigation water to the soil surface as a little stream, typically from a small-diameter tube (1–13 mm) or a commercially available emitter is mainly designed. The main application is the use of these systems in landscape irrigation systems. In this chapter, the use and application of mini-bubbler system in agriculture is mainly envisaged.

7.2 Types of Irrigation

Irrigation is the application of controlled amounts of water to plants at needed intervals. Irrigation helps to grow agricultural crops, maintain landscapes, and rejuvenate disturbed soils in dry areas and during periods of less than average rainfall. Irrigation also has other uses in crop production, including frost protection, suppressing weed growth in grain fields, and preventing soil consolidation. In contrast, agriculture that relies only on direct rainfall is referred to as rain-fed or dryland farming. There are several methods of irrigation. They vary in how the water is supplied to the plants. The goal is to apply the water to the plants as uniformly as possible, so that each plant has the amount of water it needs, neither too much nor too little.

Surface irrigation is the oldest form of irrigation and has been in use for thousands of years. In *surface* (*flood*, or *level basin*) irrigation systems, water moves across the surface of an agricultural lands, in order to wet it and infiltrate into the soil. Surface irrigation can be subdivided into furrow, *border strip or basin irrigation*. It is often called *flood irrigation* when the irrigation results in flooding or near flooding of the cultivated land. Historically, this has been the most common method of irrigating agricultural land and is still used in most parts of the world.

Micro-irrigation, sometimes called localized irrigation, low-volume irrigation, or trickle irrigation, is a system where water is distributed under low pressure through a piped network, in a predetermined pattern, and applied as a small discharge to each plant or adjacent to it. Traditional drip irrigation using individual emitters, subsurface drip irrigation, micro-spray or micro-sprinkler irrigation, and mini-bubbler irrigation all belong to this category of irrigation methods.

Drip (or micro) irrigation, also known as trickle irrigation, functions as its name suggests. In this system water falls drop by drop just at the position of roots. Water is delivered at or near the root zone of plants, drop by drop. This method can be the most water-efficient method of irrigation, and if managed properly, evaporation and runoff are minimized. The field water efficiency of drip irrigation is typically in the range of 80%–90% when managed correctly. In modern agriculture, drip irrigation is often combined with plastic mulch, further reducing evaporation, and is also the means of delivery of fertilizer. The process is known as fertigation.

Deep percolation, where water moves below the root zone, can occur if a drip system is operated for too long or if the delivery rate is too high. Drip irrigation methods range from very high-tech and computerized to low-tech and labor-intensive. Lower water pressures are usually needed than for most other types of systems, with the exception of low-energy center pivot systems and surface irrigation systems, and the system can be designed for uniformity throughout a field or for precise water delivery to individual plants in a landscape containing a mix of plant species. Although it is difficult to regulate pressure on steep slopes, pressure-compensating emitters are available, so the field does not have to be level. High-tech solutions involve precisely calibrated emitters located along lines of tubing that extend from a computerized set of valves.

In sprinkler or overhead irrigation, water is piped to one or more central locations within the field and distributed by overhead high-pressure sprinklers or guns. A system using sprinklers, sprays, or guns mounted overhead on permanently installed risers is often referred to as a solid-set irrigation system. Higher pressure sprinklers that rotate are called rotors and are driven by a ball drive, gear drive, or impact mechanism. Rotors can be designed to rotate in a full or partial circle. Guns are similar to rotors, except that they generally operate at very high pressures of 275–900 kPa and flows of 3–76 L/s, usually with nozzle diameters in the range of 10–50 mm. Guns are used not only for irrigation, but also for industrial applications such as dust suppression and logging.

Sprinklers can also be mounted on moving platforms connected to the water source by a hose. Automatically moving wheeled systems known as traveling sprinklers may irrigate areas such as small farms, sports fields, parks, pastures, and cemeteries unattended. Most of these use a length of polyethylene (PE) tubing wound on a steel drum. As the tubing is wound on the drum powered by the irrigation water or a small gas engine, the sprinkler is pulled across the field. When the sprinkler arrives back at the reel the system shuts off. This type of system is known to most people as a traveling irrigation sprinkler and they are used extensively for dust suppression, irrigation, and land application of waste water.

A lawn sprinkler system is permanently installed, as opposed to a hose-end sprinkler, which is portable. Sprinkler systems are installed in residential lawns, in commercial landscapes, for churches and schools, in public parks and cemeteries, and on golf courses. Most of the components of these irrigation systems are hidden under ground, since aesthetics are important in a landscape. A typical lawn sprinkler system will consist of one or more zones, limited in size by the capacity of the water source. Each zone will cover a designated portion of the landscape. Sections of the landscape will usually be divided by microclimate, type of plant material, and type of irrigation equipment. A landscape irrigation system may also include zones containing drip irrigation, mini-bubblers, or other types of equipment besides sprinklers.

Although manual systems are still used, most lawn sprinkler systems may be operated automatically using an irrigation controller, sometimes called a clock or timer. Most automatic systems employ electric solenoid valves. Each zone has one or more of these valves that are wired to the controller. When the controller sends power to the valve, the valve opens, allowing water to flow to the sprinklers in that zone.

There are two main types of sprinklers used in lawn irrigation, pop-up spray heads and rotors. Spray heads have a fixed spray pattern, while rotors have one or more streams that rotate. Spray heads are used to cover smaller areas, while rotors are used for larger areas. Golf course rotors are sometimes so large that a single sprinkler is combined with a valve and called a 'valve in head'. When used in a turf area, the sprinklers are installed with the top of the head flush with the ground surface. When the system is pressurized, the head will pop up out of the ground and water the desired area until the valve closes and shuts off that zone. Once there is no more pressure in the lateral line, the sprinkler head will retract back into the ground. In flower beds or shrub areas, sprinklers may be mounted on above ground risers or even taller pop-up sprinklers may be used and installed flush as in a lawn area.

Sub-irrigation has been used for many years in field crops in areas with high water tables. It is a method of artificially raising the water table to allow the soil to be moistened from below the plants' root zone. Often those systems are located on permanent grasslands in lowlands or river valleys and combined with drainage infrastructure. A system of pumping stations, canals, weirs and gates allows it to increase or decrease the water level in a network of ditches and thereby control the water table.

Sub-irrigation is also used in the commercial greenhouse production, usually for potted plants. Water is delivered from below, absorbed by upward, and the excess collected for recycling. Typically, a solution of water and nutrients floods a container or flows through a trough for a short period of time, 10–20 min, and is then pumped back into a holding tank for reuse. Sub-irrigation in greenhouses requires fairly sophisticated, expensive equipment and management. Advantages are water and nutrient conservation and labor savings through reduced system maintenance and automation. It is similar in principle and action to subsurface basin irrigation.

7.3 Mini-Bubbler Irrigation

Mini-bubblers are used to irrigate bigger areas and apply water on 'per plant' basis. Water from the mini-bubbler head either runs down from the emission device or spreads a few inches in an umbrella pattern. Mini-bubbler emitters dissipate water pressure through a variety of diaphragm material (a silicon diaphragm inside an emitter flexes to regulate water output) and deflect water through small orifices. Mini-bubbler emission devices are equipped with single or multiple port outlets. Mini-bubblers are available in adjustable flow and pressure-compensating types. Mini-bubbler irrigation is a localized, low-pressure, solid permanent installation system used in tree groves. Each tree has a round or square basin which is flooded with water during irrigation. The water infiltrates into the soil and wets the root zone. The water is applied through mini-bubblers. These are small emitters placed in the basins which discharge water at flow rates of 100–250 L/h. Each basin can have one or two mini-bubblers as required.

7.3.1 System Layout and Components

The system layout is the typical one of all pressurized systems. It consists of a simple head control unit without filters and fertilizer apparatus. The mains and the sub-mains are usually buried rigid polyvinylchloride (PVC) pipes, with hydrants rising on surface. The manifolds and laterals are also often buried rigid PVC pipes. The mini-bubblers are placed above ground, supported on a stake, and connected to the laterals with a small flexible tube rising on the surface, or they can be fitted on small PVC risers connected to the buried laterals. The difference between mini-bubbler systems and other micro-irrigation installations is that whereas in the other installations the lateral lines are small (12–32 mm), the mini-bubblers are usually 50 mm (due to the lateral high discharge). This is why the laterals need to be underground.

Mini-Bubbler Irrigation

FIGURE 7.2 Mini-bubbler emitter.

7.3.2 Mini-Bubbler Emitters

The mini-bubblers are small plastic head emitters with a threaded joint. They were originally designed for use on risers above ground for flood irrigation of small ornamental areas. In recent decades, they have been used successfully in several countries for the irrigation of fruit trees. They perform well under a wide range of pressures delivering water in the form of a fountain, small stream or tiny umbrella in the vicinity of the emitter.

The main performance characteristics are:

- Operating pressure: 1.0–3.0 bars,
- Flow rate (discharge): 100–250 L/h (adjustable), and
- No filtration is required.

There is a wide range of flow rates up to 800 L/h; this presents only low-discharge mini-bubblers (Figure 7.2).

7.3.3 Irrigation Scheduling

With mini-bubbler irrigation, the percentage of the root soil volume wetted is about 80%. Thus, there are no restrictions on the way the irrigation program is prepared. This can be either fixed depletion or fixed interval, taking into consideration the soil water holding capacity, the availability of the irrigation water, the size of flow, etc.

7.3.4 Design Criteria and Considerations

Mini-bubbler irrigation is mainly applied in fruit tree orchards. The most important criteria, apart from the routine design criteria, are the system's special features and characteristics. Mini-bubbler emitters discharge water on the same spot of ground at high rates. Thus, for a uniform distribution over the basin area, a minimum of land preparation is needed. In sandy soils, the water infiltrates at the point of application and high losses occur due to deep percolation. In fine soils with low infiltration rates, the water ponds and evaporation occurs. Mature trees always take two mini-bubbler emitters, one on each side, in order to ensure an acceptable uniformity of application. The flow rate per tree is relatively high compared with other micro-irrigation techniques at about 500 L/h. Thus, the diameter of an 80 m-long lateral for a single row of 13 trees spaced at 6 m intervals should be 50 mm. The common practice is to have one lateral per two rows of trees with small flexible tubes extended on both sides and connected to the mini-bubblers. In this way, the same size of lateral pipe (50 mm), placed (buried) between two rows, can serve 12 trees on each side (24 trees in total) spaced at 6 m intervals with 48 mini-bubbler emitters. The size of the equipment for the installation should always be able to accommodate the flows required for mature trees. For longer laterals, pressure-compensating mini-bubblers can be used, though this involves higher energy consumption and more expensive higher pressure pipes.

7.3.5 Advantages

High irrigation application efficiency, up to 75%, results in inconsiderable water savings, with absolute control of the irrigation water from the source to the tree basin. All the piping network is buried, so there are no field operation problems. The technology is simple and no highly sophisticated equipment is used. The system can be operated by unskilled farmers and laborers. No filters or fertilizer injectors are needed.

7.3.6 Disadvantages

One of the main disadvantage of the mini bubbler system is that the initial purchase cost is high. Small water flows cannot be used as in other micro-irrigation systems. In sandy soils with high infiltration rates, it is difficult to achieve a uniform water distribution over the tree basins.

7.3.7 Basic Design Considerations

The goal of mini-bubbler irrigation design is to size the diameters of the pipelines to deliver an equal volume of water to each plant while keeping the total friction and minor head losses less than the available head at the water source. To permit flexibility in determining the best design, a thorough understanding of the hydraulic design principles of mini-bubbler systems is required, and ingenuity must always be employed to keep costs low and designs flexible. Generally, the simpler the design, the less the material will cost. For example, minimizing the number of manifolds and laterals, which comprise a major portion of total material costs, will reduce costs. Corrugated PE, smooth PE, and PVC pipes are the most common pipe materials used for the laterals, and care should first be taken to design the mini-bubbler system with internal pipe diameters that are commercially available in our country. In addition, using corrugated PE pipe may be impractical for designs with very low heads on level ground, because the friction losses for corrugated PE pipes are much larger and small pipe undulations may cause air locks in the laterals. Another important material consideration is the type of connection used to connect the delivery hoses to the laterals. The type of connection chosen is often subject to the availability of materials at a particular locality because commercial connections are usually not available. Because mini-bubbler systems are based on gravity flow, the slope of the field is another crucial factor in design. The designs for systems located on level ground and those on gradual slopes differ slightly because their maximum and minimum delivery hose heights occur at different points along the lateral. Also, systems on gradual slopes will gain energy down-slope, and this allows laterals to be longer than on level ground and permits greater diversity in design for a given available head. Correspondingly, designs on level ground do not gain energy; lateral lengths cannot be as long as on sloping ground with a given head, and design flexibility is limited. Another major difference between design of mini-bubblers and other micro-irrigation systems is that design flows are assumed at the beginning of the design, and these flow rates, if run continuously, are typically much greater than the peak water requirement of any crop. In contrast, sprinkler and drip systems are typically designed to operate continuously for 90% of the time during the peak consumptive period and their design flow rates are calculated at the beginning of design. Therefore, mini-bubbler systems tend to operate less frequently than sprinkler or drip irrigation systems because their design flow rates are much greater. The design procedure developed in this study is based on a composite of earlier mini-bubbler designs, and only one friction loss equation, the Darcy–Weisbach equation, is used to simplify calculations.

Irrigation frequency and duration of water application to meet the specific crop water requirements for a particular soil and climate are determined after all pipe and hose diameters have been selected. Earlier mini-bubbler designs arbitrarily assumed delivery hose flow rates at the beginning of design, but mini-bubbler design can be improved by using the design flushing velocities as a means to prevent air locks from occurring in the delivery hoses.

7.4 Conclusions

Micro-irrigation systems are subdivided into four categories related to their difference in hydraulic design: drip, spray, mini-bubbler, and subsurface systems. The design of mini-bubbler systems differs from that of other micro-irrigation systems because they are based on gravity flow and do not require external energy or elaborate filtration systems. The fact that the dissemination of mini-bubbler design has occurred largely by site visits to existing mini-bubbler systems, probably indicates that available literature does not adequately describe the simplicity of mini-bubbler design. Steps recommended for mini-bubbler design include:

1. Establishing the field layout and elevation of pipeline profiles.
2. Designing a constant head device and establishing design elevations.
3. Selecting design criteria.
4. Calculating the length and number of pipes.
5. Calculating the design flow for each pipe.
6. Sizing initial pipe diameters by using head loss gradient charts and using midway spacings for the Christiansen reduction coefficients.
7. Calculating the delivery hose elevation by determining the friction head losses in the laterals and delivery hoses. Checking that all delivery hose elevations are between the maximum and minimum delivery hose elevations and comparing the calculated friction head losses in the lateral to the Christiansen estimated friction head loss.
8. Plotting the results of the delivery hose elevations on the lateral profile to construct an energy diagram and dynamically calibrating the delivery hose elevations after installation.
9. Determining the time of application during peak consumptive periods.
10. Installing the system and dynamically calibrating the elevation of the delivery hoses.

Bibliography

Albertson, M.L., J.R. Barton, and D.B. Simons. 1960. *Fluid Mechanics for Engineers*. Prentice-Hall, Inc. New Jersey, USA.

Boswell, M.J. 1984. *James Hardie Micro-Irrigation Design Manual*. James Hardie Irrigation Inc., El Cajon, CA, USA.

Christiansen, J.E. 1942. *Irrigation by Sprinkling. California Agriculture Experiment Station Bulletin 670*. Davis, CA.

Harrington, G.J. 1971. The Mechanics of Air Locks (Air Binding) in Small Diameter Pipes. New Zealand Agriculture Engineering Institute. Internal Report No. 2, New Zealand.

Hermsmeier, L.F. and L.S. Willardson. 1970. Friction Factors for Corrugated Plastic Tubing. *ASCE Journal of Irrigation and Drainage Division*. Vol. 96, No. IR3, pp. 265–271.

Hillel, D. 1987. The Efficient Use of Water in Irrigation. The World Bank Technical Paper No. 64. Washington DC.

Hull, P.J. 1981. A Low Pressure Irrigation System for Orchard Tree and Plantation Crops. *The Agriculture Engineer*. Summer, Vol. 36, No. 2, pp. 55–58.

Jensen, M.E. (ed.) 1980. *Design and Operation of Trickle (Drip) Systems. Design and Operation of Farm Irrigation Systems*. ASAE Monograph No. 3. St. Joseph, Michigan.

Jordan, T.D. 1984. *A Handbook of Gravity-Flow Water Systems.* Intermediate Technology Publications, London, UK.

Nicklin, D.J., et al. 1962. Two Phase Flow in Vertical Tubes. *Transactions of the Institution of Chemical Engineers,* Vol. 40, pp. 61–68.

Petersen, M.S. 1980. Recommendations for Use of SI Units in Hydraulics. *Journal of the Hydraulics Division. ASCE,* Vol. 106, No. HY12, Proceedings Paper 15931.

Rawlins, S.L. 1977. *Uniform Irrigation with a Low Head Mini-bubbler System. Agriculture and Water Management.* Elsevier Scientific Publishing Company, Amsterdam, The Netherlands. Vol. 1, pp. 166–178.

Sadeghi, S.H., Mousavi, S.F., Eslamian, S.S., Ansari, S. and F. Alemi. 2012. A Unified Approach for Computing Pressure Distribution in Multi-Outlet Irrigation Pipelines, Iranian. *Journal of Science and Technology,* Vol. 36, No. C2, pp. 209–223.

Thornton, J.R. and D. Behoteguy. 1980. Operation and Installation of a Mini-bubbler System. ASAE Technical Paper No. 80-2059. St. Joseph, Michigan, USA.

USDA. 1982. Consumptive Use of Water by Major Crops in the Southwestern United States. Conservation Research Report Number 29. Agriculture Research Service, USA.

Waheed, S.I. 1990. Design Criteria for Low Head Mini-bubbler Systems. Master's Thesis. University of Arizona, USA.

Watters, G.Z. and J. Keller. 1978. Trickle Irrigation Tubing Hydraulics. ASAE Technical Paper No. 78-2015. St. Joseph, Michigan, USA.

Worstell, R.V. 1975. An Experimental Buried Multiset Irrigation System. ASAE Technical Paper No. 75-2540. St. Joseph, Michigan, USA.

8

Measured Irrigation

8.1	Introduction	150
8.2	Manual Measured Irrigation	151
	Installing Manual Measured Irrigation	
8.3	Unpowered Evaporative Valve	153
	Instructions for Installing the Unpowered Evaporative Valve • How to Adjust the Irrigation Frequency • How to Adjust the Water Usage Rate • Key Features of the Unpowered Evaporative Valve	
8.4	Unpowered Terracotta Valve	162
	Above-Ground Installation of the Unpowered Terracotta Valve • In-Ground Installation of the Unpowered Terracotta Valve • How to Use the Unpowered Terracotta Valve • How to Adjust the Interval between Irrigation Events • How to Adjust the Water Usage Rate • When is Dripper Discharge Independent of Pressure? • Key Features of the Unpowered Terracotta Valve	
8.5	Terracotta Irrigation Controller for Latching Solenoids	171
	Installing the Terracotta Irrigation Controller for Latching Solenoids	
8.6	Automation Kit for Farm Pond Irrigation	173
	Contents of the Automation Kit for Farm Pond Irrigation • Installing the Automation Kit for Farm Pond Irrigation	
8.7	Unpowered Irrigation Controller for Solenoid Valves	178
	Installing the Unpowered Irrigation Controller for Solenoid Valves • Calibrating the Unpowered Irrigation Controller for Solenoid Valves • Weather-Based Irrigation Control • Pressure-Independent Dripper Discharge • Key Features of the Unpowered Irrigation Controller for Solenoid Valves	
8.8	Soil Moisture and Measured Irrigation Scheduling	188
	Soil Moisture Probe • Introduction to Measured Irrigation Scheduling • Root Zone Scheduling for Manual Measured Irrigation and the Unpowered Evaporative Valve	
8.9	DIY Unpowered Uniform Drip Irrigation on Sloping Land	191
	Schematic Diagram for the Omodei Method • Installation for the Omodei Method • Example of the Omodei Method • Using Sub-Drippers to Further Improve the Omodei Method • Installation Cost	
8.10	Conclusions	198
	References	199

Bernie Omodei
Measured Irrigation

8.1 Introduction

Water has become an increasingly scarce and valuable resource due to the increasing population and inefficient practices. Agriculture is the largest sector using water with low efficiency and low cost. Microirrigation, the slow and targeted application of irrigated water to prescribed soil volumes, has become synonymous with modern and efficient irrigation practices that conserve precious water and maximize plant performance [1,2]. Low-cost innovative water-saving technology is the need of the hour [3,4].

Conventional drip irrigation systems use a timer or controller to control the opening and closing of valves in order to control the duration of the irrigation event and the frequency of irrigation. The volume of water emitted by a dripper during the irrigation event is controlled by using drippers with a specified flow rate and controlling the duration of the irrigation event. The acceptance of this drip irrigation scheduling paradigm has led to the development of pressure-compensating (PC) drippers, whereby the flow rate from the dripper is relatively constant for a range of water pressures. Measured irrigation uses a totally different paradigm for controlling the volume of water emitted by a dripper during the irrigation event.

One of the effects of climate change has been a decrease in annual rainfall in sub-Saharan Africa. The livelihood on millions of people in sub-Saharan Africa (SSA) depends on effective rainwater harvesting so that crops can be irrigated throughout the year. The long-term future of many communities requires the adoption of sustainable agriculture [5]. The Billion Dollar Business Alliance for Rainwater Harvesting was established in 2016 [6]. World Food Programme, ICRAF and partners envision building an integrated farm pond support system capable of expanding and sustaining the massive upscaling of farm pond technology. In October 2016, Bernie Omodei was invited to Kenya by ICRAF (World Agroforestry Centre) to participate in a workshop in Machakos County to train farmers and extension workers in the use of measured irrigation. Measured irrigation is a low-cost approach to irrigation scheduling that has the potential to significantly improve the water efficiency of gravity-feed irrigation using stored rainwater.

Measured irrigation *is an irrigation scheduling method that satisfies the following two conditions:*

1. *Variations in the water usage rate throughout the year are controlled by the prevailing net evaporation rate (evaporation minus rain).*
2. *The volume of water emitted by each emitter during an irrigation event is controlled directly without the need to control the flow rate or the duration of the irrigation event.*

Note that the flow rate and the duration of the irrigation event adjust automatically to ensure that the required volume of water is delivered by each dripper. As the water pressure increases, the flow rate increases. However with measured irrigation, there is a corresponding decrease in the duration of the irrigation event to ensure that volume of water emitted by a dripper does not change.

The conventional volume control paradigm requires the control of two variables, namely, flow rate and time. Measured irrigation requires the control of a single variable, namely, volume. Once the focus of attention changes from flow rate and time to volume, then the design of drip irrigation systems may change accordingly.

Measured irrigation is a radical departure from the conventional drip irrigation scheduling paradigm and the implications for water efficiency and energy efficiency are significant [7–9]. Measured irrigation is a new approach to drip irrigation scheduling rather than a new irrigation technology. Existing drip irrigation installations may be upgraded to measured irrigation. However, to maximize water efficiency and energy efficiency for an irrigation application, it is preferable that the measured irrigation implementation is designed from scratch.

Gravity-feed measured irrigation is well suited to smallholders in poorer countries where access to mains power and mains water is unavailable, unreliable or too expensive. In remote locations where mains power and mains water are unavailable, gravity-feed measured irrigation can provide an irrigation

Measured Irrigation 151

TABLE 8.1 Pressurized Drip Irrigation Versus Gravity-Feed Measured Irrigation

Pressurized Drip Irrigation	Gravity-Feed Measured Irrigation
Requires access to mains water or to mains power (to operate a high-pressure pump).	Does not require access to mains water or to mains power, and hence can be installed in remote locations.
Pressure-compensating drippers are required to control of the volume of water emitted by each dripper.	The volume of water emitted by each dripper is controlled directly and is independent of the flow rate.
Water usage does not respond automatically to the prevailing on-site weather conditions.	Water usage responds automatically to the prevailing on-site weather conditions.
The irrigation scheduling is controlled by an irrigation controller or timer.	The irrigation scheduling is controlled by evaporation from and rainfall into a container.
Hose clamps are necessary.	Hose clamps are not needed due to the low pressure.
A water tank needs a high-pressure pump (for example, 500W).	A water tank may need a low-pressure pump (for example, 14W).

system that delivers a measured volume of water from each dripper. Table 8.1 compares pressurized drip irrigation with gravity-feed measured irrigation.

An irrigation controller is called "smart" when the irrigation scheduling is controlled by the prevailing weather conditions [10]. In 2011, the US Environment Protection Agency (EPA) published WaterSense specifications for weather-based irrigation controllers [11]. WaterSense is a label administered by EPA and irrigation controllers can use the WaterSense label provided that they satisfy the labelling criteria [12]. The Smart Approved WaterMark is a label used for Australian products [13]. The WaterMark labelling criteria for irrigation controllers is similar to the WaterSense criteria. Over the past 10 years, there has been significant research and development into weather-based irrigation controllers, and many new products have entered the market [3].

Measured irrigation scheduling is by definition "smart". With measured irrigation, the water usage rate is directly proportional to the net evaporation rate. In March 2019, I submitted a report [14] to Smart Approved WaterMark demonstrating a strong correlation between dripper discharge for the Unpowered Evaporative Valve and reference evapotranspiration. Research trials were conducted for 40 consecutive days from 5 February 2019 to 16 March 2019 at the Adelaide Airport Weather Station of the Bureau of Meteorology (BOM). The results demonstrated a 92% correlation. In February 2010, a report [15] was published by BOM demonstrating a similar correlation between pan evaporation and reference evapotranspiration using daily historical data from 7 BOM Weather Stations in NSW. In September 2019, the Unpowered Evaporative Valve was accepted by the Expert Panel for the Smart Approved WaterMark [14].

In addition to the measured irrigation implementations discussed in this chapter, further implementations and additional details may be downloaded from the Measured Irrigation website [16].

8.2 Manual Measured Irrigation

It is assumed that a smallholder is using drip irrigation (either pressurized or gravity feed) on a garden or a small plot of land.

To install manual measured irrigation, all that is needed is an evaporator and an adjustable dripper.

The **evaporator** is any container with vertical sides, with a surface area of at least $0.05\,m^2$, and a depth of at least $0.1\,m$ (Figure 8.1).

Any adjustable dripper may be used. An excellent example of an adjustable dripper is Claber 91214 adjustable dripper (Figure 8.2).

FIGURE 8.1 Examples of suitable evaporators.

FIGURE 8.2 Claber 91214 adjustable dripper.

8.2.1 Installing Manual Measured Irrigation

Step 1. Draw a line on the inside of the evaporator about 1.5 cm below the overflow level. This line corresponds to the high level (Figure 8.3).

Step 2. Connect the adjustable dripper to the irrigation system and position the evaporator so that the adjustable dripper drips water into the evaporator during the irrigation. The adjustable dripper should be at the same level as the irrigation drippers. The adjustable dripper is called the **control dripper**.

Step 3. Place an empty measuring container under one of the irrigation drippers.

FIGURE 8.3 Draw a line on the inside of the evaporator about 1.5 cm below the overflow level.

Step 4. Adjust the control dripper so that flow rate is about the same as the flow rate of the irrigation drippers.

Step 5. You may wish to protect the evaporator to prevent animals drinking the water, but make sure that you do not impede the evaporation (chicken wire is ideal).

Figure 8.4 shows the flowchart for using manual measured irrigation.

If your plants require less frequent watering, you may choose not to irrigate on certain evenings. For example, at sunset one day the water level is below the high level and you decide not to irrigate. At sunset the following day the water level will have fallen even further, and so when you irrigate the irrigation volume will be the sum of the irrigation volumes for both days. Changing the irrigation frequency does not affect the water usage rate.

If the garden requires more frequent watering, you may choose to irrigate at the middle of the day as well as at sunset (for example, if the weather is very hot and dry).

8.3 Unpowered Evaporative Valve

It is assumed that a smallholder is using drip irrigation (either pressurized or gravity feed) on a garden or a small plot of land. Using the Unpowered Evaporative Valve, you can upgrade your drip irrigation system so that all your plants are irrigated automatically. The water supply pressure should be at least 10 kPa (1 m head). Provided you have a continuous water supply, you can leave your garden unattended for weeks on end. This will allow you to become involved in other activities away from the garden; for example, travelling to the market to sell your produce. The Unpowered Evaporative Valve can be used for gravity-feed or pressurized systems, sprinkler or drip irrigation.

The water usage rate for the Unpowered Evaporative Valve is directly proportional to the prevailing on-site net evaporation rate. This is a unique feature of measured irrigation.

The valve in the Unpowered Evaporative Valve has a 15 mm inlet and outlet and hence, the flow rate is limited by the size of the valve. Provided that the drip irrigation system is already working effectively, you may need to use a number of Unpowered Evaporative Valves to automate the irrigation system depending on the size of the plot.

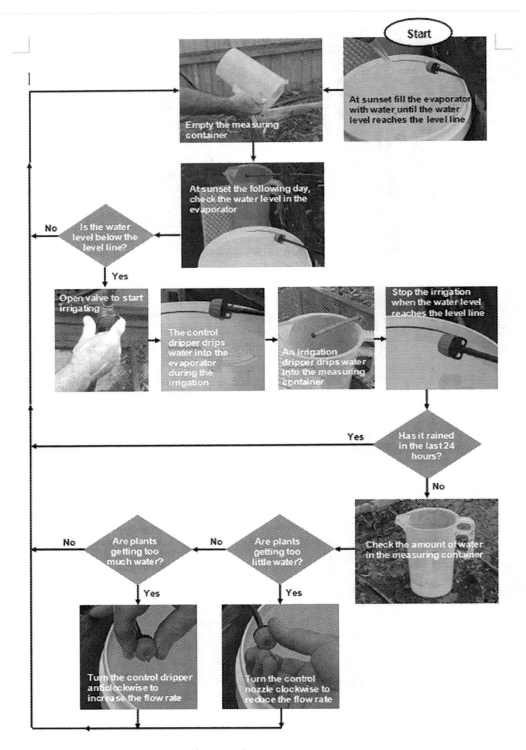

FIGURE 8.4 Flowchart for using manual measured irrigation.

8.3.1 Instructions for Installing the Unpowered Evaporative Valve

Installing the Unpowered Evaporative Valve is incredibly simple. Start with any drip irrigation application, either pressurized or gravity feed. Prior to installing the controller, it is assumed that the irrigation is operated manually by opening and closing a valve.

Step 1. Remove the Unpowered Evaporative Valve from the shipping carton. Connect the adjustable control dripper by screwing it onto the thread provided. Adjust the float shaft so that it is pointing up (be very careful when adjusting the float shaft to avoid placing any stress on the fragile plastic float shaft). Tighten the two back-nuts so that the rubber washers prevent water leaking from the evaporator.

Step 2. Position the evaporator in a suitable location so that the evaporation matches the evaporation in your garden.

Step 3. Connect the water supply to the inlet of the Unpowered Evaporative Valve (the inlet is on the opposite side to the adjustable control dripper). The water supply pressure should be between 10 and 800 kPa (Figure 8.5).

Step 4. Connect the Unpowered Evaporative Valve outlet (next to the adjustable control dripper) to the irrigation zone.

Step 5. Readjust the float shaft so that it points vertically up. Be very careful when adjusting the float shaft to avoid placing any stress on the fragile plastic float shaft. Position the adjustable control dripper so that it will drip water into the evaporator during the irrigation (Figure 8.6).

Step 6. For gravity-feed applications you may need to adjust the height of the evaporator so that the control dripper is at the same level as the irrigation drippers.

Step 7. Slide the float over the float shaft (Figure 8.7).

Step 8. Turn on the water supply and the irrigation should start. The adjustable control dripper drips water into the evaporator.

Step 9. Adjust the control dripper so that flow rate is about the same as the flow rate of the irrigation drippers (Figure 8.8).

Step 10. Fill the evaporator with water until the float jumps up and the irrigation stops (Figure 8.9).

Step 11. The float falls as water slowly evaporates from the evaporator. When the float reaches the low level the irrigation starts automatically. The float rises as the control dripper drips water into

FIGURE 8.5 Connect the water supply.

FIGURE 8.6 Float shaft must be vertical.

FIGURE 8.7 Slide the float over the float shaft.

Measured Irrigation 157

FIGURE 8.8 Adjusting the control dripper.

FIGURE 8.9 Fill the evaporator with water until the irrigation stops.

FIGURE 8.10 The irrigation starts when the float reaches the low level.

FIGURE 8.11 The irrigation stops when the float reaches the high level.

the evaporator. When the float reaches the high level the irrigation stops automatically. The cycle continues indefinitely (Figures 8.10 and 8.11).

Step 12. You may wish to protect the evaporator to prevent animals drinking the water, but make sure that you do not impede the evaporation (chicken wire is ideal). Replace the water and clean the evaporator regularly to remove algae and other contaminants.

8.3.2 How to Adjust the Irrigation Frequency

To increase the options for the irrigation frequency, the Unpowered Evaporative Valve is provided with an adjustable float consisting of a 7 cm diameter cylindrical float and 7 float rings that can slide over the cylinder to increase the outside diameter of the float (the bottom of the float ring should align with the bottom of the cylindrical float) (Figures 8.12 and 8.13).

Table 8.2 shows the irrigation frequency for various float rings. The irrigation frequency is determined by the net evaporation from the evaporator between irrigation events.

FIGURE 8.12 Cylindrical float and seven float rings.

FIGURE 8.13 Slide the float ring over the cylindrical float.

TABLE 8.2 Irrigation Frequency for the Unpowered Evaporative Valve

Outside Diameter of Float (cm)	Number of Float Rings	Net Evaporation between Irrigation Events (mm)
7	0	29.3
8	1	24.6
8	2	20.3
9	1	16
10	1	11.6
11	1	9.3
13	1	6
15	1	4

FIGURE 8.14 Push the float down to start the irrigation manually.

Provided that the water level in the evaporator is below the high level, you can start the irrigation manually at any time by pressing the float down. For example, you may wish to irrigate at sunset each day assuming that the water level is below the high level at sunset. Simply push the float down at sunset to start irrigating. You can delay the next irrigation or stop the irrigation at any time by removing the float. The irrigation cannot start again until the float is replaced (Figures 8.14 and 8.15).

It is important to realize that when you adjust the irrigation frequency by adjusting the outside diameter of the float, the water usage rate (litres per week, for example) does not change.

8.3.3 How to Adjust the Water Usage Rate

Position an empty measuring container under one of the irrigation drippers so that water drips into the container during the irrigation event. At the end of the irrigation event check the amount of water in the measuring container. You may also wish to check the moisture in the soil (Figures 8.16 and 8.17).

If your plants are not getting enough water, turn the control dripper clockwise to reduce the flow rate of the control dripper. If your plants are getting too much water, turn the control dripper anticlockwise to increase the flow rate of the control dripper.

Measured Irrigation

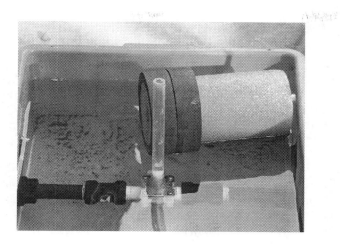

FIGURE 8.15 Remove the float to stop irrigating.

FIGURE 8.16 Measuring container under an irrigation dripper.

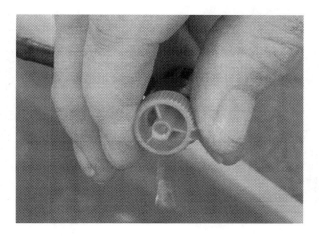

FIGURE 8.17 Adjusting the control dripper.

8.3.4 Key Features of the Unpowered Evaporative Valve

1. Completely automatic.
2. No electricity is needed (no batteries, no solar panels, no computers, no electronics and no Wi-Fi).
3. Smart irrigation controller – the irrigation is controlled by the prevailing weather conditions rather than a programme.
4. Use for both gravity-feed and pressurized irrigation.
5. Use for sprinkler or drip irrigation.
6. You can adjust the water usage rate by adjusting the control dripper.
7. You can adjust the irrigation frequency by adjusting the float.
8. Adjusting the water usage rate does not affect the interval between irrigation events, and adjusting the interval between irrigation events does not affect the water usage rate.
9. The water usage rate is directly proportional to the net evaporation rate.
10. Respond appropriately when there is an unexpected heat wave.
11. When it rains, water enters the evaporator and delays the start of the next irrigation.
12. Uses much less water without affecting the yield.
13. Simple and low tech and so there are fewer things to go wrong.
14. Provided you have a continuous water supply, you can leave your irrigation application unattended for weeks on end.

More information about the Unpowered Evaporative Valve is available in the Unpowered Evaporative Valve User Manual [17] and the DIY Unpowered Evaporative Valve Kit User Manual [18].

8.4 Unpowered Terracotta Valve

The Unpowered Evaporative Valve discussed in Section 8.3 needs to be cleaned regularly to remove algae and other contaminants. This problem can be solved by replacing the evaporator with a terracotta pot with a lid so that the water in the pot is not exposed to light.

The Unpowered Terracotta Valve is suitable for automatic sprinkler irrigation or drip irrigation, and can be installed either above ground or in ground. It includes a valve that operates with water supply pressure in the range of 10–800 kPa. The interval between irrigation events responds automatically to the on-site prevailing weather conditions (namely, evaporation and rainfall). For in-ground installation, the interval between irrigation events also responds automatically to the transpiration requirements of your plant at their current stage of growth.

Terracotta is porous, and so the water level in the pot falls as water seeps through the pot. A float inside the pot floats on the water. When the water level reaches the low level, a magnet inside the float activates the valve so that the valve opens and the irrigation starts. During the irrigation event, a control dripper drips water into the pot and the water level rises. When the water level reaches the high level, the magnet inside the float disengages from the valve so that the valve closes and the irrigation stops (Figures 8.18 and 8.19).

This remarkable low-cost invention may enable poor smallholders in remote locations to grow higher-valued crops cost-effectively.

The valve under the Unpowered Terracotta Valve has a 15 mm inlet and outlet, and so it is not suitable for large irrigation applications that require a bigger valve. If the flow rate through the valve is inadequate, you may wish to subdivide the irrigation application into zones with one Unpowered Terracotta Valve for each zone (Figure 8.20).

Measured Irrigation

FIGURE 8.18 Unpowered Terracotta Valve showing the float and the water level.

FIGURE 8.19 Float showing the ring magnet at the bottom of the float.

FIGURE 8.20 Unpowered Terracotta Valve showing the valve and the 15 mm inlet and outlet.

FIGURE 8.21 Connect the water supply to the inlet pipe.

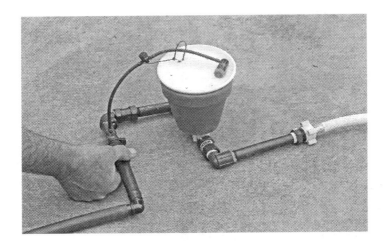

FIGURE 8.22 Connect the irrigation application to the outlet pipe.

8.4.1 Above-Ground Installation of the Unpowered Terracotta Valve

For above-ground installation, position the Unpowered Terracotta Valve in a suitable location in your garden so that the evaporation matches the evaporation at your plants.

Connect the water supply to the inlet pipe and connect the irrigation application to the outlet pipe (note that the control dripper is connected to the outlet pipe) (Figures 8.21 and 8.22).

8.4.2 In-Ground Installation of the Unpowered Terracotta Valve

In-ground installation is ideal for drip irrigation of row crops. Follow the installation steps below (Figures 8.23–8.25).

Because the terracotta pot is in the ground near the roots of plants, the Unpowered Terracotta Valve responds to changes in plant transpiration. As the crop grows and requires more water, the irrigation frequency increases automatically.

Measured Irrigation 165

FIGURE 8.23 Step 1. Dig a hole midway between two adjacent plants. There should be no irrigation drippers near these two plants.

FIGURE 8.24 Step 2. Adjust the inlet and outlet pipes so that they are vertical. Position the controller in the hole so that rim of the pot is above-ground level. Back fill the soil around the pot.

FIGURE 8.25 Step 3. Connect the water supply to the inlet pipe and connect the irrigation application to the outlet pipe.

8.4.3 How to Use the Unpowered Terracotta Valve

Turn on the water supply and the irrigation starts immediately. The control dripper drips water into the terracotta pot during the irrigation. The irrigation stops automatically after the control volume of water has dripped into the pot. The **control volume** is defined as the volume of water that seeps through the terracotta pot between irrigation events.

The irrigation starts again automatically after the control volume of water has seeped through the pot. The cycle continues indefinitely and so you can leave your garden unattended for months on end. A saucer sits on top of the pot so that the water in the pot is protected from debris, algae, mosquitoes and thirsty animals. There are six small drain holes in the saucer.

When using a conventional irrigation controller, you need to set the start time and the end time for each irrigation event. However, with the Unpowered Terracotta Valve you don't need a timer. The duration of the irrigation event is the time it takes for the control volume of water to drip into the pot, and the interval between irrigation events is the time it takes for the control volume of water to seep through the terracotta pot.

It is important to note here that the control dripper is adjustable. If you reduce the flow rate of the control dripper, it takes longer for the control volume of water to drip into the pot and so the duration of the irrigation event increases and your plants get more water. On the other hand, if you increase the flow rate of the control dripper, the control volume of water drips into the pot more quickly and so the duration of the irrigation event decreases and your plants get less water. Adjust the control dripper so that the irrigation delivers the appropriate amount of water to your plants at their current stage of growth (Figure 8.26).

The time it takes for the control volume of water to seep through the pot depends on the prevailing on-site weather conditions. When it is hot and dry, the water seeps more quickly and so the interval between irrigation events is shorter. When it is cool and overcast, the water seeps more slowly and so the interval between irrigation events is longer.

If it rains, rainwater collects in the saucer and drains into the pot. This means that the start of the next irrigation event is delayed. In addition to the control volume of water that needs to seep through the pot between irrigation events, any rainwater that has entered the pot between irrigation events also needs to seep through the pot.

To avoid irrigating during the heat of the day, you can turn off the water supply. Alternatively, a tap timer can be used so that water is only available between sunset and sunrise.

FIGURE 8.26 The control dripper is adjustable.

Measured Irrigation 167

The Unpowered Terracotta Valve uses on-site weather data (namely, evaporation and rainfall). Most smart irrigation controllers do not use on-site weather data. Instead they use weather data from the Bureau of Meteorology.

You can irrigate directly from a rainwater tank by gravity feed without using a pump provided that the water level in the tank is at least 1 m higher than the valve at the bottom of the Unpowered Terracotta Valve.

8.4.4 How to Adjust the Interval between Irrigation Events

You can adjust the interval between irrigation events by adjusting the gap between the upper and lower floats. The interval between irrigation events is the time it takes for the control volume of water to seep through the porous terracotta pot. To adjust the gap by 4 mm, rotate the upper float by two and a quarter turns (Figure 8.27).

Adjusting the interval between irrigation events does not change the water usage rate. For example, if you increase the interval between irrigation events by increasing the gap between the upper and lower floats, the amount of water used during the irrigation event increases automatically to ensure that the water usage rate (litres per week, for example) remains the same (Table 8.3).

FIGURE 8.27 To adjust the interval between irrigation events, adjust the gap between the upper and lower floats.

TABLE 8.3 Control Volume for Various Values of the Gap between the Upper and Lower Floats

Gap between the Upper and Lower Floats (mm)	Control Volume (mL)
Zero gap	77
4	109
8	141
12	173
16	205
20	237
24	269
28	300

8.4.5 How to Adjust the Water Usage Rate

Position an empty measuring container under one of the drippers so that water drips into the container during the irrigation event. At the end of the irrigation event check the amount of water in the measuring container. You should also check the moisture in the soil.

If your plants are not getting enough water, reduce the flow rate of the control dripper. Reducing the flow rate of the control dripper increases the duration of the irrigation event and so your plants get more water. If your plants are getting too much water, increase the flow rate of the control dripper.

Adjusting the water usage rate does not affect the interval between irrigation events.

8.4.6 When is Dripper Discharge Independent of Pressure?

Conventional drip irrigation systems control the volume of water discharged by a dripper by using PC drippers to control the flow rate and an irrigation controller to control the time. In a domestic garden with mains water supply, many zones are usually required to ensure that the pressure in each zone does not fall below the lower limit for pressure compensation. The irrigation controller is programmed so that each zone is irrigated at a different time.

With identical NPC (nonpressure-compensating) drippers (including the control dripper) at approximately the same level and negligible variations in the pressure within the zone due to frictional head loss, the Unpowered Terracotta Valve ensures that the volume of water discharged by each dripper during the irrigation event is approximately the same and independent of the pressure. If the water supply pressure decreases, the flow rate of the NPC drippers also decreases. However, the duration of the irrigation event increases automatically to ensure that the control volume of water is discharged by each dripper. For domestic gardens on level ground, the irrigation system can usually be designed so that variations in pressure within the zone due to frictional head loss are negligible.

By using the Unpowered Terracotta Valve in this way, many zones with PC drippers can be combined into a single zone with NPC drippers and a single Unpowered Terracotta Valve, and so the cost of the irrigation system can be reduced dramatically (Figures 8.28 and 8.29).

FIGURE 8.28 The adjustable control dripper has been replaced by an online irrigation dripper.

Measured Irrigation 169

FIGURE 8.29 The adjustable control dripper has been replaced by an in-line irrigation dripper.

FIGURE 8.30 With two additional pots the discharge during the irrigation event is trebled.

The terracotta pot is quite small and so the control volume is restricted to the range of 77–300 mL. Because each irrigation dripper discharges the control volume of water during the irrigation event, the water usage rate is often insufficient for your plants at their current stage of growth.

The pressure-independent dripper discharge can be increased by connecting additional terracotta pots to the original terracotta pot so that the water level is the same in all pots. With one additional pot the discharge is doubled, and with two additional pots the discharge is trebled. Continuing in this way the pressure-independent dripper discharge continues to increase.

One way of connecting the pots is to drill a 10 mm hole in the bottom of the pots and to use 6 mm ID (10 mm OD) rubber grommets, 6 mm barbed elbows and 6 mm flexible tubing. The drain hole in the additional pots should be sealed (Figures 8.30 and 8.31).

FIGURE 8.31 The pots are connected by 6 mm flexible tubing so that the water level is the same in all the pots.

8.4.7 Key Features of the Unpowered Terracotta Valve

1. Unpowered (no batteries, no solar panels, no electronics, no computers, and no Wi-Fi)
2. Water supply pressure 10–800 kPa
3. Use for sprinkler irrigation or drip irrigation
4. Use for gravity-feed or pressurized irrigation
5. Adjust the water usage rate by adjusting the control dripper
6. Adjust the interval between irrigation events by adjusting the float
7. Adjusting the water usage rate does not affect the interval between irrigation events, and adjusting the interval between irrigation events does not affect the water usage rate
8. Responds automatically to on-site evaporation and rainfall
9. The irrigation frequency increases significantly during a heat wave
10. Install above ground or in ground
11. For in-ground installation, the controller responds automatically to plant transpiration
12. Provided the same drippers are used throughout the irrigation application (including the control dripper), the water usage rate is independent of the water supply pressure
13. Irrigate directly from a rainwater tank without using a pump
14. Water in the terracotta pot is protected from debris, algae, mosquitoes and thirsty animals
15. Simple, unpowered and low tech, and therefore fewer things can go wrong
16. Leave your irrigation application unattended for months on end.

The Unpowered Terracotta Valve is a game-changer for automated irrigation on level ground from a rainwater tank. If you have a conventional irrigation system using pressurized irrigation with PC drippers, the following items are required.

- Pump for the rainwater tank
- Additional solenoid valves (one needed for each additional zone)
- Conventional irrigation controller
- Hose clamps

None of these items are required when you use the Unpowered Terracotta Valve, and so the cost of installing and running the irrigation systems can be reduced dramatically.

More information about the Unpowered Terracotta Valve is available in the User Manual [19].

8.5 Terracotta Irrigation Controller for Latching Solenoids

The outlet from the Unpowered Terracotta Valve is used to operate a float switch which is connected to a control box. When the valve opens, the water level in the outlet attachment rises and closes the float switch and the control box sends a pulse to the latching solenoid causing it to open and start the irrigation event. When the valve closes, the water level in the outlet attachment falls and opens the float switch and the control box sends a pulse to the latching solenoid causing it to close and stop the irrigation event. All the power required is provided by a small 9 V battery inside the control box (Figures 8.32 and 8.33).

8.5.1 Installing the Terracotta Irrigation Controller for Latching Solenoids

Step 1 (for above-ground installation). Position the Terracotta Irrigation Controller in a suitable location in your garden so that the evaporation at the controller matches the evaporation at your plants.

Step 1 (for in-ground installation). Dig a hole midway between two adjacent plants and position the Terracotta Irrigation Controller in the hole so that rim of the pot is above ground level. Back fill the soil around the pot. There should be no irrigation drippers near the two plants.

Step 2. Connect a water supply to the irrigation controller. The water pressure should be at least 10 kPa during the irrigation event.

FIGURE 8.32 In-ground installation of the Terracotta Irrigation Controller for Latching Solenoids.

FIGURE 8.33 Above-ground installation of the Terracotta Irrigation Controller for Latching Solenoids.

Step 3. Connect the adjustable control dripper to the irrigation zone so that it drips water into the terracotta saucer during the irrigation event.

Step 4. The control box has four colour-coded wires that need to be connected to the latching solenoid and the float switch (Figures 8.34 and 8.35).

Connect the **yellow** wire to one of the wires from the float switch.
Connect the **white** wire to the other wire from the float switch.
Connect the **blue** wire to the red wire from the latching solenoid.
Connect the **green** wire to the black wire from the latching solenoid.

If there are 2 latching solenoid valves, connect the second latching solenoid in parallel with the first latching solenoid.

8.5.1.1 Possible Questions

How to use the Terracotta Irrigation Controller for Latching Solenoids (see Section 8.4.3)
How to adjust the interval between irrigation events (see Section 8.4.4)
How to adjust the water usage rate (see Section 8.4.5)
When is dripper discharge independent of pressure? (see Section 8.4.6)

More information about the Terracotta Irrigation Controller for Latching Solenoids is available in the User Manual [20].

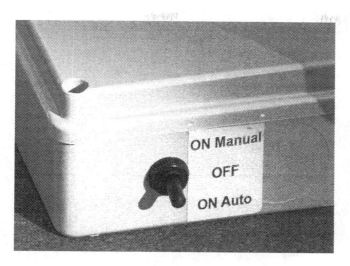

FIGURE 8.34 Control box.

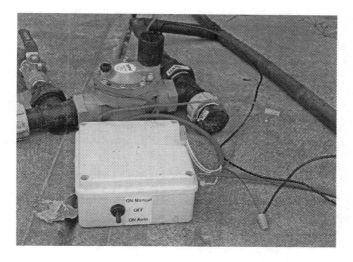

FIGURE 8.35 Four colour-coded wires connected to the latching solenoid and the float switch.

8.6 Automation Kit for Farm Pond Irrigation

The Automation Kit for Farm Pond Irrigation is for smallholders with a farm pond or dam and gravity-feed drip irrigation from a header tank [5]. Water is pumped automatically from the farm pond to the header tank and your small farm is irrigated automatically by an Unpowered Terracotta Valve. You can leave your farm unattended for weeks on end, and so you can become involved in other activities away from the farm; for example, travelling to the market to sell your produce.

It is assumed that the bottom of the farm pond is no more than 4 m lower than the irrigation drippers. The water supply pressure from the header tank should be at least 1 m head (Figure 8.36).

The kit includes an Unpowered Terracotta Valve to supply water to approximately 300 NPC drippers on flat land (assuming that each dripper has a flow rate of 2 L/H at 100 kPa). If your farm requires more than 300 drippers, you can subdivide your farm into zones with up to 300 drippers per zone (each zone has a separate controller).

8.6.1 Contents of the Automation Kit for Farm Pond Irrigation

As well as the User Manual, the kit includes the following four components (Figures 8.37–8.40). The following items will be required to install the kit and may be purchased locally:

- 12V battery
- 20W solar panel
- 2-strand electrical cable
- Wire connectors.

FIGURE 8.36 Farm pond in Kenya for gravity-feed drip irrigation.

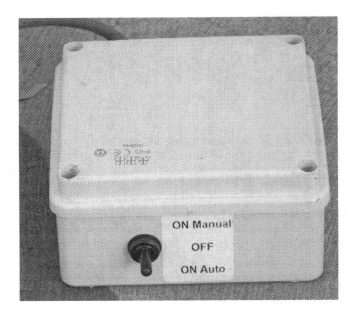

FIGURE 8.37 Waterproof pump controller with solar charge controller inside.

Measured Irrigation

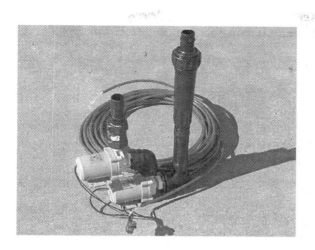

FIGURE 8.38 Double pump (two pumps connected in series) with an inlet filter, fittings to connect to 19 mm poly pipe and 9 m of waterproof electrical cable.

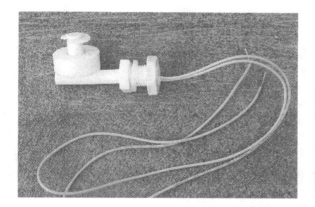

FIGURE 8.39 Float switch.

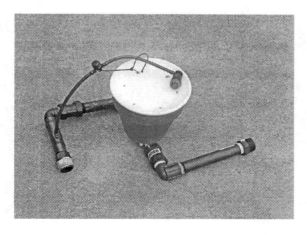

FIGURE 8.40 Unpowered Terracotta Valve.

8.6.2 Installing the Automation Kit for Farm Pond Irrigation

Step 1. Connect the pump.

Remove the header tank inlet pipe from the farm pond and connect it to the outlet from the pump.

Step 2. Install the float switch on the header tank.

Drill a 13 mm (half-inch) hole in the side of the header tank so that the hole is about 5 cm lower than the inlet to the header tank. Install the float switch on the inside of the header tank so that the float shaft points up (Figure 8.41).

Step 3. Purchase and install a solar panel (not in kit).

A 12V 20W solar panel should provide all the power required.

Step 4. Purchase a battery (not in kit).

A rechargeable 12V lead acid battery is required. You may be able to find a used car battery in good condition. A new sealed lead acid battery should have a capacity of at least 7 amp hours and a standby voltage of at least 13.5V.

Note that the solar panel and the battery may be replaced by a 12V 5A power adaptor. If you are using a power adaptor, the charge controller inside the pump controller is not required. Remove the charge controller, connect the two red wires, and connect the two black wires.

Step 5. Connect the pump controller.

The pump controller has eight colour-coded wires to be connected to the components as follows:
Connect the **red** wire to the positive terminal on the battery.
Connect the **black** wire to the negative terminal on the battery.
Connect the **blue** wire to the positive wire form the solar panel.
Connect the **green** wire to the negative wire form the solar panel.
Connect the **yellow** wire to one of the wires from the float switch on the header tank.
Connect the **white** wire to the other wire from the float switch on the header tank.
Connect the **brown** wire to the yellow (+) wire from the pump.
Connect the **purple** wire to the black (−) wire from the pump.

If you are using a power adaptor instead of a battery, connect the **red** wire to the positive wire from the power adaptor, and connect the **black** wire to the negative wire from the power adaptor.

The switch on the pump controller has three positions: ON Manual, OFF and ON Auto. For automatic filling of the header tank the switch should be in the ON Auto position.

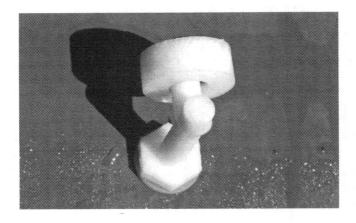

FIGURE 8.41 Float switch on the inside of the header tank with the float shaft pointing up.

FIGURE 8.42 Pump controller with the switch in the ON Auto position.

FIGURE 8.43 Connect the water supply to the inlet pipe.

There is a 30 s delay between the float switch turning off and the stopping of the pump (Figure 8.42).

Step 6. Submerge the pump in the farm pond.

The pump inlet should be at least 15 cm above the bottom of the pond to avoid pumping sediment from the bottom of the pond and clogging the inlet filter. If clogging of the filter becomes a problem, you may wish to install a larger filter.

The two pumps provided in the kit are connected in series. If the top of the header tank is less than 3.5 m higher than the water level in the farm pond, then the flow rate of the pumps may be increased by connecting the two pumps in parallel rather than in series.

Step 7. Above-ground installation of the Unpowered Terracotta Valve:

For above-ground installation, position the Unpowered Terracotta Valve in a suitable location in your farm so that the evaporation matches the evaporation at your plants.

Connect the water supply from the header tank to the inlet pipe and connect the irrigation application to the outlet pipe (note that the control dripper is connected to the outlet pipe) (Figures 8.43 and 8.44).

FIGURE 8.44 Connect the irrigation application to the outlet pipe.

FIGURE 8.45 Unpowered irrigation controller for solenoid valves.

8.6.2.1 Possible Questions

- How to use the Unpowered Terracotta Valve (see Section 8.4.3)
- How to adjust the interval between irrigation events (see Section 8.4.4)
- How to adjust the water usage rate (see Section 8.4.5)
- **When is dripper discharge independent of pressure? (see Section 8.4.6)**

More information about the Automation Kit for Farm Pond Irrigation is available in the User Manual [21].

8.7 Unpowered Irrigation Controller for Solenoid Valves

The Unpowered Irrigation Controller for Solenoid Valves (UICSV) uses evaporation as a gauge to control irrigation rather than the static timer intervals in conventional devices. It works with ordinary irrigation solenoid valves of any size, one solenoid per controller and with sprinkler or dripper systems (Figure 8.45).

After irrigation and as water evaporates from the soil, water also evaporates from the controller's container. The water in the container eventually reaches a low level corresponding to the soil drying out.

Measured Irrigation 179

FIGURE 8.46 Ring magnet and float at the low level at the start of the irrigation event.

FIGURE 8.47 Ring magnet and float at the high level at the end of the irrigation event.

The controller opens the solenoid and irrigation begins. A dripper attached to the outflow pipe slowly refills the container while the irrigation system "refills" the soil. When the water in the container reaches a high level corresponding to the required watering of the soil, the controller shuts the water off and the cycle restarts (Figures 8.46 and 8.47).

Once correctly calibrated, the UICSV only sends water when plants need it and does not overwater. It responds to the same local weather conditions as the soil. Deep or shallow watering, frequent or delayed

FIGURE 8.48 The plastic saucer sits on top of the container and it is covered with a blue polyester cloth that wicks water from inside the container to outside the container.

watering – all can be accommodated. You don't have to "turn it off" over winter as rain and cooler temperatures keep the container from drying out. You can leave your irrigation system unattended for weeks on end.

A plastic saucer sits on top of the container to protect the water in the container from debris, algae, mosquitoes and thirsty animals. Any water in the saucer drains into the container. A polyester cloth wicks water from inside the container to outside the container so that the cloth is always wet (Figure 8.48).

An irrigation system with a conventional irrigation controller can be upgraded to an unpowered system where each solenoid valve is controlled by a UICSV. The conventional irrigation controller and the associated wiring become redundant.

8.7.1 Installing the Unpowered Irrigation Controller for Solenoid Valves

Step 1. The ring magnet has an outer diameter of 72 mm and an inner diameter of 32 mm. Check that the ring magnet can activate the solenoid by sliding the magnet over the solenoid. You should hear a click as you lower the magnet over the solenoid. Irritrol, Orbit and Toro (Pope) solenoid valves all work with one ring magnet.

If the ring magnet does not activate the solenoid, two ring magnets joined together should be strong enough to activate the solenoid. For example, Hunter solenoid valves require two ring magnets to activate the solenoid (Figures 8.49 and 8.50).

It is assumed that the solenoid is non-latching.

Step 2. As power is no longer required to operate the solenoid, you may wish to cut off the two wires connected to the solenoid. It is preferable that the solenoid valve is in a location where the evaporation at the valve matches the evaporation from the soil. For example, if the garden is in full sun, then the valve should also be in full sun (Figures 8.51 and 8.52).

Step 3. The water supply should be connected to the valve inlet and the irrigation application to the valve outlet. Turn on the water supply.

Step 4. Choose the appropriate combination of shaft extensions and position the UICSV so that the solenoid is directly below the ring magnet. Rotate the magnet to progressively lower the magnet over the solenoid until the magnet activates the solenoid and the irrigation starts. An additional two full rotations of the magnet are recommended (Figures 8.53 and 8.54).

FIGURE 8.49 Check that the ring magnet can activate the solenoid.

FIGURE 8.50 Hunter solenoids require two ring magnets.

FIGURE 8.51 Solenoid valve with the wires removed.

FIGURE 8.52 Solenoid valve in a location where the evaporation at the valve matches the evaporation from the soil.

FIGURE 8.53 Position the UICSV so that the solenoid is directly below the ring magnet.

FIGURE 8.54 Rotate the magnet to progressively lower the magnet over the solenoid until the magnet activates the solenoid and the irrigation starts.

Measured Irrigation

FIGURE 8.55 Position the control dripper so that it drips water into the container via the barbed reducing joiner.

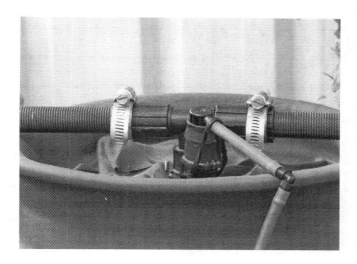

FIGURE 8.56 A rubber band is provided to enable the control dripper to be securely attached.

Step 5. Connect the adjustable control dripper to the irrigation system and position the control dripper so that it drips water into the container via the barbed reducing joiner. A rubber band is provided to enable the control dripper to be securely attached (Figures 8.55 and 8.56).
Step 6. Make sure that the polyester cloth is wet.
Step 7. You may wish to protect the UICSV with a cage or a tree guard, but make sure that the evaporation is not impeded.

8.7.2 Calibrating the Unpowered Irrigation Controller for Solenoid Valves

The UICSV controls the irrigation in two ways:

- The volume of water applied during each watering, and
- The length of time between each watering.

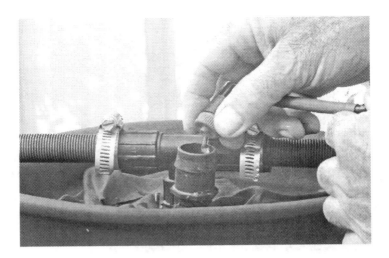

FIGURE 8.57 Adjusting the control dripper.

Both these factors are determined by the water level in the UICSV container. Water evaporates from the container and when the water level reaches the low level, irrigation starts and the control dripper then delivers water into the container. Eventually, the rising water level lifts the float and magnet away from the solenoid below and irrigation stops. The amount of water that the control dripper delivers to the container is called the **control volume**. The control volume is also the amount of water that evaporates between irrigation events.

The following two steps work together to calibrate the controller:

Step 1. First, you set how much water is applied to the plants by adjusting how long it takes for the control dripper to deliver the control volume to the container. While the container is filling the plants are being watered. The time it takes to fill the container is adjusted by changing the flow rate of the control dripper (by turning the orange part). If you set the control dripper to a fast flow rate, the container fills more quickly; thus, there will be less time for watering and the plants receive less water. If you set the control dripper to a slow flow rate, the container fills more slowly and the plants receive more water (Figure 8.57).

Step 2. Next, you set the frequency of watering by adjusting how quickly water evaporates from the container. This is done by exposing more or less of the polyester cloth outside the container. The time interval between irrigation events can be from 1 day to a week or longer (Figures 8.58 and 8.59).

The control volume depends on the particular solenoid being used (Table 8.4).

It is important to realize that the control dripper is simply replacing water that has evaporated from the container. This means that an irrigation event may be started or stopped manually at any time without affecting the water usage rate (litres per week, for example). An irrigation event can be started manually by pushing the float and magnet down. An irrigation event can be stopped manually by lifting the float and magnet up.

There is a fundamental difference between an Unpowered Terracotta Valve and an UICSV. When you calibrate an Unpowered Terracotta Valve, you are independently adjusting the water usage rate (for example, the number of litres per week discharged by an irrigation dripper) and the irrigation frequency. On the other hand, when you calibrate an UICSV, you are independently adjusting the volume of water discharged by an irrigation dripper during the irrigation event (Step 1) and the irrigation frequency (Step 2).

FIGURE 8.58 Large area of polyester cloth exposed.

FIGURE 8.59 Small area of polyester cloth exposed.

TABLE 8.4 Control Volume of Solenoids from Different Manufacturers

Manufacturer	Approximate Control Volume (mL)
Irritrol	470
Hunter	270
Orbit	650

8.7.3 Weather-Based Irrigation Control

The time it takes for the control volume of water to evaporate depends on the prevailing on-site weather conditions. When it is hot and dry, the water evaporates more quickly and so the interval between irrigation events is shorter. When it is cool and overcast, the water evaporates more slowly and so the interval between irrigation events is longer.

When it rains, water enters the container via the saucer, and so the start of the next irrigation event is delayed. Any rainwater that has entered the container between irrigation events needs to evaporate before the next irrigation event can start.

To avoid irrigating during the heat of the day, you can turn off the water supply. Alternatively, a tap timer can be used so that water is only available between sunset and sunrise.

The UICSV uses on-site weather data (namely, evaporation and rainfall). Most smart irrigation controllers do not use on-site weather data. Instead they use weather data from the nearest weather station of the Bureau of Meteorology. There are many irrigation systems where on-site weather data are significantly different from weather data from the nearest weather station (for example, irrigation inside a greenhouse).

8.7.4 Pressure-Independent Dripper Discharge

If you use the UICSV and the following three conditions are satisfied, the dripper discharge is approximately the same for all drippers in the zone and independent of pressure (Figure 8.60):

- Identical NPC drippers are used throughout the zone including the control dripper.
- All drippers are at approximately the same level.
- Frictional head loss within the zone is negligible.

When these three conditions are satisfied, the pressure-independent dripper discharge is approximately the same as the control volume. Note that the pressure-independent dripper discharge is independent of the flow rate of the irrigation drippers. For example, 4 L/H (at 100 kPa) NPC drippers will deliver the same pressure-independent dripper discharge as 2 L/H (at 100 kPa) NPC drippers. In this case the duration of the irrigation event for the 2 L/H drippers will be twice as long as the duration of the irrigation event for the 4 L/H drippers.

FIGURE 8.60 The control dripper is one of the dripline irrigation drippers.

FIGURE 8.61 Measuring the pressure-independent dripper discharge.

If the water supply pressure decreases, the flow rate of the NPC drippers also decreases. However, the duration of the irrigation event increases automatically to ensure that the control volume of water is discharged by each dripper. For domestic gardens on level ground, the irrigation system can usually be designed so that variations in pressure within the zone due to frictional head loss are negligible.

By using the UICSV with pressure-independent dripper discharge, many zones with PC drippers can be combined into a single zone with NPC drippers and a single UICSV and solenoid valve, and so the cost of the irrigation system can be reduced significantly.

By using the UICSV with pressure-independent dripper discharge, solenoid valves with flow control can be adjusted to suit the pressure limitations of the irrigation system without affecting the dripper discharge. For example, if you prefer not to use hose clamps on barbed fittings, then you can use the flow control on the solenoid valve to reduce the pressure accordingly.

To determine the pressure-independent dripper discharge, position an empty container under one of the drippers so that water drips into the container during the irrigation event. At the end of the irrigation event the pressure-independent dripper discharge is the volume of water in the container (Figure 8.61).

If the pressure-independent dripper discharge is more than your plants require at their current stage of growth, the pressure-independent dripper discharge can be halved by using two irrigation drippers for the control dripper. On the other hand, if the pressure-independent dripper discharge is less than your plants require at their current stage of growth, increase the number of irrigation drippers so that your plants get more water. Alternatively, you may be able to replace the control dripper by a control dripper with a lower flow rate and with the same emitter discharge exponent [9, 19].

8.7.5 Key Features of the Unpowered Irrigation Controller for Solenoid Valves

1. Unpowered (no batteries, no wires, no solar panels, no electronics, no computers and no Wi-Fi).
2. If you upgrade to the UICSV, the conventional irrigation controller and associated wiring become redundant.
3. Use for any size irrigation application.
4. Use for sprinkler irrigation or drip irrigation.
5. Adjust the dripper discharge by adjusting the control dripper.
6. Adjust the interval between irrigation events by adjusting the exposed surface area of the polyester cloth.

7. Adjusting the dripper discharge does not affect the interval between irrigation events, and adjusting the interval between irrigation events by adjusting the polyester cloth does not affect the dripper discharge.
8. Responds automatically to on-site evaporation and rainfall.
9. The irrigation frequency increases significantly during a heat wave.
10. When it rains, water enters the container and delays the start of the next irrigation event.
11. Provided the same drippers are used throughout the irrigation application (including the control dripper), the dripper discharge is independent of the water supply pressure.
12. Water in the container is protected from debris, algae, mosquitoes and thirsty animals.
13. Simple, unpowered and low tech, and therefore fewer things can go wrong.
14. Leave your irrigation application unattended for weeks on end.

Suppose your irrigation system has a conventional irrigation controller. When one of the solenoid valves for a particular zone fails to operate, you can take the opportunity to upgrade the zone to measured irrigation by installing a UICSV. Because the controller is unpowered, the valve can still operate even when the solenoid coil has failed. The performance of the UICSV can then be compared with the performance of the conventional irrigation controller for the other zones.

More information about the UICSV is available in the User Manual [22].

8.8 Soil Moisture and Measured Irrigation Scheduling

8.8.1 Soil Moisture Probe

The amount of water that your plants need will depend on many factors in addition to the weather. For example, as the plants grow and become bigger they will need more water. Plants growing in sandy soil will need more water than plants growing in heavy soil.

To take account of all these additional factors, you may need a soil moisture probe to check the moisture level in the soil at various depths [23]. A very simple soil moisture probe is a length of steel pipe with a long slot. A suitable diameter of the pipe is between 30 and 40 mm. An angle grinder can be used to cut a long slot in the steel pipe to that you can inspect the soil inside the pipe. A suitable width for the slot is about 15 mm. You can also use the angle grinder to sharpen the edge of the end of the soil moisture probe (Figure 8.62a and b).

FIGURE 8.62 (a) An angle grinder can be used. (b) Making a long slot in a length of steel pipe.

Measured Irrigation

By checking the moisture level in the soil through the slot in the steel pipe, you can decide whether your plants have been irrigated with too much or not enough water. An adjustable control dripper may be used to adjust the water usage rate.

Hammer the steel pipe into the soil near a dripper so that the slot faces the dripper. Remove the steel pipe from the soil and use the slot to inspect the moisture level in the soil and the position of the wetting front.

You may wish to use the slot to remove some soil from the pipe and to squeeze the soil sample between your fingers. (Figures 8.63 and 8.64).

FIGURE 8.63 Hammer the steel pipe into the soil near a dripper so that the slot faces the dripper.

FIGURE 8.64 Remove the steel pipe from the soil and use the slot to inspect the moisture level in the soil and the position of the wetting front.

8.8.2 Introduction to Measured Irrigation Scheduling

When you use manual measured irrigation, you check the water level in the evaporator at sunset each day, and if the water level is below the high level, you start irrigating. You stop irrigating when the water level reaches the high level. This method of irrigation scheduling is called **sunset scheduling**. One advantage of irrigating at sunset is that there are less evaporative losses compared with irrigating during the heat of the day.

For plants with deep roots or for plants in clay soils, it is preferable to irrigate with more water less frequently to enable the water to reach the bottom of the root zone. Between irrigation events, the soil near the surface is allowed to dry out, but there should still be moisture in the root zone. If you decide that your plants need irrigating less frequently than daily (for example, once a week), then **root zone scheduling** is recommended. Root zone scheduling takes account of evapotranspiration, the stage of growth of the crop, the soil type and the depth of the root zone

8.8.3 Root Zone Scheduling for Manual Measured Irrigation and the Unpowered Evaporative Valve

Allow the soil to dry out over several days until the soil is dry between the surface and the bottom of the root zone (use the soil moisture probe).

Place a measuring container under one of the drippers to collect the water and start irrigating just before sunset. While irrigating, check the moisture level in the soil by hammering the soil moisture probe into the soil near a dripper. Stop irrigating when the position of the wetting front is near the bottom of the root zone (or when the wetting front has reached an appropriate depth).

The volume of water in the measuring container is the **dripper control volume** and it is the amount of water that each dripper should deliver during the irrigation event to moisten the soil from the surface to the bottom of the root zone (Figure 8.65).

FIGURE 8.65 Dripper control volume for root zone scheduling.

8.9 DIY Unpowered Uniform Drip Irrigation on Sloping Land

This section is designed to assist smallholders to install a state-of-the-art irrigation system on sloping or uneven land at an affordable cost. Most drip irrigation applications on sloping land-use PC drippers. A high-pressure pump is usually required to ensure that all drippers within each zone are within the pressure range recommended by the manufacturer for pressure compensation. Furthermore, many zones are usually required to ensure that the drippers within each zone are within the pressure range recommended by the manufacturer for pressure compensation.

All of the above requirements for PC drippers make drip irrigation on sloping land from a header tank (or elevated water supply) very expensive.

This section demonstrates how drip irrigation on sloping land can use NPC drippers without compromising irrigation uniformity. It is assumed that the same drippers are used throughout the irrigation application. The irrigation controller is an Unpowered Terracotta Valve.

The irrigation method used will be referred to as the **Omodei method**.

Some key features of the Omodei method are listed below.

1. The cost of installing and running the irrigation system may be a lot less that the cost of an equivalent conventional irrigation system with PC drippers, especially for small to medium size applications.
2. Excluding the refilling of the header tank, the irrigation system is completely unpowered (no batteries, no solar panels, no electronics, no computers and no Wi-Fi).
3. A small low-pressure water transfer pump may be required to refill the header tank between irrigation events.
4. Provided that the irrigation system is designed so that frictional head loss along each lateral is negligible, then a desired irrigation uniformity for the entire irrigation system can be achieved.
5. Provided that the control dripper on the Unpowered Terracotta Valve has the same emitter discharge exponent as the irrigation drippers, the water discharged from each irrigation dripper during the irrigation event is independent of pressure.
6. You can adjust the water usage rate by adjusting the control dripper on the Unpowered Terracotta Valve.
7. You can adjust the interval between irrigation events by adjusting the float in the Unpowered Terracotta Valve.
8. The irrigation system responds automatically to on-site evaporation and rainfall.
9. Simple, unpowered and low tech, and therefore fewer things can go wrong.
10. Provided that you have a continuous water supply, you can leave your irrigation application unattended for weeks on end.

The Unpowered Terracotta Valve incorporates a small valve with a half-inch inlet and outlet. The size of the irrigation zone is limited by the size of the valve. For large irrigation applications, the Terracotta Irrigation Controller for Latching Solenoids should be used.

8.9.1 Schematic Diagram for the Omodei Method

Figure 8.66 shows a schematic diagram for the Omodei method. Figure 8.67 also shows a dripper assembly with 20 online drippers.

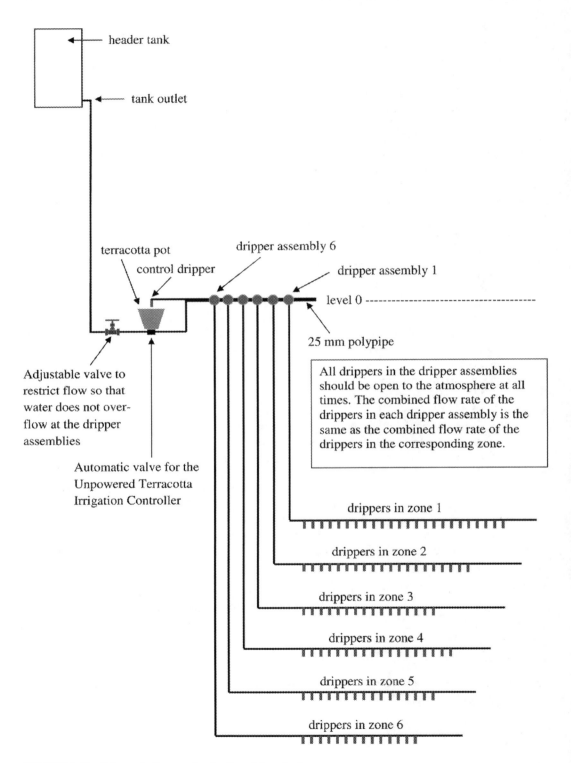

FIGURE 8.66 Schematic diagram for the Omodei method.

FIGURE 8.67 Dripper assembly with 20 online drippers.

8.9.2 Installation for the Omodei Method

Step 1. Arrange laterals so that each lateral follows a different contour level. Use enough laterals so that every plant is at approximately the same level as one of the laterals. The laterals may be NPC dripline or 19 mm polypipe with online NPC drippers. Restrict the length of the laterals so that frictional head loss along the laterals is negligible.

Step 2 (using a header tank). Connect the header tank outlet to the inlet of an Unpowered Terracotta Valve so the height of the lowest operational water level above the control dripper is about the same as the height of the control dripper above the highest lateral. Connect the outlet from the irrigation controller to a horizontal length of 25 mm polypipe at the same level as the control dripper (this level is called level 0).

Step 2 (using mains pressure). Connect mains pressure water supply to the inlet of an Unpowered Terracotta Valve (for maximum irrigation uniformity the control dripper should be as high as possible). Connect the outlet from the irrigation controller to a horizontal length of 25 mm polypipe at the same level as the control dripper (this level is called level 0).

Step 3. Let h be the height of the 25 mm polypipe above the highest lateral, let P be the desired irrigation uniformity (95%, for example) and let C be the manufacturer's coefficient of variation for the drippers.

Subdivide the slope (or uneven land) into zones as follows:

Zone 1 (Z_1) is from level $(-h)$ to level $(-h)*F$ where $F=1/(P+C)2$,
Zone 2 (Z_2) is from level $(-h)*(1.1)$ to level $(-h)*F^2$,
Zone 3 (Z_3) is from level $(-h)*(1.1)^2$ to level $(-h)*F^3$,

and so on, until all the laterals have been allocated to a zone. If the land is sufficiently level, only one zone may be required.

If $P=0.95$ and $C=0.01$, then $F=1.085$;
If $P=0.95$ and $C=0.02$, then $F=1.068$;
If $P=0.90$ and $C=0.03$, then $F=1.156$.

Alternative Step 3. To achieve maximum irrigation uniformity, do the planting in rows so that each row (or swale) follows a contour level. The laterals for each zone should be at the same level as the row.

Step 4. For each zone Z_i, count the number of irrigation drippers N_i in the zone. Construct a dripper assembly using N_i irrigation drippers. The drippers used for the dripper assembly and the drippers used in the corresponding irrigation zone should be identical. The flow rate of the dripper assembly should be the same as the combined flow rate of the corresponding drippers in the zone. Attach the dripper assembly to the 25 mm polypipe in Step 2 so that all the drippers in the dripper assembly are at level 0.

Step 5. Polypipe can used to deliver water from each dripper assembly to the laterals in the corresponding zone. The diameter of the polypipe will depend on the slope of the land. For example, on steep sloping land 13 mm polypipe can be used. The drippers in the dripper assembly should be open to the atmosphere at all times (in other words, the dripper outlets should be at atmospheric pressure). If water starts overflowing at any of the dripper assemblies, you will need to use an adjustable valve on the inlet side of the Unpowered Terracotta Valve to restrict the flow just enough to stop the overflow.

To minimize frictional head loss along very long laterals you may need to deliver water from the dripper assembly to more than one point on the lateral. For example, deliver water from the dripper assembly to three points on the lateral, namely, at one-sixth of the way along the lateral, at half way along the lateral and at five-sixths of the way along the lateral.

Provided that the irrigation system is designed so that frictional head loss along each lateral is negligible, then the desired irrigation uniformity for the entire irrigation system should be achieved. The irrigation uniformity can be further improved by increasing the number of zones.

There is no upper limit on the vertical gap between the dripper assembly and the irrigation drippers (for example, the irrigation drippers may be 50 m lower than the dripper assembly).

If you want the water discharged by each irrigation dripper during the irrigation event to be independent of pressure, replace the adjustable control dripper by a control dripper with the same emitter discharge exponent as the irrigation drippers. To increase the discharge of the irrigation drippers during the irrigation event, use a control dripper with a lower flow rate. However, it is often difficult to find a control dripper with a suitable flow rate and the same emitter discharge exponent.

It is preferable that the control dripper is identical to the irrigation drippers so that the pressure-independent dripper discharge can be increased by connecting additional terracotta pots to the original terracotta pot with the same water level in all pots. With one additional pot, the pressure-independent dripper discharge is doubled. With two additional pots, the discharge is trebled. Continuing in this way the pressure-independent dripper discharge continues to increase. See Section 8.4.6 for the details.

8.9.3 Example of the Omodei Method

This example is used to obtain some preliminary results to test the following two claims in relation to the Omodei method (Figures 8.68–8.72).

Claim 1. Provided that the irrigation system is designed so that frictional head loss along each lateral is negligible, then a desired irrigation uniformity for the entire irrigation system can be achieved.

Claim 2. Provided that the control dripper on the Unpowered Terracotta Valve has the same emitter discharge exponent as the irrigation drippers, the water discharged from each irrigation dripper during the irrigation event is approximately the same and is independent of pressure.

The drippers used for this example (including the control dripper) are all Antelco Agri Drip Classic 8 L/H drippers (at 100 kPa). The water supply is from a mains water tap. The control dripper is at level 0, zone 1 has 2 drippers at level −1.74 m and zone 2 has 3 drippers at level −2.77 m.

These results support both Claims 1 and 2 for the Omodei method (Table 8.5).

Measured Irrigation 195

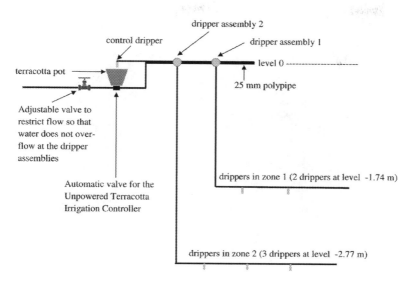

FIGURE 8.68 Schematic diagram for this example.

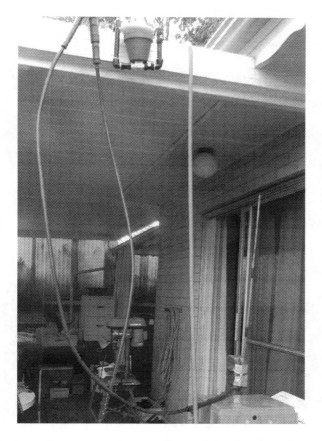

FIGURE 8.69 The Unpowered Terracotta Valve is at the top of the photo with the control dripper at level 0. A garden hose delivers water at mains pressure to an adjustable valve connected to the inlet of the controller. Polypipe delivers water from the dripper assemblies to the corresponding zones.

FIGURE 8.70 Dripper assemblies for Zones 1 and 2 at level 0 on the roof. The drippers are open to the atmosphere.

FIGURE 8.71 Zone 1 with 2 drippers at level −1.74 m.

FIGURE 8.72 Zone 2 with 3 drippers at level −2.77 m.

Measured Irrigation

TABLE 8.5 Results from 8.9.3 Example of the Omodei Method

	Zone 1 Dripper 1	Zone 1 Dripper 2	Zone 2 Dripper 1	Zone 2 Dripper 2	Zone 2 Dripper 3	Standard Deviation (SD)	Mean	Coefficient of Variation	Uniformity
Flow rate (lph)	2.240	2.230	2.210	2.160	2.210	0.0308	2.210	0.0139	0.986
Discharge (L)	0.307	0.314	0.315	0.326	0.322	0.0074	0.317	0.0233	0.977
Pressure (m)	0.830		0.750						
Flow rate (lph)	1.642	1.635	1.747	1.785	1.777	0.0733	1.717	0.0427	0.957
Discharge (L)	0.301	0.303	0.338	0.320	0.314	0.0150	0.315	0.0475	0.953
Pressure (m)	0.460		0.500						
Flow rate (lph)	1.230	1.267	1.133	1.148	1.288	0.0087	0.309	0.0282	0.972
Discharge (L)	0.299	0.306	0.315	0.305	0.321				
Pressure (m)	0.350		0.370						
Flow rate (lph)	0.743	0.768	0.817	0.840	0.855	0.0476	0.805	0.0592	0.941
Discharge (L)	0.250	0.259	0.277	0.282	0.270	0.0131	0.268	0.0489	0.951
Pressure (m)	0.125		0.105						
Discharge (SD)	0.026	0.025	0.025	0.020	0.025				
Discharge (mean)	0.289	0.296	0.311	0.308	0.307				
Discharge (CV)	0.091	0.084	0.081	0.064	0.081				
Discharge (uniformity)	0.909	0.916	0.919	0.936	0.919				
Discharge Statistics When the Lowest Pressure Is Excluded									
Discharge (SD)	0.004	0.006	0.013	0.011	0.004				
Discharge (mean)	0.302	0.308	0.323	0.317	0.319				
Discharge (CV)	0.014	0.018	0.041	0.034	0.014				
Discharge (uniformity)	0.986	0.982	0.959	0.966	0.986				

8.9.4 Using Sub-Drippers to Further Improve the Omodei Method

In Step 4 of the installation of the Omodei method, N_i irrigation drippers are used in zone Z_i. Instead of using each irrigation dripper to directly irrigate plants, suppose that the discharge from each irrigation dripper delivers water to a length of dripline with n identical sub-drippers. All the sub-drippers should be at the same level as the corresponding irrigation dripper without causing overflow at the dripper assembly. The irrigation dripper together with the n sub-drippers form a closed system and this motivates the following claim.

Claim 3. Provided that the control dripper on the Unpowered Terracotta Valve has the same emitter discharge exponent as the irrigation drippers, the water discharged from each irrigation sub-dripper during the irrigation event is approximately the same and is independent of pressure. Furthermore, pressure-independent dripper discharge = $n \times$ (pressure-independent sub-dripper discharge).

For the example above, the irrigation drippers used (including the control dripper) are all Antelco Agri Drip Classic 8 L/H drippers (at 100 kPa). Suppose that each irrigation dripper delivers water to a length of Netafim Miniscape (Landline 8) dripline with 16 irrigation sub-drippers. Then the pressure-independent sub-dripper discharge is one-sixteenth of the pressure-independent dripper discharge.

8.9.5 Installation Cost

The cost of installing the DIY drip irrigation system on sloping land from a header tank using the Omodei method is a fraction of the cost of installing and running an equivalent PC system.

- NPC drippers (or dripline) are less expensive than PC drippers (or dripline).
- A header tank does not normally provide enough pressure for PC drippers and so a high-pressure pump is needed.
- The pressure requirement for PC drippers means that hose clamps are usually required for barbed connectors. For a header tank and NPC drippers, hose clamps are not needed due to the low pressure.
- The range for pressure compensation for a PC system means that many zones are often required. Each zone will require its own solenoid valve and so the cost of the solenoid valves is likely to be significant. The Omodei method is unpowered and so there are no solenoid valves.
- The PC system requires a conventional irrigation controller and electrical wiring to each of the solenoid valves. The Omodei method requires an Unpowered Terracotta Valve and hence no wiring.
- A careful choice of irrigation drippers and irrigation sub-drippers can also reduce the installation cost for the Omodei method.

More information about the Omodei method is available in DIY Unpowered Uniform Drip Irrigation on Sloping Land [9].

8.10 Conclusions

Measured irrigation provides a radically different approach to irrigation scheduling. Conventional irrigation systems *indirectly* control the volume of water discharged by a dripper by using PC drippers with a specified flow rate and an irrigation controller to control the duration of an irrigation event. However, measured irrigation *directly* controls the volume of water discharged by a dripper rather than controlling flow rate and time. Because it is no longer necessary to control the flow rate, one can use NPC drippers.

Measured irrigation controllers use on-site weather information rather than data from the Bureau of Meteorology, and so they are ideal for applications where evaporation and rainfall in your garden are different from BOM data from the nearest weather station (greenhouses, for example).

Suppose your irrigation system has a conventional irrigation controller and one of your solenoid valves for a particular zone stops working. You can take this opportunity to upgrade the zone to measured irrigation by installing an UICSV. Because the controller is unpowered, the valve can still operate even if the solenoid coil has failed.

Conventional smart irrigation controllers use data from the nearest weather station to calculate daily evapotranspiration and adjust the irrigation schedule accordingly. The evapotranspiration is calculated by multiplying the reference evapotranspiration by the crop coefficient [15,25]. Reference evapotranspiration is based on a formula that uses weather data excluding daily evaporation. Furthermore, the crop coefficient is a theoretical value that depends upon the stage of growth of the crop. Measured irrigation responds to changes in the actual on-site evaporation and rainfall. This approach to irrigation scheduling is more appropriate than using a theoretical formula based on off-site weather data. Research done by the BOM has demonstrated a strong correlation (about 90%) between pan evaporation and reference evapotranspiration [15].

The Unpowered Terracotta Valve and the Terracotta Irrigation Controller for Latching Solenoids are game-changers for automated irrigation from a rainwater tank. Conventional automated irrigation from a rainwater tank using PC drippers requires:

- a pump,
- solenoid valves (one for each additional zone),
- conventional irrigation controller, and
- hose clamps.

None of these items are required if you use an Unpowered Terracotta Valve or a Terracotta Irrigation Controller for Latching Solenoids. PC drippers can be replaced with less-expensive NPC drippers. Hence, the cost of installing and running the irrigation system can be reduced dramatically.

The main reason for not using gravity-feed irrigation on sloping or uneven land is that the plants at the bottom of the slope get more water than those at the top. The Omodei method [9] may enable gravity-feed measured irrigation to be implemented effectively on sloping ground using a single Unpowered Terracotta Valve. For these implementations there may be many zones with each zone at a different contour level.

Since the development of PC drippers in Israel in 1974, the global sales of pressurized drip irrigation systems over gravity-feed drip irrigation systems have grown dramatically. The market share for PC dripline versus NPC dripline continues to grow. As the cost of electricity rises, many farmers with drip irrigated areas greater than 1 ha are complaining that they can no longer afford the cost of electricity to operate the pumps needed for PC drippers. With installation and running costs growing, drip irrigation is becoming less viable. The installation and running costs for gravity-feed drip irrigation systems may potentially be much less than the current costs. Since the development of PC drippers, research and development on gravity-feed drip irrigation systems have been seriously neglected. Measured irrigation provides a radically different paradigm for irrigation control, and further research and development may lead to more water-efficient and energy-efficient solutions, driving a revival in sales for NPC dripline.

A smallholder in Kenya may be growing vegetables on a small plot of land (less than 0.1 ha) using gravity-feed drip irrigation from a header tank. Irrigation scheduling is simply the smallholder opening and closing the valve on the header tank. For the cost of a bucket and an adjustable dripper, any gravity-feed drip irrigation application can be upgraded to manual measured irrigation. The water saved by this simple upgrade could enable the smallholder to grow more vegetables.

References

1. Hezarjaribi, A, Dehghani, A.A., Meftah Helghi, M., and Kiani, A. 2008. Hydraulic performances of various trickle irrigation emitters. *Journal of Agronomy,* 7(3): 265–271.
2. Lamm, F.R., Ayars, J.E., and Nakayama, F.S. (eds.) 2007. *Microirrigation for Crop Production.* Elsevier, Amsterdam, Netherlands.
3. Hema, N. and Kant, K. 2019. Cost-effective smart irrigation controller using automatic weather stations. *International Journal of Hydrology Science and Technology,* 9(1): 1–27.

4. Omodei, B. and Koech, R. 2016. Gravity-Feed Drip Irrigation for Agricultural Crops. In M.R. Goyal and M.K. Ghosal (eds.) *Innovations and Challenges in Micro Irrigation: Potential Use of Solar Energy and Emerging Technologies in Micro Irrigation.* Apple Academic Press, Waretown, NJ.
5. Mujere, N. and Isaac, R.N. 2017. Chapter 15: Rainfall Management for Sustainable Agriculture. In S. Eslamian and F.A. Eslamian (eds.) *Handbook of Drought and Water Scarcity: Principles of Drought and Water Scarcity.* Taylor & Francis Group, CRC Press, Boca Raton, FL, 15 pages.
6. UN Sustainable Development Goals. 2018. *Billion Dollar Business Alliance for Rainwater Harvesting.* Available at: https://sustainabledevelopment.un.org/partnership/?p=11904. Accessed 1 March 2022.
7. Omodei, B. 2013. Measured Irrigation: A Significant Development in Water Efficient Irrigation. In C.A. Brebbia (ed.) *Water Resources Management VII.* WIT Press, Southampton.
8. Omodei, B. 2015. Accuracy and uniformity of a gravity feed method of irrigation. *Irrigation Science,* 33(2): 121–130.
9. Omodei, B. 2022. *DIY Unpowered Uniform Drip Irrigation on Sloping Land.* Available at: https://www.measuredirrigation.com/. Accessed 1 March 2022.
10. Dukes, M.D. 2010. Water conservation potential of landscape irrigation smart controllers. *Transactions of the American Society of Hydrological and Biological Engineers,* 55(2): 563–569.
11. EPA. 2011. *WaterSense Specifications for Weather-Based Irrigation Controllers.* Available at: https://www.epa.gov/sites/production/files/2017-01/documents/ws-products-spec-irrigation-controllers.pdf. Accessed 1 March 2022.
12. EPA. 2017. *WaterSense Labelled Irrigation Controllers,* Fact Sheet. Available at: https://www.epa.gov/sites/production/files/2017-01/documents/ws-products-factsheet-irrigation-controllers.pdf. Accessed 1 March 2022.
13. Smart Approved WaterMark. 2019. *Welcome to Smart Approved WaterMark.* Available at: https://www.smartwatermark.org/. Accessed 1 March 2019.
14. Omodei, B. 2019. *Evapotranspiration and Measured Irrigation - Report for Smart Approved Watermark.* Available at: https://www.measuredirrigation.com/. Accessed 1 March 2022.
15. Webb, P.C. 2010. *BOM Reference Evapotranspiration Calculations.* Available at http://www.bom.gov.au/watl/eto/reference-evapotranspiration-report.pdf. Accessed 1 March 2022.
16. Omodei, B. 2022. *Measured Irrigation website.* Available at: https://www.measuredirrigation.com/. Accessed 1 March 2022.
17. Omodei, B. 2022. *Unpowered Evaporative Valve User Manual.* Available at: https://www.measuredirrigation.com/. Accessed 1 March 2022.
18. Omodei, B. 2022. *DIY Unpowered Evaporative Valve Kit User Manual.* Available at: https://www.measuredirrigation.com/. Accessed 1 March 2022.
19. Omodei, B. 2022. *Unpowered Terracotta Valve User Manual.* Available at: https://www.measuredirrigation.com/. Accessed 1 March 2022.
20. Omodei, B. 2022. *Terracotta Irrigation Controller for Latching Solenoids User Manual.* Available at: https://www.measuredirrigation.com/. Accessed 1 March 2022.
21. Omodei, B. 2022. *Automation Kit for Farm Pond Irrigation.* Available at: https://www.measuredirrigation.com/. Accessed 1 March 2022.
22. Omodei, B. 2022. *Unpowered Irrigation Controller for Solenoid Valves User Manual.* Available at: https://www.measuredirrigation.com/. Accessed 1 March 2022.
23. Dobriyal, P., Qureshi, A., Badola, R., and Hussain, S.A. 2012. A review of the methods available for estimating soil moisture and its implications for water resource management. *Journal of Hydrology,* 458–459: 110–117.
24. Omodei, B. 2022. *Measured Irrigation Training Manual for Smallholders.* Available at: https://www.measuredirrigation.com/. Accessed 1 March 2022.
25. Albaji, M., Eslamian, S., and Eslamian, F. 2020. *Handbook of Irrigation System Selection for Semi-Arid Regions.* Taylor and Francis, CRC Group, Boca Raton, FL, 317 Pages.

IV

Subsurface Irrigation

9
Qanat Irrigation Systems

Naser Valizadeh and Dariush Hayati
Shiraz University

Hamid Karimi
University of Zabol

Saeid Eslamian
Isfahan University of Technology

Mohammad Mohammadzadeh
Shiraz University

9.1	Introduction	203
9.2	Techno-Physical Aspects of Qanat Irrigation System	205
	The Structure of a Qanat System • Executive Operations for Maintenance of Qanat	
9.3	Social-Cultural System of Qanat Irrigation System	209
	Varieties of the Buneh • Characteristics of Buneh	
9.4	How Do the Qanats Contribute to the Sustainability of Water Resources?	213
	Energy Saving • Life Cycle Cost • Water Conservation • Reduction of Salinity • Social Development • Further Stabilization of Agriculture • Taking the Issues of Water Justice in to Account • Strengthening Moral Commitment of the Water Users	
9.5	Conclusions	215
	References	216

9.1 Introduction

Water scarcity is one of the most important issues facing environmental sustainability (Shahroudi & Chizari, 2009; Valizadeh et al., 2016; Salehi et al., 2017; Bijani et al., 2019). Water is of great importance because of its impact on ecological functions, socio-economic development programs and cultural, religious and aesthetic values (Valizadeh et al., 2020). In addition, its limited amount makes it more important (Yazdanpanah et al., 2012). However, the demand for water in some developed and developing countries has increased so far that it exceeds the capacity and potential of water supply sources (Valizadeh et al., 2021; Haji et al., 2020). On the one hand, this has led to some water management issues and problems in some areas, and on the other hand, it has caused serious reactions in various international statements (Hurlimann et al., 2009). These factors have led to more in-depth study of water management systems in the past, since the evidence suggests that the past water resources management systems have been more sustainable and did not lead to degradation of water resources and excessive exploitation (Mokhtari, 2012). Various activities such as drilling simple wells and creating complex irrigation networks have been undertaken to manage water resources (Manuel et al., 2018). But, Qanat irrigation system may be one of the most famous and amazing irrigation systems in ancient times (Madani, 2014; Mokhtari, 2012; Manuel et al., 2018). This irrigation system is still in use in many parts of the world such as Iran, North Africa and China (Habashiani, 2011; Balali, 2009).

Qanat or Karez is a waterway or canal dug underground to allow water to flow to the surface (Eslamian et al., 2016). This underground channel or waterway is created for connecting wells originating from the well mother (Hayati & Valizadeh, 2021; Semsar Yazdi & Labbaf Khaneiki, 2017). Qanats are used for transferring water and its conservation for agriculture and other needs (Manuel et al., 2018; Khan & Eslamian, 2016; Mostafaeipour, 2010). The length of Qanats is usually more than 8 km (Mokhtari, 2012;

Semsar Yazdi & Labbaf Khaneiki, 2017). Qanat is one of the major sources of water supply, especially in the countries of the Middle East, Central Asia and North Africa (Wilkinson et al., 2012). With the advent of new technologies, Qanats replaced with deep wells and the use of water pumps became increasingly common. Construction of deep wells without planning led to drying up of Qanats. So, many of them can no longer be revived, because digging deep wells has greatly reduced the water level (Balali, 2009).

A review of Qanat history shows that the idea of Qanat is of Persian origin and dates back to more than 2,000 years. It is thought that water supply of Persepolis (the historic city of Iran) was carried out by Qanats about 500 BC (Cressey, 1958). Many scholars believe that Qanats were first made in Persia. Qanat irrigation system (which is known by various names) has been frequently mentioned in ancient and medieval literature. For example, the Greek historian Polybius in the second century BC describes a Qanat that had been built during the Persian ascendancy in a desert region. Qanats are characterized by various names in other regions: in Afghanistan and Pakistan, they are renowned as Karezes; in North Africa, as Foggaras; and in the United Arab Emirates, as Aflaj. During the period of 550–331 BC and when the Iranian rule expanded from the Indus to the Nile, Qanat construction technology was spread throughout the empire. The Achaemenid Kings provided many benefits and incentives for the builders of Qanats and their heirs. As a result, thousands of new settlements were created and developed (Mostafaeipour, 2010). To the west, Qanats were constructed from Mesopotamia to the shores of the Mediterranean, as well as southward into parts of Egypt. To the east of Persia, Qanats were constructed in Afghanistan, the Silk Route oases settlements of Central Asia, and Chinese Turkistan (Mostafaeipour, 2010; Saffari, 2005).

The extraction system in the Qanats is such that water is withdrawn from the earth without the aid of electrical and mechanical means and this function is operationalized only at the expense of using the force of gravity (Mostafaeipour, 2010; Semsar Yazdi & Labbaf Khaneiki, 2017). Evidence from a comparison of existing deep wells and Qanats suggests that water extracted from Qanats is cheaper than water extracted from wells. The water of the Qanat is permanent and is always available even in emergency situations including droughts and water shortages (Nasiri & Mafakheri, 2015), although they are regularly harvested (Jome-Poor, 2009). Qanats have some other benefits as well. First, the bulk of the water channel is located underground, resulting in reduced water loss due to evaporation and infiltration into the soil. Second, the system is powered by earth's gravity and does not need a pump. And third, groundwater resources are used more sustainably. This itself has so many other advantages (Mostafaeipour, 2010; Nasiri & Mafakheri, 2015; Manuel et al., 2018).

Maintenance of Qanats in the past was based on landlord-peasant ties. Accordingly, the development of land reform policies in the mid-twenties and the introduction of new and western methods of irrigation resulted in destruction of Qanat irrigation system (Balali, 2009; Habashiani, 2011). In addition, it should be noted that the hydraulic mission was a paradigm introduced in the second half of the 20th century in some developing countries and the Middle East and North Africa. As a result of the introduction of the new water management system (in the form of deep wells and large dams), many of the values and activities/practices of the traditional water management system (Qanat irrigation system) replaced with a mechanical worldview (Kuros & Khaneiki, 2007; Habashiani, 2011). Following this paradigm shift in water resources management, governments came to this conclusion that deep wells should be considered an important strategy to make water resources more accessible to different users. With the introduction of these new technologies in the field of water resources management, the age of Qanats came to the end (Kuros & Khaneiki, 2007). As such, a mechanical paradigm emerged in this field in which communities and natural resources were considered to be unrelated components (Balali, 2009). However, in Qanat irrigation system, these two factors were interrelated. In other words, in Qanat irrigation system, water resources (and natural resources in general) did not conflict with human (communities). In contrast, in mechanical worldviews and modernized water resources management system, human societies are in line with natural resources (Valizadeh et al., 2016; Bijani, 2012). The basic view of such a mechanical worldview was that natural resources (such as water resources) should be exploited as much as possible by humans, science and technology (Balali, 2009; Habashiani, 2011).

In order to modernize the water management system, governments sought to change the traditional irrigation system both technically and socially. To this end, many countries have adopted land reform policies to change their land and water structures. Therefore, widespread propaganda was carried out in the context of Qanats' weaknesses. At the same time, the strengths of the deep wells were magnified. Since then, various extension and enlightenment programs have been designed and implemented for farmers. The purpose of these programs was to encourage farmers to use new methods of irrigation such as deep wells instead of traditional systems (including Qanat irrigation system) (Mokhtari, 2012). These programs and advertisements quickly resulted in the desired outcome of the sponsors of new water management system and consequently many of the Qanats were replaced by deep wells. In Iran, for example, there are currently over 350,000 semi-deep wells, many of which lack official licenses (Balali, 2009; Habashiani, 2011). Therefore, Qanats which are used to supply more than 70% of agricultural water, currently supply less than 10% of the water needs of this sector (Mokhtari, 2012). In addition to the degradation and over-exploitation of water resources, disappearance of Qanats has had other negative effects on local communities. For example, with the disappearance of Qanats, villagers and farmers whose livelihoods were dependent on Qanat irrigation system (especially in hot and dry areas) were forced to migrate. Furthermore, increasing energy consumption for groundwater extraction and transmission, increasing water loss and evaporation, increasing salinity of downstream soils, increasing unemployment, decreasing life expectancy in local and agricultural communities, threatening food security of communities, increasing inequality in distribution of water resources and decreasing the moral commitment of farmers and stakeholders to water conservation are other negative consequences of Qanats' destruction.

Given that Qanat is still a successful experience of water resources management in some areas, many of experts believe that its construction and/or reconstruction can be one of best strategies for getting out of current water crisis. However, since such a goal cannot be achieved without sufficient understanding of the Qanats' structure and operation, the main objective of the present study was to introduce different aspects of Qanat irrigation system. In studying the literature on Qanat irrigation system, it can be seen that this technical system operated in a social and cultural context. Therefore, three specific goals were identified in order to achieve the main goal. The first specific goal was to introduce the techno-physical dimensions of Qanat irrigation system. The second specific goal describes the social-cultural dimensions of Qanat. The third specific goal also emphasizes the fact that how Qanats can contribute to the sustainability of water resources.

9.2 Techno-Physical Aspects of Qanat Irrigation System

9.2.1 The Structure of a Qanat System

In terms of physical function and structure, Qanats can be considered as gently sloping subterranean tunnels that link wells over long distances. In other words, the physical system of the Qanat consists of a series of vertical shafts or chains of wells that are connected using steep horizontal tunnels. Qanat collects groundwater in a very efficient way and delivers large quantities of it to the surface without pumping. The water in these canals moves to low levels and agricultural or residential areas using gravity. The Qanats allow water to travel very long distances in arid and dry climates, without losing a large volume of water due to seepage and evaporation. Qanats are largely insensitive to precipitation levels. The reason for this is that they are usually of constant flow, with only slight and gradual changes in the dry and wet years. Therefore, their water flow variations are only related to seepage into the tunnels. The fact is that Qanats are a sustainable way of exploiting groundwater resources. Construction of the physical structure of the Qanat is a highly skilled operation. So, the trade of Qanat-makers (Muqannis) has mainly been hereditary (Lambton, 1953)

According to Habashiani (2011), the following list shows the main elements of a Qanat (Figure 9.1):
Mother well: The latest and deepest shaft in the Gallery is called Mother well.

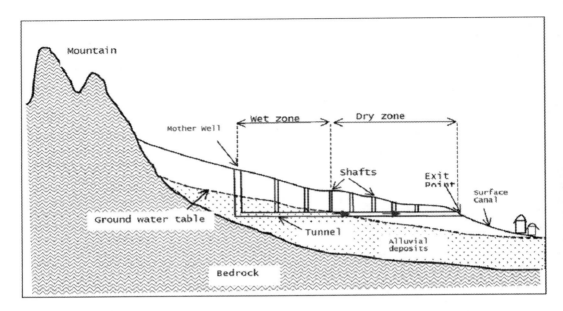

FIGURE 9.1 The structure of a Qanat system (Habashiani, 2011).

Shaft: Refers to the wells that are drilled along the gallery path and perpendicular to it. They are usually used for entry and exit of Qanat-makers, equipment and tools needed, gallery ventilation and eventually dredging. The depth of the shafts usually increase from the exit points to the mother wells.

Dry zone: It is a part of the gallery in the Qanat that is solely responsible for directing and transporting water to the surface of the earth.

Wet zone: It is a part of the gallery that is always wet and has the task of watering through the aquifer. The length of this section is subject to water table fluctuations. Also, in some cases, the full length of a Qanat may be wet zone.

Exit point: The place where the Qanat appears on the surface of the ground is called the Exit point or Mazhar.

Surface canal: After the water drains out of the underground canal, there is usually a canal to transfer it to agricultural lands and residential areas which is called the surface canal. This channel is called the surface channel because it is not underground.

Gallery: The underground conduit that connects the exit point to the mother well of the Qanat is called gallery. The gallery has an elliptical cross section and a flat or oval shape and dimensions are designed so that Qanat-makers can perform their tasks. The gallery sometimes has divisions that eventually become interconnected. So, the task of the gallery is to drain the aquifer and collect and transfer water to the exit point of the Qanat through gravity. It consists of two parts: dry zone and wet zone.

9.2.2 Executive Operations for Maintenance of Qanat

One of the most important techno-physical aspects of Qanat irrigation system is the executive operations that are employed for maintenance of the Qanat. One of the invaluable texts about introducing Qanat irrigation system (especially techno-physical aspects of this system) is the book written by Semsar Yazdi and Labbaf Khaneiki (2017). They introduce eight pivotal executive operations for maintenance construction of Qanat irrigation system.

9.2.2.1 Tunnel Checking

Tunnel checking refers to going through the tunnel from the Qanat exit toward the mother well for controlling the pace of water flow. This action may take the Qanat workers several days, depending on the length of the Qanat. This practice is carried out for controlling the tunnel and checking whether any barrier is hindering the flow of water all over the tunnel. If the barrier is small, the Qanat worker will eliminate it at once; otherwise he marks the barrier's place so he can come back and eliminate it with more suitable instruments. In the seasons like winters and falls, when the precipitation level is higher than other seasons, Qanats are more subject to breaking and collapse. In this regard, they need more "flow easing". In general, tunnel checking which is also named Ab-Harzi refers to controlling or checking the Qanat gallery. The shaft wells applied for this aim are healthy with suitable footholds through which the Qanat workers can easily climb up and down the well. These wells are usually 6 km away from each other.

9.2.2.2 Canal Cleaning

In some cases, Qanats require to be cleaned or dredged once in a while and some others do not. For instance, Qanats which run through the soils with clay formations should be cleaned some times, but Qanats constructed in the soils with sand and gravel formations are not cleaned that often. It is enough to ease the water flow in the Qanats which are not crumbling, but the Qanats with any type of crumbling should be cleaned out at least once a year. Cleaning the Qanat is described as eliminating whatever blocking factor/object that closes up the tunnel of the Qanat. The hindrance may be result from tunnel collapse or erosion which leads to the soil particles build up in the Qanat system tunnels. In addition, it should be mentioned that the tree roots can block the tunnel. This case can happen when the tree roots penetrate the tunnel and get tangled there in large number. In such cases, the Qanat workers should cut off all the tree roots and drag them out of the tunnel. The roots of the trees have also the potential to break apart the soil by growing into the porous structure of the soil and therefore can break down the tunnel if not eliminated. In this regard, on the surface of the earth and where a Qanat is constructed, the Qanat experts do not suggest implanting the trees which tend to get their roots to the water sources underground rapidly such as Tamarix, mulberry, walnut and Haloxylon. Cleaning the Qanat in water-transferring part results in facilitating water flow and minimizing water loss, but in water production section of the Qanat cleaning practice mainly contributes to opening the tiny holes through which groundwater seeps into the Qanat gallery.

9.2.2.3 Tunnel Extending

Tunnel extending refers to moving toward the aquifer by digging the end of the gallery. This practice is to increase the area of penetration in the tunnel and to obtain more water. This action is the most widely used method to increase water discharge in the Qanat irrigation system. However, it should be kept in mind that using this method is only logical and allowed where "a Qanat does not trespass on the bound of another Qanat". Increasing the length of water production section will result in draining out more water by the Qanat. However, this long water production section may decrease the discharge of surrounding Qanat irrigation systems. In Iran, Qanat tunnel extending is basically monitored for restraining the Qanat systems from overtaking each other and advancing into the aquifer any farther. This practice is of significant paramount, since the capability of the aquifer is confined to be discharged, and the adjoining Qanats will face with water shortage if a Qanat nearby would over-extract underground water resources. Extending the Qanat irrigation system's tunnel toward the aquifer can extend the contact region of the tunnel with water-bearing zone. This practice can also increase the height of water column above the tunnel. These two factors increase the infiltration of water into the Qanat tunnel. It is worth mentioning that performing Qanat tunnel extending in various directions and diverse sides simultaneously, can uplift the discharge of Qanat. Tunnel extending is mainly done simultaneously with constructing new wells. Thus, extending tunnel is carried out at the end of tunnel which is named Pishkar or Sineye Kar which mainly moves forward into the aquifer. Tunnel extending in Qanat irrigation system is called Pishkar kani but has some other local names such as "Now Kani" or "Now Kari".

9.2.2.4 Tunnel Branching

Adding more side branches to the Qanat system's main tunnel can uplift the infiltration area. In this regard, it can be understood that the discharge of Qanat will be increased. In fact, each of the branches acts independently and finally their waters join and flow down together. Qanat experts normally prefer the direction of a gully to orient a side tunnel. Through constructing a tunnel along a gully, Qanat masters will be able to gather the water storing in the sediments under the gully. Some of the Qanat irrigation systems enjoy more than five side branches. Each of these side branches has a specific name. Apart from the significant role of them in increasing the Qanat irrigation system discharge, they also can extend the neighborhood of Qanat system. The side branches existing around a Qanat irrigation system act as a buffer zone on which no other Qanat can trespass.

9.2.2.5 Tunnel Deepening

Tunnel deepening is also named "Kaf shekani" which refers to digging out a specific layer of the tunnel floor in Qanat irrigation system. This practice can be considered as a substitute for tunnel extending. Of course, it must be mentioned that these two practices are complementary. Sometimes it is possible that a drawdown happens in the aquifer and therefore the Qanat tunnel is left out of water. Such problems have become more frequent in past decades due to continuous droughts and over-exploitation of groundwater resources. In order to solve this issue, the Qanat irrigation system gallery must be put in contact with water table once again; otherwise the Qanat will be useless. To do so, it is not always possible to extend the Qanat tunnel into the aquifer horizontally, due to some natural and economic obstacles. For instance, a super-hard formation of soil during constructing the Qanat tunnel may rule out extending. In addition, tunnel may pass by the aquifer and reach dry zone again if being extended any farther. Also, it should be mentioned that a Qanat irrigation system has a buffer zone which should be considered in the process of extending. Therefore, some Qanat irrigation systems are preferred to be deepened (vertically) in order to avoid all these problems. In this process, the gallery of Qanat is extended vertically from its exit point to the mother well. Deepening Qanat tunnel leads to more water seepage into the tunnel. After deepening the tunnel, the amount of water may dramatically increase and in such cases the Qanat workers must widen the bottom of tunnel for easing the water flow. It should also be mentioned that the tunnel usually is not deepened more than 1.5 m. The reason is that it is no economically cost-effective to eliminate and lift out such a large amount of soil along a tunnel.

9.2.2.6 Tunnel Doubling

When a dramatic drawdown breaks out in the aquifer and the level of groundwater falls significantly below the Qanat tunnel, tunnel deepening cannot be considered as a suitable method and thus the workers have to employ the method of doubling. It is worth mentioning that Qanat experts usually would like to deepen the Qanat tunnel if water depletion is less than 1.5–2 m below the tunnel floor. Tunnel doubling is also named "Tahsoo Roosoo" in Persian which refers to digging another tunnel below the old tunnel. This new tunnel should be in parallel with the old tunnel. In other words, the new exit point turns up some distance down slope from the old one. The reason for digging new tunnel just under the old one is to use its shaft wells. The shaft wells are deepened to reach the new tunnel, so the Qanat workers do not need to sink new shaft wells from the beginning. Therefore, the workers save time and energy through using the old shaft wells already done. In some cases, this kind of Qanat is mistakenly named two-storey Qanat. However, it is not in line with the reality, since in this case the upper tunnel is phased out and the lower one only works. A two-storey Qanat enjoys two functional tunnels tapping water from two separate and specific aquifers lying at two different levels.

9.2.2.7 Removing Sediments

When the water is laden with calcium bicarbonate, physical condition of water and some chemical reactions led this soluble to settle and harden on the Qanat floor in the form of calcium carbonate. This sedimentation decreases the infiltration capacity of the gallery and diminishes its cross section, which leads to a decrease in water discharge in Qanat. Sediments settled on the tunnel floor and sides have to be eliminated and lifted out of the Qanat, otherwise, the tunnel may be blocked over the time. These sediments should be taken off the tunnel wall very carefully such that a thin layer of the sediment would be left on the wall, because it serves as a natural lining which curbs water percolation. In this regard, the workers prefer not to manipulate the sediment as long as its build-up does not interfere with the Qanat discharge. When the sediment would accumulate so much that it blocks the water, the workers manage to scrape a layer of sediment off the tunnel wall such that the water current would be eased and on the other hand, a layer would be remained to insulate the tunnel.

9.2.2.8 Tunnel Insulation

The water transfer section is insulated for stopping water percolation. The discharge of a Qanat depends on the amount of water wasted along the tunnel in water transport section. The longer the water production section, the more water would percolate. Furthermore, type of soil correlates with the amount of water loss, since the water percolation in sand and gravel is more than that in clay. In the shallow parts of water transport section, the workers sometimes dig the tunnel open like a trench and shore it up with brick and cement all the way, and then refill the trench up to the surface. In order to insulate the tunnel floor in the water transport section, the workers also used to tread clay mud into the cracks and gaps on the floor. This action is called "Lamal Kardan"; the worker spreads some clay mud on the tunnel floor and walks on it so the clay goes into the cracks through which water may escape. Also, the workers fill a sack with a mixture of straw and soil and drag it behind themselves along the tunnel, while someone else pours some soft clay down a well nearby. The sack helps the clay better mix with water and penetrates the holes and cracks on the tunnel floor. Bentonite is also used to insulate the tunnel floor. To do so, they mix Bentonite with Qanat water, 25 kg of Bentonite in every 100 m of tunnel, so that Bentonite settles on the floor and stops water percolation.

Nowadays, the water transport section is insulated sometimes by spreading the sheets of plastic on the tunnel floor all the way from the exit point to the border between water transport section (WTS) and water production section (WPS). These sheets are 120 cm wide. In the past, the workers used to coat the tunnel floor with a thin layer of clay mud and then put the sheets of canvas on it. The edges of canvas were fixed on the tunnel wall with pieces of mud, and then they coat the canvas with another layer of clay mud which could fill up its porous texture and get it waterproof. In case they use the above-mentioned methods to insulate the tunnel, they should look after the Qanat system more carefully so the Qanat would not need cleaning very often, otherwise the insulation would be damaged soon.

Eventually, the other way to solve the problem of water percolation is ceramic or cement hoops which are installed along the gallery one after another the same they are used to reinforce the tunnel against collapsing. In the water transport section, the space between the hoops is filled up with cement to make them seamless so the water finds no way to escape. It is worth noting that professional Qanat masters are usually reluctant to use such hoops for the purpose of insulation or reinforcing, unless they have to do so, and they cannot find any other solution. A Qanat tunnel should be dynamic and able to move horizontally and vertically to keep pace with water table. A tunnel lined with hoops is very difficult to be dug and deepened if groundwater drops below the tunnel some-day.

9.3 Social-Cultural System of Qanat Irrigation System

One of the most important features of Qanat irrigation system is that we can separate the technological systems of this management system from human activities and social entities. In other words, technological systems have an interwoven relationship with the social, ethical, religious and governance systems (Davarpanah, 2005; Kalantari, 2017; Balali, 2009; Lambton, 1953). Ancient Iranians had two fundamental tools in their water management approach which included valuable accumulated technical knowledge and a social–ethical–cultural system. These factors highlighted the ecological realities of the Iran's desert climate and the social imperative of conserving and distributing water in a way that ensured its availability to all individuals (Yazdanpanah et al., 2013). From this perspective, Qanat seems to be a socio-technical system (Balali, 2009). Similar to all traditional resource management systems, Qanats also have wide socio-cultural dimensions. In fact, Qanat is a substantial social phenomenon and should not be viewed only as an engineering wonder (Habashiani, 2011; Yazdanpanah et al., 2013). Qanat system has to be closely linked to the local communities' capacity to plan and manage their own water resources, especially for agriculture. The social institution on which Qanat depends to operate properly is called Buneh. It is worth mentioning that Buneh is known by different names in different areas. For example, Sahra and Harasseh are other names which refer to Buneh system (Habashiani, 2011). Buneh is defined as a social system in Qanat regions (Kalantari, 2017) and has been the product of adaptive reaction of human to water shortages in which individuals who are subject of the water scarcity share the managerial and ownership activities of water resources using a complex sharing ethics among local stakeholders (Habashiani, 2011; Balali, 2009).

There are various reasons for the emergence of cooperative groups of farmers such as Buneh. One of the most important factors is the inability of the farmer's family to purchase tillage tools (weak economy). Other factors include the need for group irrigation, climatic conditions, geographical conditions, optimal utilization of water resources, human social instincts and cultural factors (Kalantari, 2017). The interesting thing about Buneh is that its formation does not necessarily depend on water, since there are Bunehs in some areas whose agriculture is rain-fed. Safinejad (2000) believes that Bunehs in areas such as Iran are the result of lack of rain and dry nature of the country. In other words, they are the sustainable response of the villagers in coping with nature crises. Farmers and villagers have created similar forms of production systems due to their agricultural needs. Khosravi (1993) pointed to the emergence of Bunehs in Iran and emphasized that undoubtedly, they are a reminder of the old cropping system in Iranian agricultural communities. He continues that Iranian farmers were probably farming collectively in the past. In other words, production in rural and agricultural societies took place in the form of cooperatives such as Buneh.

There are different definitions for this Qanat-based cooperative Buneh system, each of which reflects slightly different functions and roles for Buneh. In Table 9.1, some of the most important definitions and functions of Buneh system are presented.

9.3.1 Varieties of the Buneh

There are many different classifications of Buneh. But, in general "the social system of Qanat" or "Buneh" can be divided into the following types.

9.3.1.1 Classification in Terms of Assistance

Regarding that the main focus of cooperation has been assistance, depending on the level of assistance among the members, three Bunehs can be distinguished from each other. The first type is called "complete Buneh", which refers to the ones whose main axis is assistance. In this Buneh, assistance between Buneh members continues throughout the year and in different stages of agriculture. The second type of Buneh from the perspective of assistance among members is "depleted Buneh" (which is called "water Buneh" is some regions). The depleted Buneh is a kind of Buneh that maintains the traditional Buneh mechanism and has a head. However, assistance between members is only in the field of irrigation and for providing water for agricultural lands. In other words, assistance between members is only for dredging the Qanat. In all other cases, agricultural activities are carried out independently by the family members of each member. The third type of Buneh in terms of assistance between members is "collapsed

TABLE 9.1 Definitions of Buneh System

Definition	Reference	Functions
The Buneh was a multi-family collective: a farming cooperative, whose main function was to reconcile the efficient exploitation of productive land with the careful use of the available water (that availability itself being, in large measure, a function of the socio-technical skills inherent in the Buneh)	Yazdanpanah et al. (2013)	To reconcile the efficient exploitation of productive land with the careful use of the available water (that availability itself being, in large measure, a function of the socio-technical skills inherent in the Buneh)
A traditional cooperative organization of sharecropping, in which people cultivated the land cooperatively	Rezaei-Moghaddam et al. (2005)	Organizing cooperative sharecropping
A communal system under which the arable lands of a village were organized into units farmed cooperatively by teams of sharecroppers	Hooglund (1982)	Organizing and dividing shared lands between villagers and farmers
Collective organization of production management in the Qanat regions	Jomehpour (2009)	Organizing production management in a collective method
An agricultural unit on which some farmers have the right to work cooperatively	Balali (2009)	Division of labor between farmers
A cooperative lifestyle more than just a management system	Farshad and Zink (1997)	As a lifestyle for villagers, not only Buneh is a resource management system, but also influences all aspects of their life
A social hierarchy which defines roles and responsibilities	Jome-Poor (2009)	Division of labor/responsibilities among individuals
As a system of collective ownership and management of Qanat water along with some participatory practices	Habashiani (2011)	Collective management of Gant water and agricultural activities
The villagers are divided into groups and select a land area for agriculture. Each group is called a Buneh. The number of Bunehs depends on the amount of water and the size of the land being cultivated	Kalantari (2017)	Using water is a collective and efficient way
Buneh is a complex social organization in the field of agricultural production. Each of the Bunehs has a specific share of water and land and their management is semi-formal. The main task of Buneh was to optimize land use and save water. Since life in the countryside was closely related to agriculture, Bunehs also affected other aspects of rural life	Vosoughi (2012)	The main task of Buneh was to optimize land use and save water
Buneh is a simple assemblage of several rural households whose members have created a solid social fabric through social and economic relationships. Membership in the Buneh offered the individual a prestigious social status. In addition, his position in the Buneh marked the extent of his influence on rural life	Lahsaeizadeh (2003)	Creating a solid social fabric through social and economic relationships Membership in the Buneh offered the individual a prestigious social status
Buneh is an independent cropping unit including some peasants with specific social statuses on one or more plots of land. They have designated amount of water, plowing force and several agricultural working tools. The division of labor in this independent farming unit was based on social status and economic privileges	Safinejad (2000)	Buneh is an independent cropping unit The division of labor based on social status and economic privileges is of most significant functions of Buneh
Buneh is a collective agricultural cooperative whose main focus is unorganized cooperation	Farhadi (1997)	To make a cooperation

(shattered) Buneh". Although we may no longer call them Buneh, they are historically significant. These Bunehs are the remnants of complete and collapsed Bunehs. In these Bunehs, the Buneh organization is lost and its functions are either deleted or faded. Assistance between members for dredging and revival of the Qanat has also been lost (Kalantari, 2017).

9.3.1.2 Classification in Terms of Irrigation

Bunehs are divided into three groups from the point of view of irrigation. The first types of Bunehs are "rain-fed Bunehs" in which irrigation is carried out naturally by means of rainfall. The second type of Buneh is "thorn-based Buneh". They also called "summer cropping Bunehs". In these Bunehs, the crops were cultivated using water and sprouts of some desert plants such as thorns. Irrigated Bunehs are the third type of Bunehs. These Bunehs were subdivided into several other types depending on the type of irrigation source. For example, spring, river and pumped-water Bunehs are considered irrigated (Safinejad, 2000).

However, it should be noted that there are other sub-categories such as indigenous/non-indigenous (Buneh is classified based on the indigenous or non-indigenous majority of the members), self-motivated/intervention-based (interference or non-interference by the government and government agencies as the criterion for segmentation), mono-proprietary/multi-proprietary (this type of categorization refers to ownership of the production factors by Buneh members) and peers/non-peers Bunehs (the role, social status, division of labor and the share of members from the produced product are the main criteria for segregation) reported in different studies (Farhadi, 1997).

9.3.2 Characteristics of Buneh

As it formerly mentioned, Buneh can be articulated as a social structure that had been created to cooperate in construction of and maintaining Qanat and also using Qanat's water for irrigating agricultural farms (Habashiani, 2011). All the Buneh members usually were peasants that belong to the same social status, but there was an exact division of labor among them. A Buneh, normally, included six peasant members: the Buneh head (or irrigator), two assistants and three sharecroppers (Yazdanpanah et al., 2013; Rezaei-Moghaddam et al., 2005). It should be mentioned that Buneh was not independent and each Buneh was tied into a broader network that included the landlord (or his representative/agent), other Bunehs within the village and many of vital specialists: the Qantas diggers (Muqanui), blacksmiths, carpenters and village-level service providers such as barbers and bath keepers. This sort of network, with all its diversity and mutuality, constituted the production system (Yazdanpanah et al., 2013).

The general characteristics of Bunehs can be summarized as follows (Kalantari, 2017):

- The social organization of Beneh has not been widespread. Rather, these types of organizations are more common in the water-scarce areas and desert margins. Water scarcity and low rainfall have been the most important causes of this type of production in the given areas. Although water shortage in hot and dry regions has led to the choice of this method of production, it is not the only method of agricultural production.
- The social organization of Qanats (Buneh) is often deeply linked to the landlord-based and feudalism system. In other words, Bunehs were mostly found in areas where there was a feudalism system. This indicates that the landlord has found this method appropriate to serve his interests.
- Only the workforce is collaborative in this social organization of production. In a way that landlord provides land and water. The seed and force of the plow are also usually provided by the master or some of the Buneh heads.
- The geographical location and agricultural land of each Buneh usually change every year. In other words, a land is not cultivated by many farmers for many years. In addition, Buneh members may also change every year. Changing Buneh members is usually done by members and master.

- The duty of Buneh members is not the same. Sar-Buneh (the head of the social organization) determines the duties and responsibilities of each member.
- The landlord and head of the Buneh are responsible for meeting the nutritional and economic needs of the members.
- Providing agricultural water is the responsibility of the landlord. This includes activities such as dredging the Qanat, drilling wells, repairing the aqueducts and Qanat preservation activities.
- The land area of all Bunehs and their water is usually equal. On the other hand, the number of members in Bunehs is also equal. The cultivation plan is determined by the landlord.
- Each member could leave the Buneh at any time. In that case, he would sell his membership to a farmer or somebody else in the village.
- Once the crop was harvested, production costs were first calculated and subtracted. Subsequently, according to local custom, the share of the landlord and the person who supplied the plow was subtracted. Finally, the residual product was evenly distributed among the Buneh members.

9.4 How Do the Qanats Contribute to the Sustainability of Water Resources?

In this section, some of the most important contributions of Qanat irrigation system to the sustainability of water resources are presented:

9.4.1 Energy Saving

Qanats use gravity force to move water through the main channel (Gallery). Therefore, there is no need for electric power, diesel and pump spare parts or petroleum products to do so. This will significantly reduce energy consumption. In comparison with diesel motor-equipped deep wells, Qanats help reduce greenhouse gas emissions as well (Nasiri & Mafakheri, 2015; Alizadeh, 2008).

9.4.2 Life Cycle Cost

In spite of the fact that Qanats require a great deal of manpower, their life span is better than the deep wells (Nasiri & Mafakheri, 2015). It can take years to build a Qanat, while a deep well can be built in less than a few months (Beaumont, 1971). Valid data on the cost and time of construction of the Qanat irrigation systems are not available, because there is no proper documentation in this area. On the other hand, differences in the Qanat systems in terms of length, depth of the mother well and gallery size have made many of these data incomparable. However, it should be noted that with the improvements made in the Qanat building tools and technologies, the costs and construction time can be significantly reduced. For example, geographic information systems and remote sensing technologies are used today to determine the location of the mother well in Qanat building (Nasiri & Mafakheri, 2015; Abdin, 2006).

9.4.3 Water Conservation

The rate of water flow in the Qanats depends on the natural flow of groundwater which prevents over-exploitation of the freshwater of the aquifer. In addition, the evaporation in the Qanats is very low, as the water transfer channel is underground. Underground water canalization also reduces the amount of water penetration in the ground. In arid and dry areas where digging deep wells is unreasonable, Qanats can be suitable irrigation systems to supply agricultural water and other water needs (Nasiri & Mafakheri, 2015). Accordingly, Qanats can be considered as sustainable and suitable water utilization systems that help conserve ground and surface water resources.

9.4.4 Reduction of Salinity

Qanats transfer fresh water from higher altitudes to downstream agricultural and residential areas in the plains. Downstream areas usually have saline soils. Therefore, transfer of subterranean water to these areas reduces soil salinity and consequently reduces desertification (Nasiri & Mafakheri, 2015; Haeri, 2003). In general, it can be said that Qanats play the role of drainage at the local and regional level.

9.4.5 Social Development

As mentioned earlier, Qanat irrigation system is based on a highly developed social system (Balali, 2009; Habashiani, 2011). In other words, Qanat system is a model of a cooperative system (Manuel et al., 2018; Yazdanpanah et al., 2013; Rezaei-Moghaddam et al., 2005). From the perspective of social development, construction, deployment and protection of the Qanat irrigation system require the involvement of local communities, participatory decision-making and policymaking. Plus, it provides local communities with many job opportunities. By building a Qanat in an area, many people are involved in the various stages of its construction and maintenance, which itself contributes to increase the employment and livelihoods of the area's residents. This irrigation system will also increase the sense of responsibility of individuals and farmers for the proper management and distribution of water resources (Nasiri & Mafakheri, 2015; Motiee et al., 2006).

9.4.6 Further Stabilization of Agriculture

The reliability of the Qanat's continuity (in terms of meeting users' water needs) is greater than that of the well system. As it was previously mentioned, the Qanat systems are less affected by factors such as drought, climate changes and so on. However, deep-well farming systems are highly volatile and unstable. For this reason, agriculture is less threatened in Qanat-oriented irrigation systems. Also, the impacts of drought or climate changes are less for Qanat-based agricultural systems (Bani-Tabe, 2009).

9.4.7 Taking the Issues of Water Justice in to Account

In many parts of the world, water is owned by a particular body or entity. In other words, water resource ownership has raised many issues regarding the distribution and fair use of water resources among different users and stakeholders. For example, in areas where government or government agencies own freshwater and agricultural water resources, farmers and other users must pay for meeting their water needs. However, many farmers may not be able to afford the water needed because of poverty and livelihood problems. In such circumstances, it is only the wealthy and big farmers who will be able to use water resources, because of their financial resources (Valizadeh, 2015). However, Qanat irrigation system distributes water resources more equitably among users and stakeholders, because it is based on a community-based social system. Qanat irrigation system entirely relies on the social system and cooperation of community members. In other words, even small-scale farmers can use its water, since Qanat irrigation system provides them with cheaper water (Bani-Tabe, 2009).

9.4.8 Strengthening Moral Commitment of the Water Users

Issues of ownership, sovereignty and agency have always caused many problems in the field of water resources management. The separation of ownership and sovereignty of water resources from its agency has created many disagreements among stakeholders. In cases where the government owns and governs agricultural water resources, farmers (who have the agency on water resources) consider the government responsible for supplying their water needs. In other words, because of the facts that the government designates itself as the main water supplier, farmers and other users have less concern and commitment to solve water scarcity problems (Valizadeh, 2015). In Qanat irrigation system, however, farmers have the ownership, sovereignty and agency over water resources. For this reason, there is no conflict of interests between them and other stakeholders (such as the government). As a result, in Qanat irrigation system, they are doing their utmost to make sustainable use of water resources.

9.5 Conclusions

The main aim of this chapter was to introduce the techno-physical and social-cultural sub-systems of Qanat irrigation system. As it was previously mentioned, from the point of view of physical function and structure, Qanat irrigation system can be regarded as gently sloping subterranean tunnels that connect wells over long distances. In other words, the physical system of the Qanat irrigation system includes a series of vertical shafts or chains of wells that are linked using steep horizontal tunnels. Qanat irrigation system also has a wide socio-cultural dimension. In fact, this irrigation system is a substantial social phenomenon and must not be considered merely as an engineering wonder. Qanat irrigation system has to be strictly connected to the local communities' capacity to plan and manage their own water resources, especially for agriculture. The social institution on which Qanat depends to operate properly is called Buneh. Buneh is defined as a social system in Qanat regions and has been a result of adaptive reaction of human to water shortages in which individuals who are subject of the water scarcity share the managerial and ownership activities of water resources using a complex sharing ethics among local stakeholders.

Despite centuries of use, Qanats are still important in some countries, since they require little technology to operate and are considered as a sustainable system of water resource utilization, especially in low-water areas. Although Qanats are considered as an example of a sustainable irrigation and water management system, they have been greatly affected by factors such as earthquakes, droughts, wars and social and technical revolutions. As a result, many of them are now destroyed. In the age of modernity, the emergence or development of electrically powered deep wells is the most prominent factor that has had the fastest and most destructive effect on the Qanat irrigation system. In other words, with the advent of advanced technologies for deep well drilling and the use of electric pumps, the groundwater level has declined substantially and the aqueducts supplying the water of Qanats have dried up.

Regarding that water scarcity is becoming a global problem in today's world, upgrading and improvement of Qanat irrigation systems can play a very important and key role in the sustainable use of water resources, because this irrigation system has been the product of centuries of experience. In other words, it has successfully passed the test of human experience. The relative ease of construction and management of this irrigation system and its greater adaptation to the socio-economic conditions of local communities can serve as milestones in encouraging local communities to use and re-operate Qanats. However, if Qanat irrigation system is to be restored and reused in hot and dry areas, a few points need to be considered.

First, it must be acknowledged that the success of Qanat irrigation system was the result of the specific characteristics of agricultural society, agricultural policies, land management and social stratification in the past. In other words, Qanat irrigation system was a combination of the techno-physical and social-cultural sub-systems. However, due to the political and structural changes (such as implementation of land reform programs) in many countries, social-cultural systems (which have been the basis of Qanat irrigation system) have disappeared. As a result, the first step to rehabilitate the Qanat irrigation system is to establish a proper social system (as it was called Buneh in the past).

The second point is that some reforms should be done in the modern and dominant water resources management system. Governments in different countries, for example, should prohibit drilling deep wells near the Qanats. Otherwise, the aquifers that feed on the Qanat will dry up. As a result, the Qanats will also dry up. On the other hand, different governments need to adopt realistic views and policies on the restoration of the Qanat irrigation system. In other words, they should not expect that the use of Qanat system immediately (in a short time) be able to result in agricultural development and improvement in the status of water resources, because the construction or restoration of a Qanat itself is time-consuming practice. On the other hand, the irrigation system of Qanat would not (at least in the early stages) develop the cultivation area as much as the deep-well system. However, these costs and inputs will yield better returns in the long run.

The third point is that in the reconstruction and restoration of Qanat irrigation systems, the factor of "technology" should not be considered as a completely opposite or negative factor to the development of this system. In the past and when deep wells and electric technologies were introduced as a substitute for Qanat, opponents of this irrigation system criticized it in terms of costs and time required for construction and maintenance. However, technological advances as well as advanced tools of present age can greatly reduce the time and cost of constructing and maintaining Qanat irrigation systems.

References

Abdin, S. (2006). Qanats a unique groundwater management tool in arid regions: the case of bam region in Iran. In: *International Symposium on Groundwater Sustainability (ISGWAS)*, January 24–27, Alicante, Spain.

Alizadeh, A. (2008). Agricultural water management in Iran: issues, challenges, and opportunities. In: *Iran-United States Workshop on Water Management*, Sacramento, CA, USA, August 18–19.

Balali, M.R. (2009). *Towards reflexive land and water management in Iran: linking technology, governance and culture*. Doctoral thesis, Wageningen University, Wageningen, Netherlands.

Bani-Tabe, M. (2009). Qanat: the most certain way of exploiting water. *Geography Education*, 18 (65), 36–43.

Beaumont, P. (1971). Qanat systems in Iran. *International Association of Scientific Hydrology*, 16, 39–50.

Bijani, M. (2012). *Water conflict: a human ecological analysis in downstream zone of Doroodzan Dam*. Doctoral Thesis, School of Agriculture, Shiraz University, Shiraz, Iran.

Bijani, M., Ghazani, E., Valizadeh, N., and Fallah Haghighi, N. (2019). Predicting and understanding of farmers' soil conservation behavior in Mazandaran Province, Iran. *Journal of Agricultural Science and Technology (JAST)*, 21 (7): 1705–1719.

Cressey, G. B. (1958). Qanats, Karez, and Foggaras. *Geographical Review*, 1, 27–44.

Davarpanah, G. (2005). Comparable survey of benefits and disadvantages from ground water to the method of wells and qanats. In: *Proceedings of the 2nd International Conference on Qanat*, Kerman, Iran.

Eslamian, S., Davari, A., and Reyhani, M. N. (2017). Iranian Qanāts: An Ancient and Sustainable Water Resources Utilization, Ch. 9, in *Underground Aqueducts Handbook*, Ed. by Angelakis A. N. et al., Taylor and Francis, CRC Group, Boca Raton, FL, 123–150.

Farhadi, M. (1997). *Culture of Assistance in Iran: An Introduction to Anthropology and Sociology of Cooperation (Fist Volume: Traditional Assistance in Irrigation and Farming)*. Academic Publishing, Tehran, Iran.

Farshad, A., and Zinck, J.A. (1997). Indigenous knowledge and agricultural sustainability: a case study in semi-arid regions of Iran. In *CD-ROM Proceedings of the International Conference on Geo-information for Sustainable Land Management: Enschede, ITC, 17–21 August*, Ed. by Beek K.J., de Bie C.A. and Driessen P, ITC, Netherlands, 10p.

Habashiani, R. (2011). *Qanat: a sustainable groundwater supply system*. Master's thesis, School of Arts and Social Science, James Cook University, Queensland, Australia.

Haeri, M.R. (2003). Kariz (Qanat): an eternal friendly system for harvesting groundwater. In: *Adaptation Workshop*, Cenesta, New Delhi, India, 12–13th November.

Haji, N., Valizadeh, N., Rezaei-Moghaddam, K., and Hayati, D. (2020). Analyzing Iranian farmers' behavioral intention towards acceptance of drip irrigation using extended technology acceptance model. *Journal of Agricultural Science and Technology*, 22 (5): 1177–1190.

Hayati, D., and Valizadeh, N. (2021). Freshwater Management and Conservation in Iran: Past, Present, and Future, in *Tigris and Euphrates Rivers: Their Environment from Headwaters to Mouth*, Springer, Cham, Switzerland, pp. 1507–1533.

Hooglund, E. (1982). *Land and Revolution in Iran*. University of Texas Press, Austin.

Hurlimann, A., Dolnicar, S., and Meyer, P. (2009). Understanding behaviour to inform water supply management in developed nations–a review of literature, conceptual model and research agenda. *Journal of Environmental Management*, 91 (1), 47–56.

Jome-Poor, M. (2009). Kariz (Qaant) as a product of indigenous knowledge and culture of desert habitats and its dependent systems and sustainable exploitation (The case of Qanats in Kashan). *Social Science Journal*, 33, 27–62 (In Persian).

Kalantari, K. (2017). *Rural Sociology*. Payame-Noor University Press, Tehran, Iran.

Khan, S., and Eslamian, S. (2017). Managing Drought through Qanāt and Water Conservation in Afghanistan, Ch. 22, in *Underground Aqueducts Handbook*, Ed. by Angelakis A.N. et al., Taylor and Francis, CRC Group, Boca Raton, FL, pp. 385–402.

Khosravi, K. (1993). *Rural Sociology in Iran*. Academic Publishing, Tehran, Iran.

Kuros, G., and Khaneiki, M. (2007). *Water and Irrigation Techniques in Ancient Iran*. Iranian National Committee on Irrigation and Drainage (IRNCID), Tehran, Iran.

Lahsaeizadeh, A. (2003). *Sociology of Rural Development*. Rose Publishing, Shiraz, Iran.

Lambton, A.K.S. (1953). *Landlord and Peasant in Persia*. Oxford University Press, London, UK, p. 459.

Madani, K. (2014). Water management in Iran: what is causing the looming crisis? *Journal of Environmental Studies and Sciences*, 4 (4), 315–328.

Manuel, M., Lightfoot, D., and Fattahi, M. (2018). The sustainability of ancient water control techniques in Iran: an overview. *Water History*, 10 (1), 13–30.

Mokhtari, D. (2012). *Participatory Management of Irrigation Water Resources in Iran: Fundamentals and Learning from Experiences*. Ilaf, Shiraz, Iran.

Mostafaeipour, A. (2010). Historical background, productivity and technical issues of qanats. *Water History*, 2 (1), 61–80.

Motiee, H., Mcbean, E., Semsar, A., Gharabaghi, B., and Ghomashchi, V. (2006). Assessment of the contributions of traditional Qanats in sustainable water resources management. *Water Resources Development*, 22, 575–588.

Nasiri, F., and Mafakheri, M.S. (2015). Qanat water supply systems: a revisit of sustainability perspectives. *Environmental Systems Research*, 4 (1), 1–5.

Rezaei-Moghaddam, K., Karami, E., and Gibson, J. (2005). Conceptualizing sustainable agriculture Iran as an illustrative case. *Journal of Sustainable Agriculture*, 27 (3), 25–56.

Saffari, M. (2005). Iran: land of qanats. In *International Conference on Qanat*, Kerman, Iran.

Safinejad, J. (2000). *Buneh: Traditional Agricultural System in Iran*. Amirkabir Press, Tehran, Iran.

Salehi, S., Chizari, M., Sadighi, H., and Bijani, M. (2018). Assessment of agricultural groundwater users in Iran: a cultural environmental bias. *Hydrogeology Journal*, 26 (1), 285–295.

Semsar Yazdi, A.A., and Labbaf Khaneiki, M. (2017). *Qanat Knowledge: Construction and Maintenance*. Springer, Dordrecht.

Shahroudi, A.A., and Chizari, M. (2009). An analysis of farmers' behavioral domains regarding optimal agricultural water management in Kharasan-Razavi Province: a comparation of participants and non-participants in water users' cooperatives. *Iranian Agricultural Extension and Education Journal*, 4 (2), 81–99 (In Persian).

Valizadeh, N. (2015). *Farmers' Participatory Behavior toward Conservation of Surface Water Resources: A Human Ecological Analysis in Southern Part of the Catchment Area of Urmia Lake*. M. Sc. Thesis, Tarbiat Modares University, Tehran, Iran

Valizadeh, N., Bijani, M., and Abbasi, E. (2016). Pro-environmental analysis of farmers' participatory behavior toward conservation of surface water resources in southern sector of Urmia Lake's catchment area. *Iranian Agricultural Extension and Education Journal*, 11 (2), 183–201 (In Persian).

Valizadeh, N., Bijani, M., and Abbasi, E. (2021). Farmers' participatory-based water conservation behaviors: evidence from Iran. *Environment, Development and Sustainability*, 23 (3), 4412–4432.

Valizadeh, N., Bijani, M., Karimi, H., Naeimi, A., Hayati, D., and Azadi, H. (2020). The effects of farmers' place attachment and identity on water conservation moral norms and intention. *Water Research*, 185, 116131.

Vosoughi, M. (2012). *Rural Sociology*. Keyhan Press, Tehran, Iran.

Wilkinson, T.J., Boucharlat, R., Ertsen, M.W., Gillmore, G., Kennet, D., Magee, P., Rezakhani, K., and De Schacht, T. (2012). From human niche construction to imperial power: long-term trends in ancient Iranian water systems. *Water History*, 4 (2), 155–176.

Yazdanpanah, M., Hayati, D., and Zamani, G. H. (2012). Application of cultural theory in analysis of attitude and activities toward water resource conversation: the case of Jihad-e Keshavarzi staffs in Bushehr Province. *Iranian Agricultural Extension and Education Journal*, 7 (2), 1–19 (In Persian).

Yazdanpanah, M., Hayati, D., Zamani, G.H., Karbalaee, F., and Hochrainer-Stigler, S. (2013) Water management from tradition to second modernity: an analysis of the water crisis in Iran. *Environment, Development and Sustainability,* 15 (6), 1605–1621.

10
Subsurface Drainage and Irrigation System in Paddy Fields

Shuichiro Yoshida
The University of Tokyo

10.1	Introduction	219
10.2	Soil Profile of Paddy Fields and Drainage	220
10.3	Increasing Needs for Subsurface Drainage in Paddy Fields	221
10.4	Basic Concept of Subsurface Drainage Design for Paddy Fields	221
10.5	Combination of Main Drainpipe and Supplementary Subsurface Drainage (Mole Drain, Subsoiling)	222
10.6	Functional Deterioration of Subsurface Drainage	224
	Permeability of Drainpipes • Flow Resistance between Outside and Inside of the Drainpipe • Degradation of Filler Materials • Change in Water Flow Resistance of Soil above Backfilled Trench	
10.7	Performance Diagnostics	231
10.8	Subirrigation in Paddy Fields	233
10.9	Conclusions	236
References		236

10.1 Introduction

Subsurface drainage is substantial for ill-drained farmland such as reclaimed land or flood plain in the world (Mirzaie et al., 2021). When the subsoil is fairly permeable and the groundwater table stays high, it is very effective to install subsurface drainage pipes deep in the subsoil and drain the water out of the field. However, when there is an impervious layer in the shallow depth, deep subsurface drainage does not contribute to collecting water from large area. Thus, in such fields, drainage pipes are installed in the shallow depth above the impervious layer and its intervals are set closer (e.g. Walker et al., 1982; Nelson, 2017). The core technologies are also effective for paddy fields which have hardpan layer at the bottom of plowed soil. However, the system for paddy fields developed in Japan has some differences from the conventional ones in the world (Maruyama and Tanji, 1997; Tabuchi and Hasegawa, 1995; Tabuchi, 2004). First, the depth of the trench is deeper than the impervious layer (hardpan), which is so shallow that tiller or plow possibly damages pipes if they were installed above it. Second, backfilled trench becomes the major pathway of the water from plowed soil to the drainpipes. Although the hydraulic conductivity of hardpan and subsoil is extremely low, the hydraulic conductivity of the backfilled soil is much greater than that of undisturbed subsoil (Fujioka and Maruyama, 1971). Water flows horizontally in the topsoil and then vertically into the drain through the backfilled part. Therefore, permeability of the backfilled part is important (Tabuchi, 2004). In Japan, rice husks have been used as the filler of the backfilled trench instead of soil to obtain high permeability. Because rice husk is organic material, farmers easily accept to use it in their fields. However, rice husk easily decomposes under alternating wet and drying cycles and lose their volume. The resultant vacated trench is replaced by soil, causing a decrease in permeability (Yoshida and Adachi, 2005). Thus, to keep drainage efficiency of paddy fields, not only building subsurface drainage system but also maintaining the water pathway to the drainpipe is highly required.

This chapter aims to present the degradation process of the backfilled trench and its effect on drainage performance. Diagnostics of the system is proposed and use of the system for subirrigation is briefly reviewed.

10.2 Soil Profile of Paddy Fields and Drainage

Because paddy rice farming requires much amount of irrigation water, paddy fields in Asia are usually located in the area where water is easily available. The typical soil types in such areas are gray soil, gray lowland soil and lowland soil which are often found in alluvial plains. More than 60% of paddy soils in Japan, for instance, are classified to gray or gray lowland soil. Because the intrinsic water table in these areas is too high to control the bearing capacity of soil for mechanized farming or to cultivate upland crops, drainage works, such as construction of drainage canals and pump stations, have been performed. Although construction of subsurface drainage is considered to be among the measures to lower the water table in the field scale, a difference lies in the meaning of water table between upland fields and paddy fields. To make the field submerged for paddy rice farming, farmers need to reduce water percolation from the plowed soil to subsoil. Two strategies are adoptable to lower the percolation: raising water table higher and reducing permeability of soil. The first measure is applicable if all the surrounding fields are planted with rice. However, in farmland blocks where multiple farmers cultivate, selection of crops depends on each farmer's preference. Therefore, the percolation control cannot be generally realized by managing water table in block scale. The alternative for the percolation control is to create impervious soil layer in shallow depth. Puddling and construction of claypan have been the substantial techniques for percolation control of paddy fields.

Puddling is the most common method of land preparation for lowland rice in Asian countries (Sharma and De Datta, 1985), and is very effective for percolation control. The operation is accomplished by stirring the soil by a rotary harrow with ponding water. Sharma and De Datta (1986) studied the effects of puddling on pore-size distribution, and observed that puddling decreased the proportion of pores larger than $30\,\mu m$ in equivalent diameter, and increased the proportion of pores less than $30\,\mu m$. Guidi et al. (1988) also observed a decrease in the proportion of pores larger than $50\,\mu m$ after puddling. Although the reduction of large pores directly decreases the permeability of the puddled layer, the effect also reaches the water pathway in subsoil. Adachi and Inoue (1988) analyzed the mechanisms of percolation control by puddling based on the measured water potential profiles and dry bulk density after puddling. They revealed that hydraulic consolidation of the puddled layer and clogging of pores in subsoil (claypan) due to the water percolation causes reduction of hydraulic conductivity of puddled paddy fields. The effect of puddling keeps on exclusively when the puddled layer is submerged or sufficiently wet. Once the puddled soil begins to desiccate, cracks are easily formed causing the drastic increase in percolation capacity. Because puddling is widely performed in paddy rice farming, vertical soil profile of paddy field is generally affected by this farming operation. Subsurface drainage for paddy fields has been took account this special feature of soil profile.

Because the compression of soil causes reduction of water permeability and increase in mechanical strength, the construction of compacted soil layer, i.e., claypan, also contributes to the percolation control and stabilizing soil bearing capacity for the stable operation of machineries even under flooded condition. For puddling, cone penetrometer resistance of claypan should exceed $0.2\,MPa$. Although puddled soil and claypan are important for the flooding and mechanized tillage, puddling and transplanting, the function hinders quick drainage of excessive water above it for harvest of rice or cultivation of upland crops.

10.3 Increasing Needs for Subsurface Drainage in Paddy Fields

Before mechanization of paddy farming in Japan, majority of paddy fields are located in lowland where underground water table is kept high due to the regional water level determined by the rivers and lakes. However, the modernization of paddy farming requires efficient drainage of the fields for introducing heavy machineries. For tillage and harvest by tractors and combined harvesters, cone penetrometer resistance of the plowed soil needs to be more than 0.2 MPa. To meet this condition, rapid drainage of excess water in the plowed soil is necessary. Another historic change in Japanese paddy farming is overproduction of rice, which induced the change in agricultural policy from rice monoculture into production of multiple crops in paddy fields around 1970. For the realization of the policy, improvement of field drainage following regional drainage is actively implemented by the national and local government. Subsurface drainage has been among the promising technologies for drainage of the rotationally cropped fields. The area of newly constructed subsurface drainage system had been kept high from 1970 till 1990s. As a result, aged and degraded subsurface drainage system began to increase from 1990. Then the maintenance of the drainage facility has become an important issue. Thus, the study on subsurface drainage for paddy fields is gradually shifting from the design of new system into the management of the aged system.

10.4 Basic Concept of Subsurface Drainage Design for Paddy Fields

The general theory of subsurface drainage assumes two-dimensional water flow over the soil layer above and a little below the drainpipe. However, in paddy fields, the thickness of the plowed soil is no more than 20 cm and impervious layer (claypan) lies below it. The water in the plowed soil hardly percolated into to the subsoil. Thus, the specially designed subsurface drainage system for paddy fields was developed in Japan to improve field drainage efficiency. The technology is also applied in the countries outside Japan, although the acreage may be not high (Ebrahimian and Noory, 2015; Ogino et al., 1993). The studies in 1960s and 1970s (Tabuchi, 1966; Negishi et al., 1972) are the basis of the present subsurface drainage system for paddy fields in Japan. Figure 10.1 shows the conventional design of the system. The depth of the drainpipe ranges between 50 and 80 cm. Although perforated plastic pipes are frequently used, ceramic pipes are also used in some regions as in Figure 10.2. Any special envelopes are not used but filler material is filled in the backfilled trench. As the filler material, rice husk (hull), crushed stones,

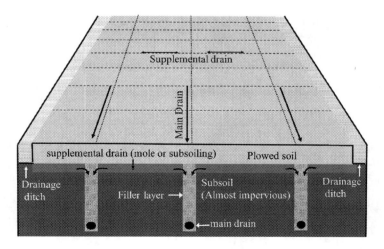

FIGURE 10.1 Conventional subsurface drainage system for paddy fields in Japan.

FIGURE 10.2 Ceramic drainpipe installed in paddy field.

wood chips and shells are used considering the cost and availability in the region. The trench is dug by a chain trencher or backhoe, leaving backfilled trench having a width of 15–30 cm. Rice husk filler is filled up to the ground surface and compressed down. However, other filler materials are filled up below the plowed layer to avoid mixing to the plowed soil. The combination of crushed stones in the lower part and rice husk for the upper part is also applied in some regions to take into account the advantage and disadvantage of the decomposable material as the filler.

10.5 Combination of Main Drainpipe and Supplementary Subsurface Drainage (Mole Drain, Subsoiling)

Subsoiling and mole drain are among the techniques of the supplementary subsurface drainage usually operated by farmers. The operation is highly recommended for the subsurface drainage of paddy fields where the drainage efficiency is mainly controlled in plowed soil and claypan. Although the upper part of the backfilled trench is broken by tillage soon after the installation, the water pathway remains in the claypan for a few years. However, the influence of tillage and puddling also reaches to the backfilled trench in the upper part of the claypan and gradually reduces the permeability of the filler layer. Thus, continual countermeasures are required to keep the vertical water pathway from the plowed layer to deeper backfilled trench. The overland drainage efficiency also depends on the horizontal mobility of the water in the plowed layer. But, because the interval of the main drainpipes cannot be less than 5 m due to economical reason, enhancement of water flow normal to the drainpipes is also an alternative way to supplement the lack of the horizontal drainage efficiency.

Horizontal water movement in plowed soil layer with supplementary drainage can be simply modeled by the analogy of Hoogoudt's Equation (Hoogoudt, 1940) as follows. Let the x axis start from the middle of two supplementary drains installed with an interval of s [m] (Figure 10.3). In this model, the direct flow into the main subsurface drainage is ignored. Assuming steady rainfall intensity R, the Darcy equation and the volumetric water balance in the right-hand half of the plowed layer ($0 < x < S/2$) are:

$$Q = -kh\frac{dh}{dx}, \qquad (10.1)$$

Subsurface Drainage/Irrigation System in Paddy Fields

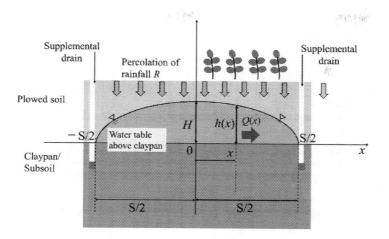

FIGURE 10.3 A simple model of horizontal distribution of water table above claypan.

$$Q = Rx, \qquad (10.2)$$

where k and h denote hydraulic conductivity of the plowed soil for horizontal water flow, and the level of the water table is measured from the bottom of the plowed soil.

The horizontal water flow is assumed to occur only in the saturated plowed soil, i.e., between the bottom of the layer and the water table above it. Combining Eqs. (10.1) and (10.2) yields the following differential equation:

$$Rx + kh\frac{dh}{dx} = 0. \qquad (10.3)$$

If the permeability of the supplementary drains and the filler layer is high enough to drain all the water that should be drained, the boundary value of h defined on the supplementary drain ($x = s/2$) can be set to zero. Then, the solution of Eq. (10.3) is:

$$Rx^2 + kh^2 = \frac{Rs^2}{4}, \qquad (10.4)$$

where h takes the maximum value H at the midpoint between two supplementary drains ($x = 0$) and H satisfies the following equation:

$$kH^2 = \frac{Rs^2}{4}. \qquad (10.5)$$

Equation (10.5) can be rewritten like:

$$s = 2H\sqrt{\frac{k}{R}}. \qquad (10.6)$$

This equation can be used to know the optimum interval of the supplementary drains to control the water table in plow layer within the acceptable level. Because R exceeds the hydraulic conductivity of plowed soil, surface flow happens and the percolation rate into the plowed soil cannot be higher than the hydraulic conductivity. Therefore, the minimum value of k/R in the right-hand term of Eq. (10.6) is 1. To prevent flooding above field surface in any rainfall intensity, the interval of supplementary drains s is required to be less than the twice of the thickness of plowed layer. This means a few tens of centimeters of the interval is necessary. However, the rainfall with intensity higher than the hydraulic conductivity of the plowed soil does not continue for hours. Practically, we should prepare for the continuous rainfall intensity around one-tenth of hydraulic conductivity. If we assume the thickness of plowed soil be 15 cm, maximum interval of supplementary drains should be 1 m to avoid flooding over the ground surface. The National Agricultural Research Organization of Japan recently developed a packaged subsurface drainage and irrigation system 'FOEAS' (NARO, 2011). This system applies supplementary drains at an interval of 1 m and has succeeded in improving drainage efficiency of paddy fields. Equation (10.6) also shows that the optimum interval can be three times wider if the hydraulic conductivity of the plowed soil is larger by an order of magnitude.

The two-dimensional solution of the steady-state water table above the claypan was also presented by Ogino and Murashima (1985), in which the horizontal distribution of water table in a rectangular drainage unit surrounded by two main drainpipes and supplementary drains was formulated.

10.6 Functional Deterioration of Subsurface Drainage

Subsurface drainage system for paddy field deteriorates with time. The hydraulic condition of the aged subsurface drainage is apart from the initial design. Figure 10.4 shows the water pathway from the ground surface to the drain canal. In the stage of design, the water flow in the drainpipe follows Manning's Law, and the water table in the filler or the supplementary drainage is below the bottom of plowed soil layer. In reality, due to the increase in the resistance of each flow sections, drainage efficiency gradually declines. The assumptions of the model in the previous section hardly hold in aged subsurface drainage system.

10.6.1 Permeability of Drainpipes

The increase in the flow resistance due to clogging of pipes and envelopes is reviewed in FAO (2005). Accumulation of mad in the drainpipe was reported to be very slow when flooded rice was cropped

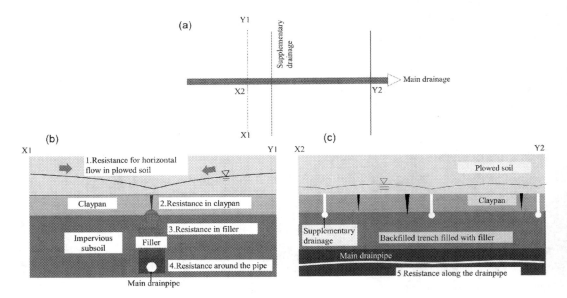

FIGURE 10.4 Water pathway from ground surface to drainpipe and its flow resistance. (a) Top view of subsurface drainage system; (b) profile normal to the main drainpipe; and (c) profile along the main drainpipe.

(Kusaka et al., 1993). However, it occurs more seriously in the drainpipes which does not discharge sufficient amount of water through the year. When mad begins to accumulate in pipes, it reduces drainage discharge and flashing function, creating a vicious circle. The accumulation is generally less near the outlet because the flow velocity is higher, although the effect on the drainage efficiency is more serious if the accumulation occurs near the outlet.

Although the tile drains up to 30 years ago, the upper end of the drainpipe was closed in the ground, the recent subsurface drainage system has inlets which can be used for subirrigation or flushing the deposition in the drainpipes. The drainpipes without such inlets, cleaning hose attached with opposing jet flushing nozzle can be applied to wash inside the pipe from the drain canals. However, there are many drawbacks of this tool. First, it can be used for the drainpipes straightly drained to the drainage canal. Second, the nozzle can only reach a few meters from the outlet. This cleaning system also does not work for branch type system, in which multiple drainpipes are gathered at the downstream of the field and drained from one outlet. For such cases, making spot holes above drainpipes to supply flushing water from the irrigation canal was proposed (Kohtani et al., 1989). This idea is taken over by the design of the subirrigation and drainpipe flushing system, i.e., water management pit in Hokkaido Pref. (Hokkaido, 2008) or irrigation and drainage box of FOEAS (NARO, 2016).

10.6.2 Flow Resistance between Outside and Inside of the Drainpipe

The path of water getting into the drainpipe differs by the type of the pipe, i.e., ceramic pipe or corrugated plastic tube. Ceramic pipes ($L=50$ cm including the flange) are put on the bottom of trench and are connected to each other. A side of the ceramic pipe has a socket which is fit to the end of another pipe. Although the pipe can absorb water, the main path of water into the pipe is the gap between the pipes. Thus, the inlet of water from the envelope (filler) to the drainpipe locates every 45 cm. In contrast, corrugated plastic tube is in rolled form and is installed on the trench by uncoiling the roll. The perforations exist very densely on the surface of the pipe, which facilitate smooth entry of water into the pipe. However, the permeability of the strainers can decrease with increase in the depositing soil particles on the filler (envelope) contacting to the strainers. Mihara et al. (1995), for example, presented that the effective strainer structure of drainpipes is the funneled strainer in which pore radius is larger in the flow direction.

10.6.3 Degradation of Filler Materials

In the most of countries, envelope is equipped around drainpipes. Envelope has the role to improve the permeability around the drainpipe, and acts as permeable constraints to impede entry of damaging quantities of soil particles and soil aggregates into drainpipes (FAO, 2005). The subsurface drainage for paddy fields, with impervious subsoil in shallow depth, requires not only the envelope but also backfilled trench filled with a material facilitating vertical water flow. Because the filler is artificial flow path through the impervious subsoil, the hydraulic conductivity is always expected to be extremely high. It should be noted that filler is different from filter that filtrates suspended particles. If the filler captured the particles very efficiently, the permeability of the layer should decrease soon after the installation although the clogging of the drainpipes would be temporally mitigated. Thus, the materials for the filler need to meet the conditions that the pore size should be large enough to permit most of soil particles suspended in the drain water to easily pass through, while the pore space is small enough to avoid entry of soil clods and large aggregates which may block the pores. Although mineral materials such as gravels widely used as filler, organic materials are favorable because they can be mixed to the soil for cultivation or farmland consolidation. However, their lifetime is limited. Among the organic filler materials, rice husk (hull) is the most popular material because it is easily available in paddy farming region and is decomposable. However, the progress in decomposition or clogging of the material sometimes causes a decrease in drainage efficiency, or formation of cavity along the backfilled trench.

Nagaishi (1977a, b, 1980) are the pioneering works investigating decomposition of rice husk filler. He proposed two indices for quantifying the degradation of rice husk filler. One is 'Decomposition Index (*DI*)', and the other is 'Filtered Soil Content (*FSC*)'. The methods to measure the *DI* are as follows:

1. Wash the rice husk sampled from the backfilled trench on a 0.42 mm meshed sieve.
2. Dry it at 70°C for 48 hours in an oven.
3. After measuring the weight (W_1), immerse the sample in 10% of sodium chloride solution for 24 h.
4. Wash it again on the same sieve.
5. Dry the residue on the sieve at 70°C for 48 hours in an oven.
6. Measure the weight (W_2).

DI is evaluated by the following formula:

$$DI = (W_1 - W_2)/W_1 \times 100(\%). \qquad (10.7)$$

The correspondence of *DI* to the state of the husk is summarized in Figure 10.5. The methods to measure the *FSC* are as follows:

1. Dry the rice husk sampled from the backfilled trench at 70°C for 48 hours in an oven.
2. Measure the weight (W_3).
3. Immerse the sample in 6% of hydrogen peroxide solution for 24 hours.
4. Wash it on a 0.42 mm meshed sieve.
5. Dry the residue on the sieve at 70°C for 48 hours in an oven.
6. Measure the weight (W_4).

FSC is evaluated by the following formula:

$$FSC = (W_3 - W_4)/W_4 \times 100(\%). \qquad (10.8)$$

The difference in lifetime of rice husk probably highly depends on the temperature and mean water table. To clarify more general nature of the rice husk filler, Yoshida et al. (2005) investigated 38 paddy fields under different crop history and the age of the subsurface drainage.

The typical vertical profiles of *FSC* and *DI* are provided in Figure 10.6. The most of the filtered soil distributes at the top of the filler layer, because the suspended soil particles come from plowed soil and clogging starts from the upper layer. Since the pore size of the rice filler is large compared to the flowing soil particles, the soil particles cannot be filtered perfectly. The filtration efficiency usually increases with the increase in retained particles, the filtered soil tended to increase at the topsoil but the middle of the filler did not retain much soil. The bottom of the filler layer also retains soil, because the vertical water flow concentrates to the perforations of the plastic drainpipes or the connections of the ceramic drainpipes. The high *FSC* can increase the water flow resistance of the filler. The profiles of *DI* show that decomposition of rice husk filler progressed from the upper layer in most of the profiles. Decomposition is generally accelerated by the oxidized condition and flowing in of fertile water, while it is depressed under the anaerobic conditions. The observed profiles of *DI* clearly show these basic understanding of the decomposition of organic filler. Because highly decomposed rice husks are losing elasticity, the bulk density of the filler decreased by decomposition, leading to depression of the filler layer or the formation of cavity at the top of the backfilled trench.

Figure 10.7 shows the interannual change in *DI* and *FSC* of the husk samples taken from 30 to 40 cm-depth layer of filler. *DI* increased with the years after installation, but it widely ranged, because the degree of decomposition depends on the wards. But statistical analysis showed *DI* did not depend on the history of cultivation, which was unexpected result. Between 10 and 12 years, most of the husks were

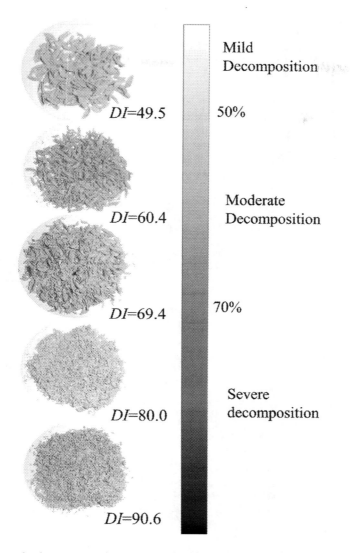

FIGURE 10.5 Relationship between visual appearance of rice husk filler samples and its decomposition index, DI.

decomposed and fragmented, and a part of them became powdery state. FSC also increased with the year after the installation of subsurface drainage. But decrease in water permeability due to the filtered soil was not observed even in the most contaminated filler.

Nagaishi (1977a, b, 1980) showed that the standard lifetime of rice husk in western Japan was 10 years, and the decomposition could progress in a few years under cropping upland crops. He also stated that severe reduction of saturated hydraulic conductivity occurred in the filler with high FSC. Seino et al. (1994) reported that the complete decomposition of rice husk requires 11–12 years in upland field, while the lifetime is three times longer in paddy fields. They also reported that decomposition of husk did not deteriorate the drainage function in all the investigated fields.

Occurrence of gapping void in the filler layer (backfilled trench) is another impact of the filler deposition on field management. The voids occurred from 8 years and were very probably made in 12 years after installation (Yoshida et al., 2005). Compression of decomposed husk causes the depression of the

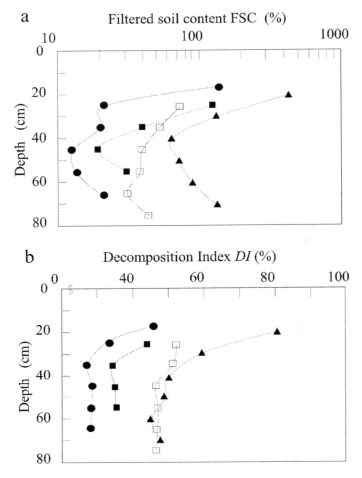

FIGURE 10.6 Examples of vertical distribution of FSC (a) and DI (b) of the rice husk filler. The symbols denote the sampling sites, for each of which the region and installation year differ (reproduced from Yoshida et al., 2005).

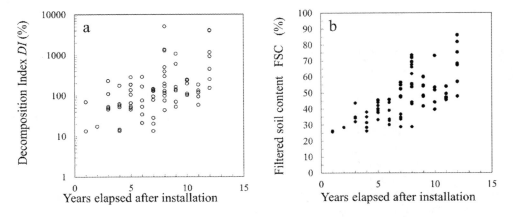

FIGURE 10.7 Change in FSC (a) and DI (b) of the rice husk filler after installation (reproduced from Yoshida et al., 2005).

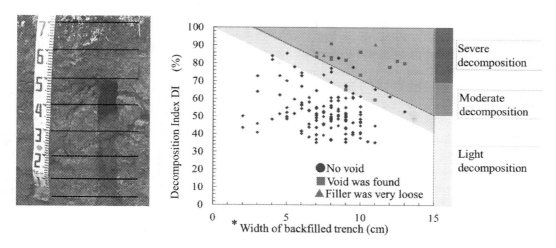

FIGURE 10.8 Condition for occurrence of gapping void: (a) picture of gapping void in the backfilled trench; (b) condition for occurrence of gapping void. Although the width of the trench is initially 15 cm, it decreases due to horizontal earth pressure (reproduced from Yoshida et al., 2005).

filler forming the gapping voids. They tended to be formed in the backfilled trench through the rigid stable subsoil, while it is not observed in the backfilled trench through soft deformable clayey subsoil, because the compression of the backfilled trench causes shrinkage of the space for the decomposed rice husk filler. Figure 10.8 shows the formation of gapping void on the relationships between the width of the trench and the DI. The wider the backfilled trench, the more probable the gapping void formed under the same DI. The gapping voids cause leakage of flooding water during wetland rice cropping. They also threaten safe operation of planting machineries having narrow wheels.

10.6.4 Change in Water Flow Resistance of Soil above Backfilled Trench

Ability of subsurface drainage system is influenced by the permeability of the soil above the filler layer. Interannual depression of the filler layer or the formation of an impermeable layer by tillage and puddling can deteriorate the accessibility of water from topsoil to the filler layer. On the other hand, shrinkage cracks induced by field desiccation contribute to the flow of subsurface water to the filler layer in clayey fields. Building mole-drains or subsoiling also has same function as the desiccation cracks. Figure 10.9 shows seasonal change in the drainage ability of the paddy fields (Yoshida et al., 2008). The discharge was measured by ponding the field once and drained only from the subsurface drainage. The drainage ability was defined as the discharge when cumulative drained water amounted to the pore volume of the backfilled trench filled with rice husk filler ($\fallingdotseq 4$ mm).

Although puddling completely ceased the percolation of water to the subsurface drainage, the midsummer drying, which induced cracks, effectively enhanced the drainage ability, while harvesting reduced drainage ability again by compressing the soil.

Figure 10.10 shows the interannual change in the maximum drainage ability of the year. The drainage ability tended to decrease in a few years after the installation. Figure 10.11 presents the depression of the filler layer. Because the mean thickness of the plowed soil in this field was 14 cm, the filler layer became a few centimeters deeper. Yoshida and Adachi (2005) pointed out a few centimeters of difference in the thickness of the soil above the filler reduced percolation rate above the drainpipe by order of magnitude. The water flow between the topsoil and the filler layer depended on narrow cracks in the claypan. The narrow cracks are so subject to clogging and closure that the discharge in the aged subsurface drainage does not have sufficient drainage efficiency.

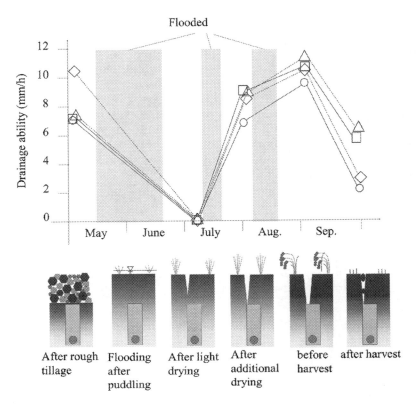

FIGURE 10.9 Seasonal change in drainage ability along with cropping calendar. ◇△: 4th year after installation; ○□: 2nd year after installation (reproduced from Yoshida et al., 2008).

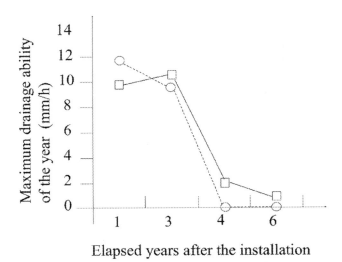

FIGURE 10.10 Interannual decrease in the annual maximum drainage ability. The measurements were performed in two experimental plots under the same management. The symbols □ and ○ represent these plots (reproduced from Yoshida et al., 2008).

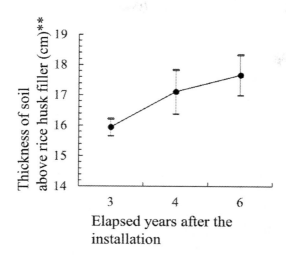

FIGURE 10.11 Change in thickness of soil above rice husk filler. *Vertical bars denote standard error. **The thickness of the plowed soil in the fields was 14 cm (reproduced from Yoshida et al. 2008).

10.7 Performance Diagnostics

Reform or reinstallation of the subsurface drainage system is required if the drainage efficiency of the field becomes deteriorated. But before performing countermeasures, diagnostics of the system is important for saving the cost. Performance of subsurface drainage system should be primarily evaluated by the maximum discharge under heavy rainfall events. Farmers usually aware of the degradation of the field drainage from the long-term flooding or high soil water content after rain. However, it is not conventional for them to measure the discharge from the subsurface drainpipe. In heavy rain events, because most of the water which cannot percolate from the ground surface goes to the surface drainage outlet, the amount of water which is required to be drained from the subsurface drainage is the water stored in the coarse pores of plowed soil or remaining on the depression of the ground surface. If 50 mm/d is adopted as the design discharge of the subsurface drainage system and a discharge around 2 mm/h is observed, the system can be evaluated to be normal. However, it gets below 1 mm/h, the drainage ability probably deteriorated. When the discharge does not meet the targeted level, the next step is to determine the bottle-neck.

Because the water flow from ground surface to the drainpipe is forced by the gradient of total hydraulic head, its distribution along flow path provides information about the permeability of each section. Figure 10.12 shows an example of measuring distribution of pressure head along the water flow path. The tensiometers must be installed just upon the drainpipe so as to measure the hydraulic head along the path. The total hydraulic head can be calculated by summing the pressure head and the height from the reference point such as the outlet of the drainpipe.

Figure 10.13 shows the vertical profiles of hydraulic head measured upon the main subsurface drainpipes. The profiles a and b were taken on the same drainpipe in a paddy field where subsurface drainage adequately functioned. The maximum discharge from this field was close to 2 mm/h. The profile c was taken in the neighboring field where drainage was very bad and being waterlogged above the claypan for many days after rain. The maximum discharge from this field was a quarter of the field of the profiles a and b. These profiles were observed at the moment when the maximum discharge was recorded during a moderate rainfall event (total rainfall was 21 mm). In the profile (a), the total head of the bottom of plowed soil (H_0) was almost consumed for the vertical water flow to the top of the filler, resulting in small total head gradient for the filler layer and for the interface between both sides of the drainpipe. The maximum total head gradient appeared between the middle of claypan and the top of the filler

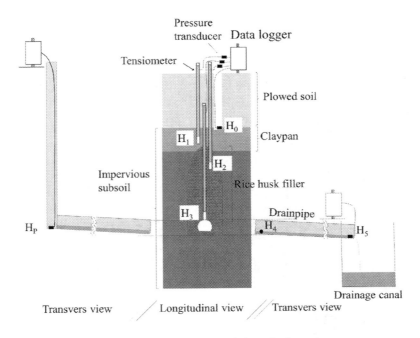

FIGURE 10.12 Measuring points of pressure head above and along the drainpipe.

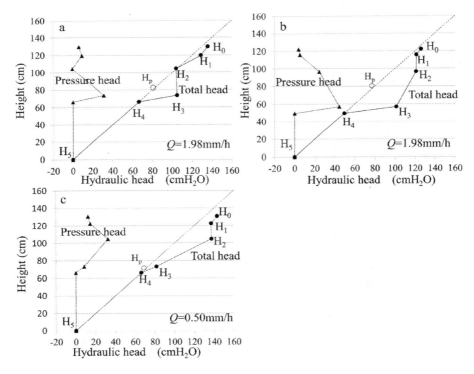

FIGURE 10.13 Vertical distribution of pressure head and total head at the peak discharge from the drainpipe after a rainfall event.

A_i and A_d were measured at 20 m from the irrigation canal and drainage canal, respectively in the field with good drainage ability. B_i was measured at 20 m from the irrigation canal in the field with poor drainage ability.

layer. This profile suggests the resistance of the soil layer above the filler controlled the water flow into the drainpipe. In contrast, profile b shows almost static pressure head between the plowed soil and the top of the filler, suggesting the resistance from the plowed soil to the drainpipe was very low, to provide hydraulic energy to the interface between the outside and inside of the drainpipe.

This field has densely constructed supplementary drain, interval of which was 2 m. Its effect clearly appeared in profile b, although the connection between the supplementary drainage and the filler of main drainage was probably little hindered in profile a. In profile c, since the highest gradient of total head appeared in the filler layer and the pressure was positive, flow resistance in the filler layer was abnormally high. Due to the loss of hydraulic head, the head difference between the outside and inside of the drainpipe was so small that the water flow into the drainpipe must have been very small at this location. The observation of the vertical section of the filler layer at this location revealed that the upper part of the backfilled trench was considerably narrowed, and pasty soil particles retained in the rice husk filler. These were the probable causes of the increase in flow resistance of the filler layer.

The performance diagnostics of the aged subsurface drainage based on the hydraulic head measurement is rational and promising method. However, data accumulation from the various fields is necessary to establish the diagnostic system and to apply it to the governmental project on the management of subsurface drainage system.

10.8 Subirrigation in Paddy Fields

Subirrigation using subsurface drainage system has been used for long years in the world. Although the cultivation of flooded rice by transplanting cropping system was not required subirrigation, direct sowing system of flooding rice or the cultivation of upland crops in paddy fields has widened the benefit of introducing subirrigation. To implement subirrigation, we need to decide the targeted depth of water table. The recommended depth of the water table for a variety of crops is available in the papers and handbooks. Kohda (1982), for example, showed that wheat and maize can normally grow when the water table was between 20 and 30 cm from the ground surface, while sweet potato, burley and spinach require the water table depth less than 50 cm. They derived this conclusion based on the experiment using a sloped lysimeter filled with coarse-grained farmland soil, which had higher unsaturated conductivity compared with fine textured soils. Although the information provides the preferences of the crops for the soil moisture condition, we should note that the relationship between the depth of the water table and the moisture condition in the root zone depends on the unsaturated hydraulic conductivity and the demand for the evapotranspiration,

Let define the vertical axes z starting from the water table in upward direction. Buckingham and Darcy's Equation is written as:

$$q = -k(\Psi)(d\Psi/dz + 1), \qquad (10.9)$$

where q denotes the upward water flux, and the unsaturated conductivity k is a function of soil matric potential Ψ. When we define the depth of the bottom of plowed soil as z_p, integration of the equation from $z = 0$ (the water table) to $z = z_p$ (the bottom of the plowed soil) yields the following equation (Gardner, 1958):

$$-z_p = \int_0^{\Psi_p} \frac{d\psi}{1 + \dfrac{q}{k(\psi)}}, \qquad (10.10)$$

where Ψ_p denotes the matric potential at the bottom of the plowed soil. If we assume steady upward water flux from the water table to the plowed soil where major evapotranspiration occurs, the equation provides the relationship between the depth of the water table and the matric potential at the depth of the plowed soil. Hasegawa (1986) compared this relationship for a clayey paddy soil and an upland volcanic ash soil (Figure 10.14). For the volcanic ash soil, if the depth of the water table from the bottom of the plowed soil is around 1 m, 5 mm/d of upward water flux can be expected. However, for the clayey paddy soil, the water table is required to be within 10 cm from the bottom of the plowed soil to keep the same upward water flux. Thus, in clayey soil, the matric potential at the bottom of the plowed soil hardly changes with the depth of the water table. These differences due to the soil type clearly reveal the weakness of the water table depth as a common criterion for the management of subirrigation in clayey paddy field. Even if the supplementary drains were densely implemented, the sufficient upward water flux from the water table below the plowed soil cannot be expected because the rate of capillary rise is not highly accelerated by such vertical macropores. Thus, in the clayey paddy soil, the water table should be kept higher than 20 cm below the plowed soil, or intermittently be risen above the bottom of the plowed soil to provide irrigation water horizontally to the plowed soil. Takahashi et al. (2018), for example, tested two subirrigation methods for cultivation of soybean in the clayey paddy field. One method was maintaining water table at −35cm from the ground, and the other was intermittent irrigation which involved raising the water table level during the irrigation period to within 16 cm of the ground surface each morning and then dropping it to −35 cm in the evening. They reported both methods could keep the matric potential at 10 cm below the ground surface more than −40 kPa.

Subirrigation can be introduced to paddy fields without building new irrigation canal. Figure 10.14 is an example of the device for supplying irrigation water from pipeline to the subsurface drainpipe (Hokkaido, 2008). The irrigation water from open channels can be also used by another type of the device. To keep the predefined water table, various devices have been developed. The simplest way is to adjust the flow rate by the irrigation valve. But because the demand temporally changes due to soil moisture conditions and weather, the relief well for the drainage usually has a function to overflow the water at the predefined height. Figure 10.15 shows a schematic chart of a relief well used in FOEAS (NARO, 2016). The system to control water table by raising the subsurface drainage outlet (Figure 10.16) is not unique in Japan but also popular in US as 'controlled drainage' (e.g. Gilliam

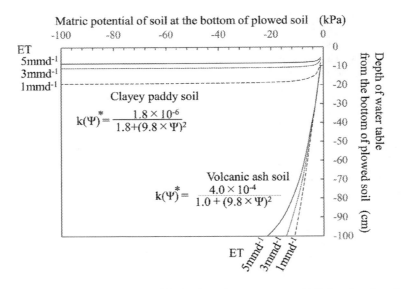

FIGURE 10.14 Relationship between water table depth and matric potential of plowed soil under steady-state evapotranspiration. *The parameters of unsaturated conductivity for the soils were cited from Hasegawa (1986).

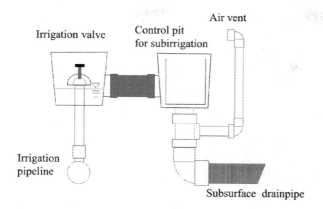

FIGURE 10.15 Example of device for supplying water from irrigation canal to subsurface drainpipe (remade from Hokkaido 2008).

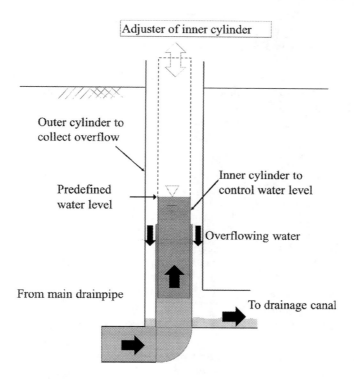

FIGURE 10.16 Design of the double-cylinder release valve for subsurface drainage controlling the water table. Drawn by the author based on the chart of 'Water table controller for FOEAS' provided by Paddy Research Company (http://www.paddy-co.jp/data/f_12.pdf).

et al., 1979; Evans et al., 1995), which targets reduction of nitrogen and phosphorous loss from upland fields. By combining control of the irrigation discharge and the height of the drainage water table, farmers can keep the water table inside the subsurface drainage system. However, if the demand is much less than the discharge, it causes waste of irrigation water. Thus, the irrigation valve having a function of switching depending on the water level in the drainpipe is also available.

10.9 Conclusions

Subsurface drainage system suitable for paddy fields with thin plowed soil and consolidated subsoil (claypan), which is uniquely developed in Japan, is presented. Such a soil profile is highly adopted to the conventional mechanized rice cropping system in which puddling, transplanting and harvesting are conducted by machineries. However, the profile causes ill-drainage for harvesting rice or cultivation of upland crops. The subsurface drainage system for this type of paddy fields has the backfilled trench above the main drainpipe filled with filler material up to the bottom of the plowed soil. Supplementary drains having the depth of approx. 40 cm are also densely constructed normal to the main drainpipe, because two-dimensional water flow toward the main drainpipe in subsoil cannot be expected. Although the design of this system has been established, the maintenance or restoration of the aged system is recently becoming the important issue.

Before the restoration or reconstruction project, the diagnostic of the aged system is required to save the cost. The drainage efficiency should be primarily evaluated by measuring the discharge from the drainpipes at the rainfall. If the discharge does not meet the standard, the cause should be analyzed. Direct survey of the profile above the drainpipe is effective for checking the filler degradation, i.e., depression, decomposition and clogging; its effect on the drainage function is not necessarily clarified. In contrast, measurement of the distribution of total hydraulic head along the water pathway from plowed soil to drainpipe is a promising method to identify the bottle-neck section reducing the drainage discharge.

To introduce subirrigation into such a profile of paddy fields, the water table must be controlled by taking into account the unsaturated conductivity, i.e., texture, of subsoil. For the clayey subsoil, the water table should be intermittently risen to plowed soil or control it approx. 10 cm below the plowed soil. Because various devises to control the water level in the subsurface drainage system have been developed, the effective use of these facilities to homogeneously manage the moisture condition of the field should be studied more.

References

Adachi, K., Inoue, H., 1988, Effects of puddling on percolation control of rotational paddy fields -studies on the water movement of rotational paddy fields-, *Transactions of Japanese Society of Irrigation, Drainage and Reclamation Engineering*, 140, 19–26. (In Japanese)

Ebrahimian, H., Noory, H., 2015, Modeling paddy field subsurface drainage using HYDRUS-2D, *Paddy and Water Environment*, 13, 477–485.

Evans, R.O., Skaggs, R.W., Gilliam, J.W., 1995, Controlled versus conventional drainage effects on water quality, *Journal of Irrigation and Drainage Engineering*, 121(4), 271–276.

FAO, 2005, Materials for subsurface land drainage systems, FAO Irrigation and Drainage Paper 60 Rev. 1, Italy.

Fujioka, Y., Maruyama, T., 1971, Studies on underdrainage of clayey paddy field 1, *Transactions of the Japanese Society of Irrigation, Drainage and Reclamation Engineering*, 35, 48–53. (In Japanese)

Gardner, W.R., 1968, Some steady-state solution of the unsaturated moisture flow equation with application to evaporation from a water table, *Soil Science*, 85, 228–232.

Gilliam, J.W., Skaggs, R.W., Weed, S.B., 1979, Drainage control to reduce nitrate loss from agricultural fields, *Journal of Environmental Quality*, 8, 137–142.

Guidi, G., Pini, R., Poggio, G., 1988, Porosity in a puddled rice soil as measured with mercury-intrusion porosimetry, *Soil Science*, 146, 455–460.

Hasegawa, S., 1986, Soil water movement in upland fields converted from paddy fields, *Soil Physical Conditions and Plant Growth*, 53, 13–19. (In Japanese)

Hokkaido, 2008, Manual of subirrigation using water managing pit, http://www.pref.hokkaido.lg.jp/ns/nts/grp/06-11tebiki.pdf (In Japanese).

Hoogoudt, 1940, Physical and mathematical theories of tile and ditch drainage and their usefulness, Research Bulletin, 436, Iowa Agricultural Experiment Station, 667–706.

Kohda, H., 1982, Study on rotational cropping of vegetables, cereals and potatoes in wetland, (1) Water table and crop growth and yield, *Bulletin of Ibaraki Prefectural Agricultural Research Station*, 22, 25–91. (In Japanese)

Kohtani, J., Kawado, Y., Adachi, T., 1989, Renewal method of filler for tile drainage and cleaning method of drainpipe by making spot hole, *Journal of Japanese Society of Irrigation, Drainage and Reclamation Engineering*, 63, 695–699.

Kusaka, T., Fukuda, M., Ono, T., Roy, K., 1993, Effectiveness of drainage and durability of, and condition of soil deposited inside clay pipes buried long term in paddy fields - Mechanism of drainage and maintenance of drain pipes (I)-, *Transactions of Japanese Society of Irrigation, Drainage and Reclamation Engineering*, 168, 97–104. (In Japanese)

Maruyama, T., Tanji, K.K., 1997, *Physical and chemical processes of soil related to paddy drainage*, Shinzansha Sci & Tech, Tokyo, Japan.

Mirzaie, Z., Fatahi, R., Eslamian, S., Azizi, A., 2021, Comparing and evaluating reaction factor and drainage water quality respect to direction of surface irrigation, *International Journal of Hydrology Science and Technology*, 12(2), 214–222.

Nagaishi, Y., 1977a, Use of rice husk for filler material of tile drainage, *Field and Soil (Hojoh to Dojo)*, 91, 10–13. (In Japanese)

Nagaishi, Y., 1977b, Durability of rice husk filler for tile drainage, *Journal of Japanese Society of Irrigation, Drainage and Reclamation Engineering*, 45, 387–390. (In Japanese)

Nagaishi, Y., 1980, Durability of rice husk filler for tile drainage (II), *Journal of Japanese Society of Irrigation, Drainage and Reclamation Engineering*, 48, 387–391. (In Japanese)

NARO, 2016, Manual for using water table control system in rotational cropping in paddy fields (revised version) https://www.naro.affrc.go.jp/publicity_report/publication/files/narc_suidenrinsaku_foeas_rr.pdf. (In Japanese)

Negishi, H., Tada, A., Furuki, T., Moriya, M., Shibuya, K., Sugawara, K., Kamimura, H., 1972, Study on subsoil improvement and irrigation and drainage system (1) General overview, *Bulletin Research Institute of Agricultural Engineering*, 10, 43–94. (In Japanese)

Nelson, K.A., 2017, Soybean yield variability of drainage and subirrigation systems in a claypan soil, *Applied Engineering in Agriculture*, 33(6), 801–809.

Ogino, Y., Murashima, K., 1985, Theoretical investigation for a pipe drainage design – design of pipe drainage for multipurpose paddy field (1), *Transactions of Japanese Society of Irrigation, Drainage and Reclamation Engineering*, 119, 1–6. (In Japanese)

Ogino, Y., Murashima, K., Kim, S.W., Kim, S.J., 1993, Present situation of paddy land consolidation and standardization of field drainage planning and design in Korea, *Journal of Japanese Society of Irrigation, Drainage and Reclamation Engineering*, 62, 23–27. (In Japanese)

Seino, M., Dokai, Y., Chikano, M., Seno, Y., 1994, Durability of rice husk as filler material for tile drain and improvement of the drainage function, *Bulletin of Yamagata Prefectural Agricultural Experiment Station*, 28, 99–114. (In Japanese)

Sharma, P.K., De Datta, S.K., 1985, Puddling influence on soil, rice development, and yield, *Soil Science Society of America Journal*, 49, 1451–1457.

Sharma, P.K., De Datta, S.K., 1986, Physical properties and processes of puddled rice soils, *Advances in Soil Science*, 5, 139–178.

Tabuchi, 1966, Study on drainage in clayey paddy fields, (1) Problems and hypothesis on drainage in clayey paddy fields, *Transactions of Agricultural Engineering Society of Japan*, 18, 7–11. (In Japanese)

Tabuchi, 2004, Improvement of paddy field drainage for mechanization, *Paddy and Water Environment*, 2, 5–10.

Tabuchi, T., Hasegawa, S. (eds), 1995, *Paddy fields in the world*, The Japanese Society of Irrigation and Drainage and Reclamation Engineering, Tokyo, Japan.

Takahashi, T., Katayama, K., Nishida, M., Namikawa, M., Tsuchiya, K., 2018, Effect of using subirrigation and slit tillage to increase soybean (Glycine max) yield in clayey soils in rice paddies converted to uplands, *Soil Science and Plant Nutrition*, 64(4), 491–502. doi: 10.1080/00380768.2018.1451226

Walker, P.N., Thorne, M.D., Benham, E.C., Sipp, S.K., 1982, Yield response of corn and soybeans to irrigation and drainage on a claypan soil, *Transactions of ASCE*, 25, 1617–1621.

Yoshida, S., Adachi, K., 2005, Spatial variability of percolation rate along subsurface drainage in paddy fields, *Transactions of Japanese Society of Irrigation, Drainage and Reclamation Engineering*, 235, 35–41. (In Japanese)

Yoshida, S., Adachi, K., Seki, M., 2005, Decomposition of rice husk filler causing gapping voids along subsurface drainage, *Transactions of Japanese Society of Irrigation, Drainage and Reclamation Engineering*, 235, 25–33. (In Japanese)

Yoshida, S., Adachi, K., Tanimoto, T., 2008, Seasonal and interannual change in under-drainage ability in clayey paddy fields, *Transactions of Japanese Society of Irrigation, Drainage and Reclamation Engineering*, 257, 51–56. (In Japanese)

11

Pitcher Irrigation as a Viable Tool of Enhancing Nutrition and Livelihood of Small-Scale Farmers: The Case Study of the Drier Areas of Katsina State, Nigeria

S. Jari, T. A. Rimi, and R. T. Nabinta
Federal University Dutsin-Ma

Saeid Eslamian
Isfahan University of Technology

11.1	Introduction	239
	Justification of the Study • Objectives of the Study	
11.2	Materials and Methods	242
	Data Collection	
11.3	Results	242
11.4	Discussion	244
11.5	Conclusions	245
11.6	Recommendations	245
	Acknowledgments	245
	References	245

11.1 Introduction

The growing demand for water in the country emphasizes the need to introduce and develop low-input and water-saving irrigation technologies for sustainable crop production, particularly in semi-arid and arid areas (Bainbridge, 2001). Irrigation is a crucial input for growing plants and accounts for the high water demand in agriculture (Albaji et al., 2020). Water conservation measures using techniques such as drip irrigation require significant investment. In this context buried clay pot irrigation, which is a cost-effective traditional technique, can be adopted for controlled irrigation (Vasudevan et al., 2006). Use of buried unglazed clay plots, often called 'pitcher irrigation', is another very ancient irrigation technology appropriate for home gardens, especially in dry areas (Bainbridge, 2001).

Pitcher irrigation, in its simplest form, entails burying an unglazed, porous clay pot next to a seedling. Water poured into pot seeps slowly into the soil, feeding the seedling's roots with a steady supply of moisture.

When a pot, filled with water and covered by a lid (wooden or clay), is buried in the soil, the water oozes out of the clay pot due to hydraulic head difference (moisture content difference) between the pot surface and the surrounding soil until it is in equilibrium with the surrounding area. The pressure gradient results from positive pressure head inside pitcher and negative pressure head at the outer surface of the pitcher which is in contact with soil. The rate of seepage of water from pitcher will depend on the type of plant and soil and climatic conditions around the pot. The movement of water is as a result of the uptake by the crops and it continues as long as the plants take it up and it evaporates. The system is therefore self-regulating. With this irrigation, deep percolation losses are negligible since water is released from smaller areas, and the rate of water loss can be controlled site to site by the amount of water put in each pitcher. Pitcher irrigation is ideal for sandy to loamy soil with good porosity (40%–60%) (Bainbridge, 2001). Small pitchers are often used because they are less expensive than large ones (Bainbridge, 2001).

Pitcher irrigation is an inexpensive small-scale irrigation method practiced in the semi-arid areas. Additionally, being easy and cheap, the installation and operation of the method are inexpensive (Bainbridge, 2001). Irrigation through buried clay pots has been reported to increase yields of annuals and establishment of perennial crops and trees compared to alternative irrigation methods, to reduce weed growth due to the selective water supply and to save water due to lower drainage especially in sandy soils (Bainbridge, 2001). Their use can be advantageous to smallholders as they are reportedly less prone to clogging than drip irrigation tubes, do not need water pressure systems and generally are sturdier and need less maintenance Bainbridge (2001). When the pot is filled with water, the natural pores in the pot's walls allow water to spread laterally in the soil, creating the moist conditions necessary for plant growth. Pitchers are filled as needed, maintaining a continuous supply of water directly to the plant root zone. Pitcher irrigation is used for small-scale irrigation, where water is either scarce or expensive, fields are difficult to level such as under uneven terrain, and more so, that the water is saline and cannot be normally used in surface methods of irrigation and in remote areas where vegetables are expensive and hard to come by Dubey et al. (1991). Water poured into pot seeps slowly into the soil, feeding the seedling's roots with a steady supply of moisture. Pitcher irrigation uses water more efficiently than other systems since it delivers water directly to plant root zones, instead of broader areas of the field. The surrounding soil is almost always at field capacity (approximately 80% of soil pores filled with water) as long as the pot is not allowed to dry up completely due to evapotranspiration (Bainbridge, 2001). Pitcher irrigation is a traditional system (Mondal, 1974, 1983; Das, 1983) of irrigating plants and considered several times more efficient than a conventional surface irrigation system.

Pitcher irrigation is claimed to be a self-regulative system with a very high water-saving potential and good capabilities for irrigation of various types of crops (Mondal, 1978, 1983; Chigura, 1994). The saturated hydraulic conductivity of clay pots is the key factor in controlling the success of this irrigation method in the field. Pitcher irrigation systems use clay pots which are baked at high temperature (800°C) to produce walls of the desired porosity. The porosity of pitcher wall depends on the mixture of clay and sand at a ratio of 4:1; it also depends on baking temperature (Daka, 1991).

Despite its apparent simplicity, factors affecting the system performance have not been well described and analyzed in the literature. Stein (1990) and Daka (1991) found that using clay pots can save up to 70% of water compared to watering with buckets and sprinkler irrigation. Pitchers have also been used to establish deserts shrubs (Bainbridge et al., 1988a). There is a need to adopt traditional methods of irrigation that could have similar efficiency to that of drip irrigation but with less cost (Batchelor et al., 1996). Developing traditional, low-input and water-saving technologies for sustainable crop production, particularly in semi-arid and arid areas, is one of the major challenges to scientists, one which has been ignored by most international developmental programs (Bainbridge, 2001). One of these neglected methods is pitcher irrigation. Alternative technologies such as the water filled buried clay pots have been used in many dry-lands of the world for thousands of years (Bainbridge, 2001). Employing locally made clay pots was found to be appropriately efficient for small-scale horticultural crop production (Bainbridge, 2001).

Thus far, little research has been carried out on the performance of pitcher irrigation systems, including the various factors affecting water seepage out of the pitchers.

For successful design, installation and operation of pitcher irrigation systems in arid and semi-arid regions, pitchers should be placed at suitable distances from each other so that the wetted areas do not needlessly overlap, while at the same time ensuring that areas of the soil root zone are not unintentionally left dry. Batchelor et al. (1996) found that the pitcher irrigation can save water up to 70% when compared to the conventional techniques of irrigation. The pitcher irrigation system saves more water because it eliminates water loss through run-off, evaporation and deep percolation. Besides increasing yield owing to adequate soil aeration, soil under this system is always under field capacity (Daka, 1991).

High-tech irrigation methods no doubt save large quantities of water, but technical, economical and socio-cultural factors obstruct the adoption of these methods. Hence, developing and introducing simple, efficient, low-input, easy to install, operate and maintain irrigation technologies, especially suitable for small-scale irrigation in arid regions, is one of the major tasks for the scientists (Bainbridge, 2001). Pitcher irrigation method is one of these efficient methods which is used for small-scale irrigation where water is scarce, fields cannot be easily leveled, soils are coarse-textured with high water infiltration rates, water is saline and cannot be normally used in surface irrigation methods and in remote areas where fresh vegetables are expensive to fetch (Bainbridge, 2001).

11.1.1 Justification of the Study

- Developing traditional, low-input and water-saving technologies for sustainable crop production, particularly in semi-arid and arid areas, is one of the major challenges to scientists, one which has been ignored by most international developmental programs.
- Drip irrigation systems are well known but are expensive to set up, but more efficient traditional methods of irrigation such as pitcher irrigation that could be of great help have not been well studied or publicized.
- Pitcher irrigation system is innovative, low-cost and easy-to-maintain technology which is operated and managed by individual to grow high-value, high-nutritious and multi-purpose crops and trees during dry periods for direct consumption and/or the local market, and therefore, has a very significant potential for enhancing food security, poverty alleviation and enhancing the livelihood of small-scale farmers in the drier areas.
- The use of pitcher irrigation for off-season small-scale irrigation in arid and semi-arid areas is not exploited sufficiently in terms of agricultural and livelihood improvements on one hand, and technological development and market/institutional adoption on the other. One of the key factors that contribute to poor adoption, replication and transfer of pitcher irrigation technology and practices is the lack of technical capacity by small-scale farmers to implement these systems in a cost-efficient manner. Thus, in order to realize the benefits of pitcher irrigation system, the systems must be designed and constructed professionally and be properly maintained. Adequate technical skills are also required to facilitate a strategic cost-efficient planning, implementation, monitoring and evaluation of pitcher irrigation systems. Therefore, there is a need to increase the capacity of small-scale farmers, in order that they can design operate and maintain pitcher irrigation systems. This can enable wide adoption of pitcher irrigation technology and ultimately increase the resilience to climate change by small-scale farmers in arid and semi-arid areas.

11.1.2 Objectives of the Study

- The objectives of this study are to foster the capacity of small-scale farmers to practically implement cost-efficient pitcher irrigation technology and practices in arid and semi-arid areas. Therefore, this study intends to provide small-scale farmers with the required information and skills to support proper planning, design and construction of cost-efficient pitcher irrigation technology and practices.

- The ultimate objective is to help small-scale farmers living in arid and semi-arid areas to use pitcher irrigation to grow high-value, high-nutritious and multi-purpose vegetable crops and trees during dry periods for direct consumption or the local market.

11.2 Materials and Methods

Ten farmers from ten villages (Agangaro, Daddara, Daga, Dannabasu, Matsai, Natsinta, Kayauki, Kukabiyar, Radi and Riko) located in Jibia, Kaita and Batagarawa local government areas of Katsina State, Nigeria, were purposely selected to participate in the study. All these villages have something in common, i.e. very limited water resources and no irrigation facilities. Three of the participants were women, three were youth and four were men. All the ten farmers have at least attended junior secondary school. Capacity building workshop was organized at the teaching and research farm of Federal University Dutsin-Ma on 20th November 2017. Each farmer was supplied with clay pots of 8–10 L capacity for the purpose of demonstration. Step-by-step procedures of establishing a pitcher irrigation system were demonstrated to the farmers by the lead researcher and farmers were assisted by the technicians to establish ten pitchers each. Pitchers were spaced 100 cm apart. Each pot was buried up to the neck, filled with water and covered with a wooden lid to reduce evapotranspiration. Top soil mixed with organic matter at the ration of 2:1 was used to bury each pot. Four seedlings of tomato per pot were transplanted 5 cm away from the pot while four seeds (stands) of okra and cucumber each were sown per pot at 5 cm away from the pot. The knowledge, skills and experience acquired by the farmers at the workshop enable them to establish vegetable garden of tomato, cucumber and okra using pitcher irrigation system. Our research team visited the farmer's garden three times in the course of this study, i.e. at the establishment, during intensive growth period and at harvest time to monitor progress and ensure compliance to the procedures and best agronomic practices. At harvest farmers were taught how to take sample (tagged) ten plants randomly from their plots for data collection/yield determination. Digital weighing balance was used to take data on fruit weight and Vernier caliper was used to measure fruit size of tomato. Ruler was used to measure pod and fruit length of okra and cucumber, respectively.

11.2.1 Data Collection

Data were collected on yield parameters such as number of fruits per plant, weight and size of fruit. *t*-Test statistical method of data analysis was employed in data processing as described by Gomez and Gomez (1984).

11.3 Results

Records of number of fruits per plant are presented in Table 11.1 and the highest number of fruits per plant in tomato was 33.3, 30.2 and 29.6, recorded at University farm Badole, Daga and Agangaro areas, respectively. The lowest number of fruits per plant in tomato was 23.0, which was recorded at Natsinta, and this number was lower by ten fruits compared with the highest; therefore, a significant difference was observed between the highest and the lowest but there was no significant difference between locations. The highest number of fruits produced per cucumber plant was eight (8.0) and the lowest was 6.4 produced at Kukabiyar and Dannabasu, respectively. But there were no significant differences between the eleven locations as per the number of fruits per plant in cucumber. The highest number of pods produced per plant in okra was 10 and the lowest was 8.4 at Kukabiyar and Dannabasu, respectively, but the difference between eleven locations was not significant.

Fruit size is an important yield parameter and is presented in Table 11.2. The results indicated that the biggest tomato fruit size was 9.08 cm and the smallest was 7.56 cm, recorded at Kukabiyar and Daga, respectively, but tomato fruit size was not significantly different between the eleven locations. The biggest fruit size of cucumber recorded was 14.2 and the lowest was 12.2 cm at Kayauki and Natsinta, respectively,

TABLE 11.1 Number of Fruits Per Plant for Vegetables Grown under Pitcher Irrigation System at 11 Locations in 2017

S. No.	Location	No. of Fruits/Plant Tomato		No. of Fruits/Plant Cucumber		No. of Pods (Plant Okra)	
		Mean	Mean Squares	Mean	Mean Squares	Mean	Mean Squares
1	Agangaro	29.6	876.1	6.9	47.61	8.8	7744
2	Daddara	29.6	876.16	7	49	10	100
3	Daga	30.2	912.04	6.7	44.89	8.8	77.44
4	Dannabasu	26.4	696.96	6.4	40.96	8.5	72.25
5	Kayauki	24.7	610.09	6.9	47.61	8.4	70.56
6	Kukabiyar	25.1	630.01	8	64	9.6	92.16
7	Matsai	27.4	750.76	7.9	62.41	9.4	88.36
8	Natsinta	23.1	533.61	6.7	44.89	9.2	84.64
9	Radi	27.6	761.76	6.6	43.56	9.1	82.81
10	Riko	29.5	870.25	7.3	53.29	8.7	75.69
11	University farm Badole	33.3	1,108.89	8.1	65.61	9.5	90.25

TABLE 11.2 Fruit Size of Vegetables Grown under Pitcher Irrigation System at 11 Locations in 2017

S. No.	Location	Fruit Size of Tomato (cm)		Fruit Size of Cucumber (cm)		Fruit Size of Okra (cm)	
		Mean	Mean Squares	Mean	Mean Squares	Mean	Mean Squares
1	Agangaro	8.13	66.1	12.5	156.25	6.9	47.61
2	Daddara	7.67	65.12	13	169	7	49
3	Daga	7.56	54.2	13.6	184.96	6.7	44.89
4	Dannabasu	8.9	70.56	13.5	182.25	6.9	47.61
5	Kayauki	7.98	63.68	14.3	204.49	7.3	53.29
6	Kukabiyar	9.08	82.26	13.8	190.44	8.5	72.25
7	Matsai	8.67	75.17	12.8	163.84	8.2	67.24
8	Natsinta	8.78	77.09	12.2	148.84	7	49
9	Radi	7.93	66.1	12.5	156.25	7.1	50.41
10	Riko	8.05	64.8	13.4	179.56	7.3	53.29
11	University farm Badole	8.72	59.6	13.3	151.29	8.1	65.61

but the difference between the eleven locations was not significant. The longest pod recorded in okra was 8.5 and the lowest was 6.7 cm at Kukabiyar and Daga, respectively, but the differences between eleven locations were not significant.

The average weight of one fruit and total fruit yield per plant of tomato, cucumber and okra are presented in Table 11.3; the heaviest tomato fruit weight recorded was 40.2 and the lowest was 32.5 g, and the difference between them is significant but there were no significant differences between the 11 locations. The recorded highest weight of one cucumber fruit was 136 g, while the lowest was 120.5 g; therefore, the difference was significant between the highest and the lowest fruit weight of cucumber. However, we observed that there were no significant differences between locations in this parameter. In okra we recorded 18.4 g as the highest and 16.0 g as the lowest weight of one fruit, but the difference was not significant between locations in this parameter.

Results of total fruit yield per plant of vegetables tomato, cucumber and okra using pitcher irrigation technology are presented in Table 11.3. The highest fruit yield per plant of tomato was 1.27 kg, while the lowest was 0.88 kg, their differences were not significant and we observed no significant difference in total fruit yield per plant between locations. The highest total fruit yield per plant of cucumber recorded

TABLE 11.3 Yield of Vegetables Grown under Pitcher Irrigation System at 11 Locations in 2017

S. No.	Location	Weight of One Tomato Fruit (g)		Yield of Tomato/ Plant (kg)	Weight of One Cucumber Fruit (g)		Yield of Cucumber Plant (kg)	Weight of One Okra Pod (g)		Yield of Okra/ Plant (kg)
		Mean	Mean Squares		Mean	Mean Squares		Mean	Mean Squares	
1	Agangaro	32.5	1,056.2	0.96	128	16,384	0.88	17.2	295.84	0.15
2	Daddara	36	1,296	1.07	134.5	18,090.25	0.94	17.1	292.41	0.17
3	Daga	37.8	1,428.84	1.14	136	18,496	0.91	16.3	265.67	0.14
4	Dannabasu	33.3	1,108.89	0.88	132.5	17,566.25	0.85	16.1	259.21	0.14
5	Kayauki	35.8	1,281.64	0.88	133	17,689	0.92	17	289	0.14
6	Kukabiyar	38.6	1,489.96	0.97	129	16,641	1.03	17.2	295.84	0.17
7	Matsai	36.3	1,317.16	0.99	123	15,129	0.97	16.5	292.25	0.16
8	Natsinta	40.2	1,616.04	0.93	120.5	14,400	0.81	16	256	0.15
9	Radi	37.6	1,413.76	1.04	127.5	16,256.25	0.84	16.8	282.24	0.15
10	Riko	35.1	1,232.01	1.04	130	16,900	0.95	16.2	262.44	0.14
11	University farm Badole	38.1	1,451.61	1.27	133.5	178,822.3	1.08	18.4	338.56	0.17

was 1.08 kg and the lowest 0.81 kg, which indicated no significant difference between the highest and the lowest total fruit yield per plant; similarly, no significant difference was observed between locations. In okra, we recorded 0.17 kg and 0.14 g as the highest and lowest total pod yields per plant, respectively, and the difference between them was not significant; similarly, we observed no significant difference in total pod yield per plant between locations.

11.4 Discussion

Achieving high fruit yield per plant (1.27 kg) of tomato by using pitcher irrigation method could be attributed to favorable growing condition created by the pitchers. Pitcher irrigation ensures continued supply of moisture throughout the growing period, which positively affects all the yield parameters. Similar observation was reported in the works of Jari and Muttaka (2018), Bainbridge (1987, 1988a, b) and Daka (2001). The lowest yield recorded per plant could be attributed to having fewer and lighter fruits per plant. Achieving high tomato yield in both the farmers' field and at the university farm could be due to having higher number of and heavier fruits per plant. Therefore, pitcher irrigation creates favorable growing conditions, which guarantees high fruit yield of tomato in the drier environment. Daka (1991) and Bainbridge et al. (2001) reported a similar observation.

Higher yield per plant of cucumber obtained at the university farm and at farmers' field could be attributed to favorable growing condition created by the pitchers, which leads to higher number of fruits per plant and heavier fruits. Pitcher irrigation assures adequate supply of moisture and adequate aeration, especially at critical stages of flowering and fruit setting which leads to formation of high-quality fruits that command premium price in the vegetable market. Pitcher irrigation assures farmers of higher cucumber yields of good quality that attract premium price, thus more income to farmers. The works of Bainbridge (1987, 1988a, b, 2001) confirmed our observations.

Our study on pitcher irrigation demonstrated that farmers can achieve high yield of good-quality tomato, cucumber and okra due to favorable growing conditions created by the pitchers, especially steady supply of moisture at critical growth stages, adequate soil aeration and less completion from weeds which ensure good performance of all the growth and yield parameters. A similar conclusion was reported by Jari and Muttaka (2018), Daka (1991, 2001), Dubey et al. (1991) and Anon (1979).

The highest yield of tomato, cucumber and okra were all recorded at University farm Badole. This might be attributed to adoption of best agronomic practices as well as knowledge and skills of the research team on the establishment and management of pitcher irrigation system.

High yield, ease of establishment and high water use efficiency of pitcher irrigation technology would guarantee wider adoption and sustainability. Several authors (Jari and Muttaka, 2018; Bainbridge, 2001; Dubey et al., 1991; Daka, 1991; Anon, 1979) confirmed our observation.

11.5 Conclusions

The results of our study suggest that pitcher irrigation is suitable for small-scale farmers in the arid and semi-arid environments because the cost of establishing and operating the system is within the capabilities of small-scale farmer and it is profitable.

Pitcher irrigation (clay pot irrigation) can contribute to significant improvement in food production and enhancement of family food security and rural livelihood of small-scale farmers living on the drier areas but farmers lack the basic skills and information to establish and operate this system efficiently. Therefore, capacity building of farmers is the key that could unlock the great potentials of this water-saving and water-efficient system of irrigation.

Pitcher irrigation creates employment opportunities for all the stakeholders along the value chain (farmers, pot makers, lid makers and marketers of vegetables).

This study demonstrated that building the capacity of small-scale farmers living in the drier areas to use pitcher irrigation technology to produce vegetables is a viable tool that could be used to enhance family food security and livelihood of small-scale farmers because it is technically, socially and economically acceptable to them.

11.6 Recommendations

This technology pitcher (clay pot irrigation) should be scaled up to cover wider drier areas of Katsina State, Nigeria, and elsewhere with similar ecological condition.

To ensure wider adoption and sustainability of pitcher irrigation technology in the drier areas we need to intensify effort to build the capacity of the major stakeholders along the value chain to establish and operate this technology efficiently and profitably.

We recommend that the government and NGO should support local clay pot makers to set up small-scale clay pot manufacturing industries to supply pots in the required quantity and quality to support the efforts of saving available water for agriculture in the drier areas.

Acknowledgments

The authors wish to acknowledge the financial support by *Tertiary Education Trust Fund (TETFUND)* to carry out this study through Institution Based Research (IBR) Project.

References

Albaji, M., Eslamian, S., and Eslamian, F. (2020) *Handbook of Irrigation System Selection for Semi-Arid Regions*, Taylor and Francis, CRC Group, Boca Raton, FL, 317 Pages.
Anon. (1979) Irrigation with buried clay pots. *African Environment* 3(3/4):378–380.
Bainbridge, D.A. (1987) Pitcher irrigation. *Cookstove News* 7(4):7.
Bainbridge, D.A. (1988a) Clay pots for efficient irrigation. *International Ag-Sieve* 1(1):7.
Bainbridge, D.A. (1988b) Pitcher irrigation. *Dry Lander* 2(1):3.
Bainbridge, D.A. (2001) Buried clay pot, a little known but very efficient traditional method of irrigation. *Agriculture and Water Management* 48:79–88.
Batchelor, C., Lovell, C., and Murata, M. (1996) Simple micro-irrigation techniques for improving irrigation efficiency on vegetable gardens. *Agricultural Water Management* 32:37–48.
Chigura, P.K. (1994) Application of pitcher design in predicting pitcher performance. Unpublished MSc Thesis, UK. Cranfield Institute of Technology, Silsoe College, UK.

Daka, A.E. (1991) *Conservation irrigation using ceramic pitchers as ancillary media for water conservation.* Greece Belkema Press, Thessaloniki: Greece, p. 5.

Daka, A.E. (2001) Clay pot sub-surface irrigation as water-saving technology for small-farmer irrigation. PhD Thesis, University of Pretoria, South Africa. uped/up.ac.za/thesis/available/td-

Das, K.G. (1983) *Controlled release technology bioengineering aspects.* John Wiley and Sons, New York, pp. 15–61.

Dubey, S.K., Gupta, S.K., and Mandal, R.C. (1991) Pitcher irrigation technique for arid and semi-arid Zones, *Advances in Dry Land Resources and Technology* 6:137–177.

Gomez, K.A. and Gomez, A.A. (1984) *Statistical procedures for agricultural research.* 2nd Edition, John Willey and Sons, New York.

Jari, S. and Muttaka, M. (2018) Productivity of tomato (*Lycopersicum esculentum L*) grown under pitcher irrigation as affected by number of plants and rates of poultry manure per pot at Sudano-Sahelian Region of Nigeria. *FUDMA Journal of Science* 2(3):112–115.

Mondal, R.C. (1974) Farming with pitcher: a technique of water conservation. *World Crops* 26(2):91–97.

Mondal, R.C. (1978) Pitcher farming is economical. *World Crops* 303:124–127.

Mondal, R.C. (1983) Salt tolerance of tomato grown around earthen pitchers. *Indian Journal of Agricultural Science* 53(5):380–382.

Stein, T.M. (1990) Development of design criteria for pitcher irrigation. Unpublished M.Sc. Thesis. Cranfield Institute of Technology, Silsoe College, Silsoe, UK.

Vasudevan, P., Dastidar, M.G., Thapliyal, A., and Sen, P.K. (2006) Water conservation and management in arid and semi arid areas: a review: *Proc. National Seminar on Technological Options for Improving Water Productivity in Agriculture*, JNKVV, Jabalpur, pp. 233–241.

12
Pot Irrigation

12.1	Introduction	247
	Background History of Subsurface Irrigation Systems • Specification of Pot Irrigation: Installation and Performance • Pot Irrigation Components	
12.2	Fundamentals and Criteria of Irrigation System Design	254
	Flow Uniformity of Pottery (C_V) • Flow or Flow–Pressure Relationship in Pottery (Q–H) • Moisture Soil Wetting Pattern under Different Soil Textures and Different Operating Pressures of the System • Hydraulic Characteristics: Flow Regime • Pressure Loss in the System • The Background History of Design Criteria and Hydraulics Assessments • Moisture and Salinity Distribution in the Soil Profile	
12.3	Pot Irrigation Performance versus Other Systems	265
12.4	Pottery Mechanism of Preparing and Firing	267
12.5	Economic, Social and Environmental Issues	268
12.6	Conclusions	269
	References	269

Hamideh Faridi
University of Manitoba

Ghasem Zarei
Agricultural Engineering Research Institute, Agricultural Research, Education and Extension Organization

12.1 Introduction

Increased population, limited water resources, increased demand for water and its transfer for other uses have created serious problems for the development of agriculture. For many years, the emphasis on sustainable agriculture, the protection of water and soil resources and the observance of environmental issues have been the focus of agricultural activities (Pacheco, 2020). However, with the advent of the 21st century, in addition to the above objectives, improving water productivity and optimal use of water have also been focused. One of the ways to adapt to drought and water scarcity is to make optimal use of available water resources (Malekian et al., 2012). Figure 12.1 shows the climate of different parts of the world. Today, agricultural sector is the largest consumer of water in the world (Figure 12.2) and therefore, ensuring the most efficient use of water is critical in this sector, but we will not achieve this unless the efficiency of water consumption increases (Wang et al., 2019). Water efficiency is defined as the amount of product produced per volume unit of water consumed (Wang et al., 2016; Yang et al., 2017). The purpose of this chapter is to get acquainted with the details of the pot irrigation system and to explain the advantages and disadvantages of this irrigation system.

12.1.1 Background History of Subsurface Irrigation Systems

The history of subsurface irrigation in the world dates back to more than 70 years (Mondal et al., 1987; Zarei et al., 2003). Studies on subsurface irrigation of plants began in Europe around 1934 by Robey and

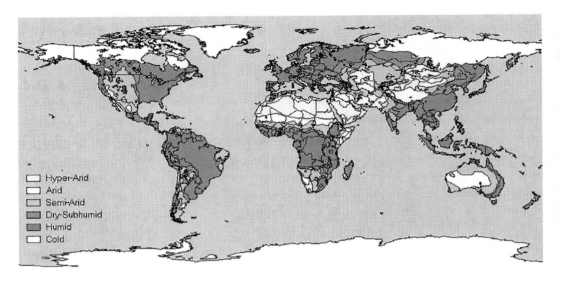

FIGURE 12.1 The climate of different parts of the world.

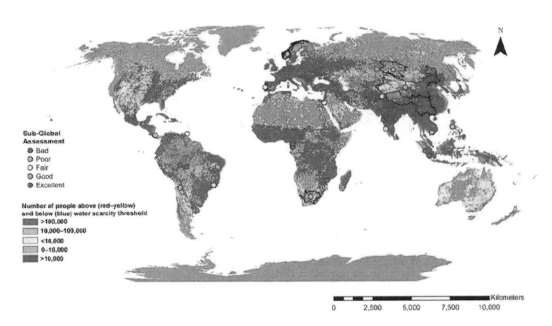

FIGURE 12.2 The world water resources.

led to the construction of canvas pipes with numerous small holes (Bainbridge, 2001). After the end of the Second World War, rubber pipes with special water drips were commercially marketed due to the special problems of these types of pipes. The pipes were used by Blass in Palestine in 1962 as an underground irrigation method that was not successful due to the short-term clogging of the holes and the entry of plant roots into them (Sohrabi and Gazori, 1997; Zarei et al., 2003). Hence, the researchers found a tendency toward micro-irrigation systems because of the following advantages: having the ability to settle on the ground, increasing the irrigation efficiency, crop yield and water efficiency (Bastani, 1975; Ashrafi et al., 2002; Siyal, 2009). Therefore, surface drip irrigation systems have evolved and been developed more rapidly in this period of time. In recent years, the invention of various methods of subsurface irrigation systems

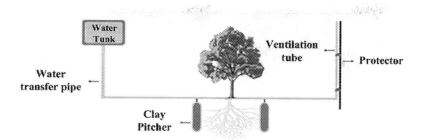

FIGURE 12.3 Schematic view of a clay subsurface irrigation.

has received serious attention due to the unique advantages of these systems and the lack of water resources for consumption in the agricultural sector of most countries (Majidi et al., 2009; Siyal et al., 2013).

In subsurface irrigation methods, the area of plant root development is moistened by water leakage from all types of pipes and permeable and perforated surfaces. These methods include pot irrigation, underground perforated pipe irrigation and subsurface drip irrigation (SDI). The use of pottery to save water and provide water for the cultivation of annual plants in the world has a history of several thousand years. Pot irrigation (Figure 12.3) is a traditional irrigation method that was first adopted in North Africa and Iran (Mondal, 1974; Aghaii et al., 2015).

This method is an efficient irrigation system to save water and irrigation (Stein, 1990). In one of the first texts on agriculture, Fan Sheng-Chi Shu's book, the use of pottery buried is described in China about 2,000 years ago (Sheng, 1974). This irrigation method was probably used long before this text was published. Pot irrigation is used for vegetables and fruit trees including stone fruit trees and shrubs in dry lands such as India, Pakistan, Iran, the Middle East and Latin America (Bainbridge, 2001). This method of irrigation has been used in Iran for centuries in desert cities such as Yazd and Ardakan. But this method gradually became abolished and did not find the necessary development among farmers due to the clogging of pottery pitchers over time, the problem of filling them by hand and creating obstacles to mechanized cultivations (Bastani, 1975; Ashrafi et al., 2002; Zarei et al., 2003; Dehghanisanij and Zarei, 2012; Aghaii et al., 2015).

The SDI system was formed based on the characteristics of the pot irrigation system (Bainbridge, 2001). In the last decade, farmers used the SDI for pistachios for the first time in Iran (Aghaii et al., 2015). In arid and semi-arid regions such as Iran, this system has many benefits. Compared to the pot irrigation system, the SDI method is suitable for any agricultural scale. Subsurface irrigation reduces water use at the farm level and raises the water use efficiency (WUE). It also has the ability to control water use and the uniformity of water distribution. This system increases food security, sustainable irrigated agriculture and greater economic benefits in agriculture in these areas (Ashrafi et al., 2002; Zarei et al., 2003; Dehghanisanij and Zarei, 2012).

12.1.2 Specification of Pot Irrigation: Installation and Performance

In this irrigation method, the pitcher is placed in the soil and its mouth is placed a little above the ground and water is poured into the pitchers at regular intervals. The number of plants and their cultivated location around the pitcher depend on the type of plant and its growth characteristics (Mondal, 1984; Gupta, 1999). Practically, about 530 pits with a width of 70 cm and a depth of 28 cm are usually dug per hectare, and about 18 kg of animal manure is added to each pit. This amount of fertilizer mixes well with the same amount of soil. A pottery pitcher with a capacity of 6 L is placed at the center of the hole and the surrounding area is filled with the prepared mixture. The opening of the pitcher should be level or slightly higher than the ground and the pitcher is filled with water. Four vegetable seeds such as melons are planted around the pitcher (Figure 12.4). The opening of the pitcher is usually blocked with a tile or lid. The pitcher should always be filled to the brim with water (Bainbridge, 2001).

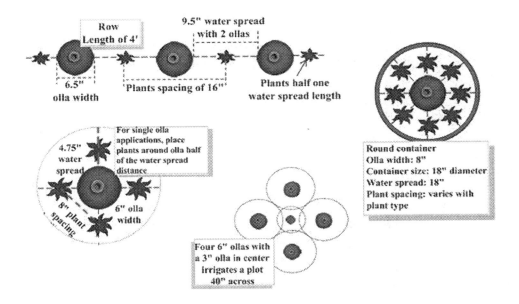

FIGURE 12.4 Schematic of the pitcher mounting in the soil and the arrangement of cultivated plants around it (clay pots called ollas).

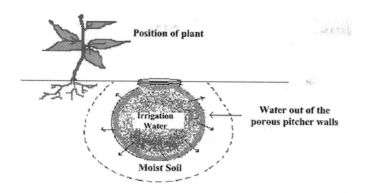

FIGURE 12.5 Schematic of installation and pot irrigation operation.

The establishment of the system includes the following points (Tanwar, 2003):

- Select a pit with a depth of 60 cm and a diameter of 90 cm.
- Mix agricultural soil with some chemical fertilizer and pour it to a depth of 30 cm.
- Place the pitcher in the middle of the pit and fill it with the rest of the soil and manure mixture up to the throat of the pitcher.
- Fill the pitcher with irrigation water.
- Put 6-8 seeds or seedlings of vegetables and summer seeds evenly around the pitcher after 2 days from the first filling of the pitcher.
- Refill the pitcher with water on alternate days to provide proper moisture in the root development area of the plants.

Basically, pottery pitchers will be porous if cooked at a mild temperature and not polished. This means that there are small holes in the wall from which water slowly leaks out. The crop is planted near the clay pot and filled with water (Figure 12.5). The water inside the pitcher seeps into the surrounding soil where the plant roots are located. When the plant uses water, more water seeps out of the pitcher. In this case, the pitcher provides the exact amount of water the plant needs and thus, WUE increases. The farmer only regularly inspects the pitchers for fractures and fills them when needed (Bainbridge, 2001).

According to historical findings, pot irrigation is the oldest form of topical irrigation used in several countries around the world. They used unglazed earthenware pitchers with a capacity of 5–8 L that were made on site. Disadvantages of this method include the following (Daka, 2001):

- The high cost of labor to install and fill pitchers on a large scale.
- Impossible mechanized cultivation of agricultural products due to obstruction of pots on machinery and agricultural implements.
- Impossibility of its development in large scales.
- The possibility of obstruction of the pitchers due to the passage of time and the use of poor-quality water.
- Pots should be moved during plowing. They should be handled with caution during installation or removal to prevent breakage.
- Buried earthenware pitchers can be clogged, especially if they are dry for a long time. If this happens, they should be removed from the soil and rubbed or soaked or reheated to clean the cavities.
- The mixture of clay, time and temperature of heating and selection of clay should be done correctly to make sure that the pitcher is sufficiently porous for this method. Fortunately, it is easy to test the pitchers, prepare the right mixture and observe the timing and amount of heating.

- If muddy water (silty) is used, the small holes in the earthenware pitcher will gradually clog and will not work well.
- In some samples of clayey (heavy) soils, this system does not work well. The use of a mixture of sand or organic matter with the soil during planting in the vicinity of the pottery helps its efficiency.

The benefits of pot irrigation include the following (Masri, 2002; Daka, 2001):

- It is more than 10 times more effective than watering plants by surface irrigation. In Mexico, two corn crops have been harvested in 1 year in a place where only one was previously cultivated.
- This method works with direct germination and good seedling and helps germination and seed establishment even in very hot and dry conditions.
- The farmer's work is less, i.e. less water is moved and less weeds are produced because the potted water is given to the plants and not to the weeds.
- It is useful for soil construction because water does not spill on the soil, and the growth bed is free and more air flows.
- Soil modifiers such as fertilizer, compost or chemical fertilizer are placed where the plant needs them, not weeds.
- It is cheaper and more reliable than many high-tech drip irrigation systems that are more likely to be trapped by insects or destroyed by animals and usually require a flat surface.
- The entire system can be built with materials and skills available on site and does not require a pump to operate. As long as the farmer controls the pitchers, the operation of the system will not be a problem.
- It is useful for job creation. Individuals can be used to make pitchers and lids.
- Useful and direct water enters the root zone and less water is lost due to evaporation.
- It is cheaper due to the use of local materials.
- This system does not require water pressure and water filter.

One of the distinctive features of pot irrigation is that it can be used on any tiny, irrigated field, regardless of its shape or size. It is also a very effective irrigation instrument that can generate a variety of goods utilising salt water, and it can be made and utilised on-site. (Stein, 1990). Also in this method, the trend of daily leakage changes from the same pitcher is in line with the trend of evaporation and transpiration changes (Kisi and Cimen, 2009).

Zarei et al. (2003) stated that clay pipes must be combined with a polyethylene water supply network in order to have proper leakage and performance (Figure 12.6). In this irrigation method, local materials are used and launched at low pressure. Therefore, it is suitable for irrigating gardens, vegetables and greenhouses.

According to Stein (1990), the following aspects should be considered in pot irrigation:

- The size of the pitchers and the way they are placed at the feet of the plants
- Pot filling programs for different products and water qualities
- Fertilization with irrigation water (irrigation fertilizer)
- Pot life and system costs
- Depth and distance of pots for different plants.

For placing the pot in the soil, a pit is dug about 3 times the width and 2 times the depth of the pot. The bottom of the pit is softened with a fork. Compost or perennial manure is mixed with the soil that comes out of the pit in a ratio of 1–2. The clumps soften. If the soil is too heavy, sand is added. If the soil is too salty or alkaline, some gypsum is added. The mixed soil is then poured into the pit. The top of the pit should be placed 2 cm above the ground. The pitcher is placed in its place and the lid is put. Then the area around the pitcher is filled with mixed soil and the pitcher is filled with water and the lid is put. After placing the pot, it is time to plant seedlings or seeds. Before planting, it can be seen how far from the pot the soil is soaked. In many soils, seeds or seedlings should be placed 5–10 cm from the outside edge of the buried clay pot. The distance between the pitchers depends on the type of crops and the soil. The

Pot Irrigation

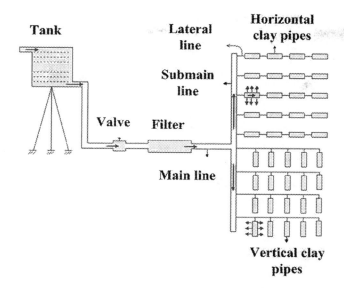

FIGURE 12.6 View of a clay subsurface irrigation system.

most common mistake in this irrigation system is to plant seedlings or seeds far from the pot, outside the wetting front. Crops are usually planted symmetrically around the pitcher. The higher the opening of the pitcher, the easier it will be to refill the pitcher when the plants are fully grown. Pitchers should be monitored regularly to prevent them from drying out. The time between refilling the pitchers varies during the growing season. Small pitchers usually need to be filled every few days, while larger pitchers are filled every 2–3 weeks. Also, if chemical fertilizers and soluble compost are added to the potting water, it should be diluted with excess water (Daka, 2001).

12.1.3 Pot Irrigation Components

Clay subsurface irrigation can be classified as a micro-irrigation system. According to Figure 12.6, in this irrigation system, pottery pieces with the right porosity, controllability, and requirement are used.

The water leaked from these parts is continuously replaced by polyethylene-made water transmission pipes. This method is designed to allow automatic installation, commissioning and operation.

Components of the clay subsurface irrigation system are (Zarei et al., 2003):

- Water tank (with a capacity commensurate with the surface under cover, for example, from barrels for greenhouses to tankers for gardens) with a height of 2–4 m to provide the required pressure of the system
- Water filter for physical and biological water purification
- Main and lateral polyethylene pipes for transferring water from the tank
- Polyethylene fittings to connect the various components of the system
- Shut-off valve for irrigation control and management
- Polyethylene lateral pipes for transferring water to pottery pieces
- Pottery pieces located at depths of 5–20 cm from the soil surface as a water distributor in the soil profile and plant root development zone.

In this irrigation system, depending on the type of cultivation, the distance of plants on a row of cultivation and also the management of water distribution, sub-pipes are connected to tubular pottery or vertical pottery (aquifer) at certain intervals. If the purpose of irrigation is to distribute water locally at regular intervals in the soil, according to Figure 12.7a, vertical potteries are used, and if the purpose

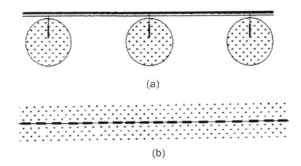

FIGURE 12.7 Condition of moisture distribution by pitchers.

of irrigation is to create a continuous moisture strip, according to Figure 12.7b, horizontal potteries are used. Important and practical features of clay parts used in pot irrigation system are geometric shape, dimensions, specific gravity and porosity.

12.2 Fundamentals and Criteria of Irrigation System Design

12.2.1 Flow Uniformity of Pottery (C_V)

One of the important criteria in designing and applying a subsurface irrigation system in farms, orchards and greenhouses is to create a balance between the discharge of the emitters of the irrigation system and the water needs of the plants so that the plants in different parts of a farm equal water. The most important factor in increasing the uniformity of water outflow from emitters is Manufacturing Coefficient of Variation. The coefficient of variation is an indicator to quantify the variations of emitters. Theoretically, all emitters of a certain type should have the same flow rate at a constant operating pressure. However, this is not achieved due to the quality and mechanism of their manufacture in the factory. Usually, designers and operators of such irrigation systems must ensure the uniformity of water emitters. The coefficient of variation of water emitters in the pot subsurface irrigation system has been calculated using Eq. (12.1).

$$C_v = \frac{1}{q_{avg}} \sqrt{\frac{\sum_{i=1}^{n}(q_i - q_{avg})^2}{n-1}} \times 100, \tag{12.1}$$

where:

C_v: Manufacture coefficient of variation (%)
q_{avg}: Average discharge of tested pottery pieces (L/h)
q_i: Discharge of each tested pottery pieces (L/h)
n: Number of water emitters tested.

The standard and recommended values for the coefficient of variation of micro-irrigation methods (ASAE EP-405) are in accordance with Table 12.1. According to this table, the coefficient of variation of the construction of the tested pottery parts is acceptable and is unobstructed in terms of application in the subsurface irrigation system.

12.2.2 Flow or Flow–Pressure Relationship in Pottery (Q–H)

In general, the discharge of an emitter is a function of its operating pressure. In sprinklers and drippers, the relationship between pressure and outlet flow is as follows (Zarei, 2003):

$$q = k.H^X, \tag{12.2}$$

TABLE 12.1 Standard Classification of Emitters for Micro-irrigation Systems Based on Manufacture Coefficient of Variation

Classification	Point Source Emitters	Line Source Emitters
Good	$C_V \leq 5$	$C_V \leq 10$
Average	$5 < C_V < 10$	$10 < C_V \leq 15$
Acceptable	$10 < C_V \leq 15$	$15 < C_V \leq 20$
Unacceptable	$15 < C_V$	$20 < C_V$

TABLE 12.2 Flow–Pressure Relationship in Three Different Soils

		Regression Analysis	
Soil Texture	Pressure Variation Trend	Equation[a]	Coefficient of Determination
Sandy loam	Increase	$y = 26.419x^{0.775}$	$R^2 = 0.86$
	Decrease	$y = 18.415x^{0.886}$	$R^2 = 0.86$
Silty clay loam	Increase	$y = 22.698x^{0.845}$	$R^2 = 0.85$
	Decrease	$y = 13.572x^{1.058}$	$R^2 = 0.95$
Silty clay	Increase	$y = 19.464x^{0.808}$	$R^2 = 0.90$
	Decrease	$y = 19.130x^{0.773}$	$R^2 = 0.62$

[a] y = clay capsule discharge (l/day) and x = operating pressure (m).

where:

q: Flow rate of sprinkler or dripper (L^3/T)
k: Coefficient of sprinkler or dripper flow rate (–)
H: Sprinkler or dropper operating pressure (L)
X: Coefficient determining the type of flow inside the sprinkler or dropper (–).

Behnia and Arabfard (2005) have reported the existence of a linear relationship between water pressure and discharge from the jar in the open air. Also, Gupta et al. (2009) have reported a linear relationship between the discharge of clay capsules and their operating pressure in the open (outside the soil) and a nonlinear relationship between these two parameters in two fine sandy and sandy loam soils. Ghorbani-Vaghei et al. (2010) also have reported a flow–pressure relationship in a clay capsule made with clay loam soil linearly, but with an increase in the percentage of sand in the mixture used to make the capsule, this nonlinear (power) relationship was reported. Zarei and Shahpari (2014) have reported flow–pressure relationship for three soils (Table 12.2):

12.2.3 Moisture Soil Wetting Pattern under Different Soil Textures and Different Operating Pressures of the System

In general, the growth and development of plant roots and their progress and development in the soil profile strongly depend on the amount of soil moisture (Albaji et al., 2020). The wetting pattern of the soil profile is called the wet bulb or the volume of the soaked soil. On the other hand, the formation of moisture bulbs due to the operation of an irrigation system is the result of the effect of two forces of gravity and capillary in the soil profile. Therefore, knowledge of the dimensions of the wet bulb created in different soils for different operating hours of the system is necessary and inevitable. In this regard, in addition to the texture and structure of the soil and the operating time of the irrigation system, water pressure within the system is also effective as a variable in the formation of moisture bulbs and should be considered (Siyal and Skaggs, 2009). Measured and predicted water contents in a loamy soil under porous clay pipe subsurface irrigation after 5 days of irrigation with four different constant pipe pressure heads are shown in Figure 12.8 (Siyal and Skaggs, 2009).

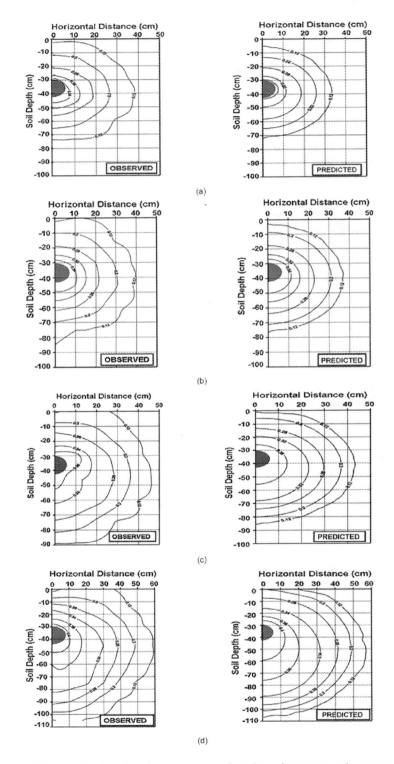

FIGURE 12.8 (a–d) Measured and predicted water contents after 5 days of irrigation with a constant pipe pressure head of 25, 50, 100 and 200 cm, respectively.

Pot Irrigation

The ability to manage it during the system's design and operation stages is made possible by having complete knowledge of and awareness of all relevant aspects of the subsurface irrigation system. The formation of moisture bulbs around the clay pipes (line source and point source) is directly related to the amount of water pressure inside the clay pipe, hydraulic conductivity of the pottery body, hydraulic conductivity of the soil around the pottery and finally, the duration of irrigation. Obviously, only if you are aware of the above, you can expect a certain volume of soil around the roots to reach the desired moisture.

12.2.4 Hydraulic Characteristics: Flow Regime

Basically, the flow regime in pottery pieces must be known and calculated in order to control the amount of water flow. In fluid mechanics, the flow regime is determined by an indicator called the Reynolds number (R_e). The Reynolds number is defined as follows:

$$R_e = \frac{d.V}{\nu}, \tag{12.3}$$

where Re is the Reynolds number (dimensionless), d is the diameter of the water conductor (cm), V is the flow velocity (cm/s) and ν is the kinematic viscosity of the water (cm²/s). Depending on the value of the Reynolds number, the flow regime may be classified as laminar, unstable, semi-turbulent or full-turbulent. It should be noted that in each of the above cases, special relations are used to calculate the coefficient of friction. The types of flow regimes are divided according to the Reynolds number (Table 12.3).

Since the pores of the wall of the clay capsule are small and the flow velocity of them is low (seepage rate), the fractional form of the above equation is small and as a result the fraction value decreases which makes the resulting flow linear (Zarei and Shahpari, 2014). In order to calculate the Reynolds number during the passage of water through the porous environment of the pottery, the results and data obtained from experiments related to determining the coefficient of structural change and the flow–pressure relationship can be used.

12.2.5 Pressure Loss in the System

12.2.5.1 Pressure Loss in Pipes

The amount of water pressure inside the clay parts depends on the water pressure in the central station of the system and the frictional and localized losses in the path of the water transmission line of the system. As this system can operate with very low pressures, so the source of pressure supply can be the potential of gravity or a source of water that is located at a very low height. Also, in this regard, and considering the possibility of reducing the discharge of pottery parts over time and after years of operation, the necessary measures should be taken to increase the operating pressure of the system to keep the amount of seepage from the body of pottery parts constant. It's vital to remember that the clay subsurface irrigation system's components resemble drip irrigation system components (apart from the water emitter), therefore you should determine the diameter of the pipes and calculate the friction drop of the main, semi-main, and lateral pipes. The same common equations used in micro-irrigation can be

TABLE 12.3 Classification of Flow Regime Types According to Reynolds Number

Flow Regime	Reynolds Number (Re)
Laminar	$Re \leq 2{,}000$
Unstable	$2{,}000 < Re \leq 4{,}000$
Semi-turbulent	$4{,}000 < Re \leq 10{,}000$
Full-turbulent	$Re > 10{,}000$

used. In this regard, the most common equation that has become universal and general is the Hazen–Williams relation as follows:

$$H_f = 163,021.8 L \left(\frac{Q}{C_{HW}}\right)^{1.852} D^{-4.87}, \qquad (12.4)$$

where:

H_f: Pressure loss along the pipe (m)
L: Pipe length (m)
Q: Flow through the pipe (L/s)
C_{HW}: Hazen –William's equation (dimensionless)
D: Pipe diameter (cm).

The Heisen–William's equation (C_{HW}) in Eq. (12.4) depends on the material and life of the pipe used. The value of this coefficient can be estimated based on Table 12.4.

It should be noted that Eq. (12.4) is valid only for the main and semi-main water transfer pipes that do not have multiple outlets (inlet flow to the pipe is fixed from the beginning to the end of the pipe). Basically, the hydraulic flow in the sub-pipe is different from ordinary pipes because the flow rate of the pipe is continuously reduced during transfer (water leaving the droppers or clay parts). In this case, due to the gradual decrease of the flow rate (Q), the pressure loss along the branch pipe (H_f) is also less than normal (pipe without outlet). Therefore, the pressure *loss* due to friction in the sub-pipes (P_f) is obtained from the following equation:

$$P_f = F.H_f, \qquad (12.5)$$

where:

H_f: Pressure loss due to friction in the pipe without outlet
P_f: Pressure loss due to friction in the branch pipe with multiple outlets
F: The correction factor.

The amount of correction factor (F) depends on the number and condition of the outlets, as well as the formula for calculating the pressure loss. The value of F is for pipes with one output equal to 1 ($F=1$) and for pipes with 15 or more outlets is 0.36. Considering that in subsurface clay irrigation, the number of branched clay pieces from each sub-pipe is usually more than 15, and for this reason, this coefficient can be considered equal to 0.36.

TABLE 12.4 Coefficient Values of Heisen–William's Equation (C_{HW}) for Different Types of Pipes

Type of Pipe	Heisen–William's Equation (C_{HW}) Coefficient
Aluminum pipe with connection	120–130
New steel pipe	130
Steel pipe with a life of more than 15 years	100
Cement and concrete pipes	140
Plastic pipe with a diameter of 10 cm and more	150
Plastic pipe with a diameter of less than 10 cm	140

12.2.5.2 Pressure Loss in Pipe Fittings

In addition to the pressure loss due to friction in the main and secondary pipes, there are also losses in cases where the flow changes in the direction or pipe size. These losses are called minor losses. In pressurized irrigation systems (including subsurface irrigation systems), these types of losses occur in connections, valves, branches, elbows, etc. Knowing the extent of these losses is important in calculating the pressure required for the operation of the subsurface irrigation system. The amount of these losses is obtained from the following equation:

$$h_{fm} = K \frac{V^2}{2g}, \tag{12.6}$$

where:

H_{fm}: Minor loss (m)
V: Flow velocity in the pipe (m/s)
K: Resistance coefficient against the flow (–).

The resistance coefficient against the flow (K) is calculated for most cases or obtained experimentally. There are several tables and curves to calculate the K values for different connections according to the pipe diameter. Also, tables have been developed to calculate the partial loss with flow rate (Q) and pipe diameter (D). Sometimes, for ease of operation, the pressure loss due to the presence of transformers and attachments can be considered about 10% pressure loss in the network.

12.2.6 The Background History of Design Criteria and Hydraulics Assessments

Stein (1990) identified the design criteria for this irrigation system. The results of this study showed that the hydraulic conductivity of pitcher materials is due to three other factors: (1) the movement of water in the pitcher wall, (2) the size of the side surface and (3) the thickness of the pitcher wall, which are more important in the design.

Studies by the Central Soil Salinity Research Institute showed that the amount of water emitted from each pot depends on the type of pottery (pottery) and the number of plants irrigated by each pot depends on the type of soil, the pores of the pot wall and the shape of the pot (Chandrasekaran et al., 2010). This research also shows that soil moisture and salt distribution in the plant root development zone are more appropriate than other systems in potted irrigation method and, the salts accumulate outside the root development zone and at the soil surface under the conditions of using the subsurface (clay) irrigation system. Also, in these studies, it was concluded that the amount of salt transferred from inside the pitcher to the outside and in the environment of plant root development is less than the water inside the pitcher. This makes it possible to use low-quality water. In the institute's specialized experiments, tomato yield was 29 t/ha at saline level of 12 dS/m with the usage of 5,000 pots per hectare. Watermelon and melon also employed this technique to tolerate at salinity level of 12 dS/m. In these studies, 7–10 L pitchers have been identified as suitable for irrigation and cultivation of vegetables and suitable economic products.

Siyal and Skaggs (2009) stated that subsurface irrigation using clay pipes could be a useful and efficient way to save water in most arid and semi-arid regions if the design and application guidelines have already been determined.

Bahrami et al. (2011) investigated using a fuzzy model to estimate the vertical and radial depth of wetting caused by irrigation with a porous clay capsule buried. The fuzzy logic algorithm was used to define the relationship between saturated hydraulic conductivity, flow rate, porosity, water volume and height of hydrostatic pressure and surface wet radius and vertical distance. 200, 400, 600 and 800 g of fine sand were mixed in each kilogram of clay loam and then baked in an oven at 900°C to prepare porous clay capsules with different juices. The water flow rates of porous capsules were 200, 400, 800, 1,200 and 2,000 cm³/h at the hydrostatic pressure of 2.5 m. The radius and wet depth values of this model were compared with the analysis of errors remaining on the farm. The results showed a significant relationship between estimating and measuring the radius and depth of wet and shallow depths. The results also showed the wetted radius and vertical depth values in 2.5 meter pressure with low discharge are about 13.5 and 22 cm, respectively. These values were much higher at 14 and 55 cm, respectively. This means that increasing the fluidity of the clay capsule increases the depth of wetting. Based on the results of this research, the fuzzy logic algorithm is found to be suitable for accurate calculation of the depth and radius of wetting.

The unique feature of pottery is basically the continuous and optimal distribution of moisture in the area of plant root development according to their water needs (climate conditions) and its role in increasing WUE. This method works with direct germination and helps germination even in very hot and dry conditions (Masri, 2002, Daka, 2001). Also in this method, the trend of daily leakage changes from the pitcher is equal to and in line with the process of evaporation and transpiration changes (Kisi and Cimen, 2009).

A series of experiments and observations were carried out in Iran for the first time to investigate the possibility of using pots in irrigation by the Institute of Soil Science and Fertility in 1975, in which the method of water leakage from pots to soils and plant growth in the vicinity were studied. Also, the effects of shape, volume and material of the pitcher, pressure and quality of water inside the pitcher, field soil texture and the amount of free water evaporation in the amount of water leakage from the pitcher and the amount of radiant wetting front and soil moisture percentage were studied (Javaheri, 1976). These studies have shown that pot use is successful for irrigation. In this study, many issues of pitcher irrigation were investigated including distribution of moisture in light and heavy soils, passing salts through the pottery and the possibility of using salt water in pot irrigation. The researcher concluded that increasing the water pressure from 40 to 140 cm doubled the radius of moisture distribution. Also it was reported that in the old method of pitcher irrigation, the pitchers were filled with water and refilled every few days without any special and logical rules. In this case, some of the pitchers may be empty and others may still be half full, or the plant may not be given enough water. In addition, due to the limited development of the moisture radius, it was necessary to plant the crop at the shortest distance from the pitcher, which in turn exacerbated the problem of root progress to the pitcher, eventually causing clogging or breaking of the pitcher. It would be very easy to control the irrigation system if there was a relatively clear relationship between the amount of water passing through and the time the water was drained from the pitcher. In addition, applying pressure to the water in the pitcher can develop a moisture radius. For this reason, the crop can be planted at a distance farther from the pottery (compared to the old method) by applying pressure and eventually developing a moisture radius.

Bastani (1975) has studied subsurface irrigation with potted tubes. This researcher has conducted experiments to provide a suitable soil formula for the strong construction of high-water pipes. The results showed that the pitcher tubes were able to deliver enough water to the root environment by leaking from the wall and had good resistance to the movement of agricultural machinery and plowing operations. In subsurface irrigation with potted pipes, the highest amount of crop is produced due to the prevention of surface evaporation with the least amount of water. This method is economically viable as the low cost of production.

Vasudevan et al. (2007) have considered pottery irrigation as a suitable method to maintain and optimize water consumption and have recommended that all the characteristics of this system such as water leakage rate, water permeability coefficient and moisture bulb created should be determined before implementation.

Abu-Zreig and Atoum (2004) examined the hydraulic characteristics and the rate of water leakage from pottery made in Jordan and stated that hydraulic conductivity is the most important characteristic that should be obtained for pottery used in pot irrigation.

Pot Irrigation

Stein (1990) conducted a study on pot irrigation in two parts. In the first part, a comprehensive study of the principles and techniques of pot irrigation was performed and the experimental part (second part) was based on 2 years of greenhouse experiments and based on several years of laboratory experiments implemented at the Faculty of Rural Engineering and Conservation of Natural Resources at Castle University in Germany. The main importance of practical work is to study the characteristics of pitcher materials and its changes over time. Then, the close interaction between the pitchers and the environment around them was tested and other influential factors were carefully examined. In the last section, there was a special emphasis on the production of agricultural products with pot irrigation and its effect on increasing the WUE. The results of this study showed that the hydraulic conductivity of pottery materials is due to three other factors; The movement of water in the pitcher wall, the size of the side surface and the thickness of the pitcher wall are more important in the design. A mixture of sand and clay was prepared as the raw material for the pitchers to be a practical and effective way to increase the hydraulic conductivity of the saturation. The results showed a very significant exponentially increase in the hydraulic conductivity of saturated pitchers of heat-treated materials with an increase in sand crumbs. In general, it was observed that in addition to the composition of the materials, the production method and the heating process also have a significant effect on the permeability (permeability) of the pitcher. Also, the hydraulic conductivity of the saturation is not constant, but changes with time when the pitchers are used continuously.

Despite the other results, it is clear that pots used in this type of irrigation can regulate their leakage rate based on changes in evaporation or the water requirement of crops. In some cases, an increase in leakage rate can be more than 200% of the observed rate. The degree and capability of the setting can be introduced as the "reactivity" of the pitcher, which interacts with the three main factors: the hydraulic conductivity of the pitcher wall, the size of the side surface and the thickness of the pitcher wall. System optimization with reference to the amount of leakage and its "reactivity" can be done using the interaction diagram shown in Figure 12.9 (Stein, 1990). This interaction diagram makes possible only the qualitative evaluation of the interaction of the characteristics of the pitcher and its quantitative evaluation is not possible. Referring to the interaction of pitcher and soil, it is concluded that the interaction is not a fixed value. Increasing the growth of the roots around the pitchers reduces the ability to interact over time. Therefore, the rate of leakage and "reactivity" decrease with the growth of the irrigated crop. In practical irrigation conditions, the rate of leakage and "reactivity" should be much less than what is potentially possible. It should be noted that the design process may not be based on standardized

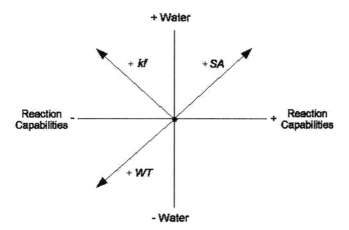

FIGURE 12.9 Interaction diagram of leakage rate (+/− water) and "reaction capability" (+/− reaction capability) in the reaction to the characteristics of the pitcher. Kf, hydraulic conductivity of saturated pitcher materials; SA, surface area of the pitcher and WT, pitcher thickness.

pitchers for transferring results to practical applications for irrigation. This can be problematic due to the very complexity of pitcher production and, as a result, the unpredictability and re-production of the amount of leakage and the "reactivity" of the pitchers. The pitchers are precisely connected to the environment in terms of performance compared to other existing irrigation systems. Therefore, they are affected by environmental factors (soil and climate) and it is recommended that an experimental (cross-testing) method be used to design, implement and optimize pot irrigation systems. This creates an optimal adaptation for the system based on the crop and soil conditions under cultivation. Consistency with the characteristics of the pottery should be considered based on the interaction diagram (Figure 12.9) and product optimization, for example, by adjusting the number of plants grown to design and optimize the pot irrigation system and to improve WUE.

Ashrafi et al. (2002) conducted a study to simulate the penetration of porous clay pipes in subsurface irrigation and stated that the use of clay irrigation was as an efficient irrigation method for storage water due to reduced evaporation losses from the soil surface and deep penetration into the soil. They used porous clay pipes to simulate penetration and to predict the geometry of the soil in the VS2D computer model. Experimental studies on soil samples have shown that two different soil textures were needed in a particular design space to understand the flow phenomenon and the accuracy of the model. In a soil texture, the depth of pipe placement and the volume of water used in the soil are the main factors affecting the created moisture bulbs. In their opinion, different parameters such as side distance, installation depth, irrigation time, hydraulic conductivity of the clay pipe and hydraulic height of the system are among the design criteria of the pot irrigation system. Two different textures of fine sandy loam and sandy loam, with 1.5 and 1.3 g/cm^3 bulk density, respectively, were used in moisture weight percentage of 2% and 10%. The hydraulic conductivity of soil saturation was determined using the fixed height method. Dry and wet retention curves were determined for the tested soils using the Pressure Chamber method and the Hanging Column. VS2D computer program was used to simulate penetration of porous piping in subsurface irrigation. The possibility of using the model for use in subsurface irrigation was confirmed by the experiments. A method was developed to design a porous subsurface irrigation system for row cultivation. The results of computer simulations showed that in the existing soil texture, the geometry of the wet area is sensitive to the depth of establishment and the volume of water used in the soil. Unsaturated hydraulic conductivity of soils is an important factor for creating a moist area in the soil. For a certain volume of water, soil with higher unsaturated hydraulic conductivity creates a larger wet area. The results showed that there was an inverse relationship between the depth of deployment and the lateral distance of the pipes. For hydraulic drop and specific irrigation time, when the installation depth is high, the lateral distance of the pipes is reduced. The use of high-transmission pipes in subsurface irrigation is very useful. In a certain volume of water, reducing irrigation time or using a lower hydraulic drop in the irrigation system is one of the benefits of using high-permeability pipes.

Abu-Zreig et al., (2009) conducted a field study to determine the effect of pottery materials, potential evaporation and soil on clay leakage rate. Eleven pitchers with different hydraulic shape and guidance were used in these experiments. The amount of leakage from the pitchers in the open air was measured after the pitchers were buried up to the neck in silty loamy soil. The average leakage from pitchers when buried in the ground under constant pressure (5,500 mL/day) was higher than in variable pressure conditions (2,700 mL/day) or atmospheric leakage (1,700 mL/day). The rate of permeability was directly related to the permeability coefficient of the pitchers and the evaporation potential of the medium. The linear relationship between the rate of leakage and the potential evaporation in all pitchers indicated that they could be adjusted automatically. The direct relationship between the amount of pot leakage and the potential of the soil matrix potential around the wall of the pitchers showed that the behavior of the pitcher was affected by soil moisture conditions. In this study, a mathematical model was used to predict the amount of leakage of buried pitchers in the soil under constant and variable operating pressure conditions. The correlation between the predicted leak rate in the model and the experimental data was good, and the coefficient of explanation for the variable and constant pressure tests was 0.74 and 0.62, respectively.

Gupta et al. (2009) conducted laboratory and numerical studies to analyze the effect of soil texture on the diffusion characteristics of porous clay used in subsurface irrigation. The results showed that the amount of leakage from the clay pipe has a linear relationship with the hydraulic head used in the open environment. With the pipe placed in the soil, the amount of functional leakage of soil texture and hydraulic pressure has been used. For a given soil texture, when the used hydraulic pressure increases, the moisture content in the vicinity of the pipe increases and the effect of capillary force on the amount of pipe emission decreases. In saturated conditions, in the vicinity of the pipe, the effect of capillary forces on the flow area and the amount of leakage from the pipe is negligible. The results are suitable for designing the lateral distance and depth of pipeline deployment, length and period of irrigation and hydraulic head required for the use of porous clay pipes for subsurface irrigation.

Abu-Zreig and Atoum (2004) modeled the hydraulic characteristics and leakage of pottery pitchers produced in Jordan. The hydraulic conductivity of the saturated pitchers is the most important factor influencing the amount of junction of the pitchers. Fourteen pitchers of different shapes, sizes and production temperatures were selected. The falling-head method and the constant-head method were used to measure their hydraulic conductivity. Both methods were suitable for calculating the K_s of the pitcher, although the constant-head method was faster and easier. The hydraulic conductivity of the pitchers varied between 0.219 and 2.37 mm/day. The amount of K_s increases with increasing production temperature. Sanding the surface of the pitcher wall increased the K_s about 30%. The amount of leakage from the pitchers was between 600 and 3,700 mL/day and had a strong relationship with the hydraulic conductivity of the pitchers. A mathematical model was developed to predict the extent of leakage from pitchers using measurements and observations of pot geometry. The estimated leak rate and the laboratory findings matched well ($R^2=0.97$). The prediction of the amount of leakage was successful due to the volume of the pitcher, the operating pressure, the hydraulic conductivity and the cooking temperature of the pottery ($R^2=0.56$) so that the average absorption error was 23%. In this study, in laboratory conditions, hydraulic of the pot irrigation in saline water was investigated considering the behavior of flow in the pitcher, soil moisture distribution, moisture front progression and soil salt distribution using pots made with different materials.

Ajit et al. (2007) used two modified falling-head and constant-head methods to measure the hydraulic conductivity of eight selected pitchers. Both methods were accurate for calculating K_s, although the falling-head method was faster and easier. The hydraulic conductivity of the pitchers varied between 360.5 and 480.04 mm/day using the falling-head method, and in the constant-head method this value varied between 398.6 and 574.4 mm/day. The calculated conductance of the pitchers varied between 13,705,837.54 and 17,983,403.85 mm^2/day in the falling-head method and between 15,640,326.92 and 21,082,408.74 mm^2/day in the constant-head method.

Ghorbani-Vaghei et al. (2010) argued that misunderstandings about how to distribute soil water, which is affected by the unstable hydraulic properties of the soil, and the hydraulic and physical properties of the porous capsule have, in some cases, led to mismanagement. They improved the physical and hydraulic properties of porous clay capsules made from Gorgan Calcareous soils with a loamy clay texture. Porous clay capsules were made with a different discharge from the soil and added to correct 200, 400, 600 and 800 g of fine sand per kilogram, and were baked at two temperatures of about 600°C and 980°C in the furnace. Water leakage and moisture distribution in the soil of porous clay capsules at 10, 25, 50, 80 and 100 kPa hydrostatic pressures were measured by automatic flow-pressure device. The results showed that there was a significant relationship between the permeability of porous clay capsules and the percentage of sand mixed with the primary soil. The water leakage of porous clay capsules cooked at 980°C was greater than that of porous clay capsules cooked at 600°C.

Behnia and Arabfard (2005) evaluated the form of statistical design of crushed plots with 4 different levels of water pressure (20, 50, 100 and 200 cm water column). They performed 3 pot sizes (0.5, 5 and 15 L) in the laboratory to investigate the effect of water pressure and the volume of the pitcher on the amount of leakage from the pitcher. The leakage of the pitchers was measured after 4 hours. The results showed that in all pressures there is a direct linear relationship between the volume of the pitcher and the flow rate. Therefore, at a constant water pressure, the larger the volume of the pitcher, the higher the discharge from the pitcher. Therefore, in areas where there is no possibility of increasing water pressure in the pot irrigation system, larger pots can be used to increase the irrigation flow. Also, in pitchers of the same volume, there is a linear relationship between the water pressure and the discharge flow of the pitcher, and with increasing water pressure, the discharge flow also increases. Hence, in areas where a larger pitcher cannot be used, the irrigation flow rate can be increased by increasing the irrigation pressure. Also, increasing the level of seepage of the pitcher will increase the discharge flow.

12.2.7 Moisture and Salinity Distribution in the Soil Profile

Measurement or prediction of soil moisture distribution pattern in surface and SDI has been investigated by a number of researchers (Lunt et al., 2005; Cote et al., 2003; Lazarovitch et al., 2006).

Sohrabi and Gazori (1997) examined the efficiency of using semi-permeable pipes in underground irrigation. The results showed that the radius of moisture development depended on the physical properties of the soil and the hydraulic conditions of the system. In fact, in proportion to the increase in pressure in the network, the soil moisture range expands more.

This was performed for the pressures of 0.5, 1 and 1.5 bars and was statistically significant at the level of 1%. Also, along the pipe, with the exception of the beginning, the amount of moisture given to the soil did not differ significantly, and also after the ascending transfer of water in the soil and its evaporation from the soil surface, salts accumulated in the soil surface layer and caused soil salinity to a depth of 10–15 cm. Irrigation efficiency has increased by 50% on average compared to gravity irrigation method.

Alemi (1980) conducted a study to determine the distribution of water and salt in the pot and drip irrigation systems. For this purpose, an elliptical earthenware pottery was buried vertically in the center of a lysimeter as a means of supplying water to the soil. The distribution of water and salt in the soil was compared with the source of pot irrigation and drip irrigation. 500 ml of calcium chloride solution was used in both methods in soil. The calcium chloride was then rinsed using 50 L of drinking water. Soil solution was periodically sampled using suction cups. Soil samples were also taken to determine the water content and chloride ion concentration. Using the pot method, 130 mL/h of water was used to shift the salt to a radius of 41.5 cm in 390 hours, however using drip irrigation, 1 L/h of water relocated the salt to 42 cm in 52.5 hours. In equal amounts of used water, the salt moves to a greater depth in the soil profile at a lower water intensity. More salt was released from the source of the droplets, with higher water intensity. After 72 hours of redistribution, the moistened volumes were almost equal in pot and drip irrigation methods.

Since pot irrigation is an old and efficient method in arid and semi-arid regions and small pitchers are used in this method because they are cheaper than larger ones, Siyal et al. (2009) studied the soil moisture from the pitchers and compared small pitchers with larger ones. Three pitchers in different sizes, small (11 L), medium (15 L) and large (20 L), were studied and simulated. The hydraulic conductivity of the saturated material of the pitcher was determined using the constant-head method. Measured values ranged from 0.07 cm/day for large pitchers to 0.14 cm/day for smaller sizes. The pitchers were buried up to their necks in sandy loam soils to determine the soaked soil profile and filled with water. Moisture distribution was measured after 1 and 10 days at a distance of 20, 40 and 60 cm from the center of the pitcher and at depths of 0, 20, 40 and 60 cm of soil. Moisture distribution was also simulated by HYDRUS-2D model. An acceptable correspondence was founded between the distribution of measured and simulated moisture using the HYDRUS-2D model. The mean square error was between 0.0044 and 0.023. This close correlation showed that HYDRUS-2D model is a good tool for evaluating and designing pot irrigation systems. Also, numerical and experimental results showed that the small pitcher, which is half the size of a large pitcher, has two times the hydraulic guidance and produces a moisture front almost equal to the large pitcher. The simulation of the large pitcher also showed that, as expected, the horizontal diffusion of water in a fine-textured soil was greater than that of coarse-textured soil.

Saleh and Setiawan (2010) conducted numerical and experimental studies to study the flow of water in the soil adjacent to a pitcher and to determine the availability of products to provide the necessary moisture to the soil. In this study, Darcy's and Richard's equations of water flow were used in a cylindrical coordinate system, and the finite element method was used to describe soil moisture profiles. Two types of silicate and sandy clay soils were studied. The hydraulic conductivity of the pitcher was 6–10 cm/s, which was 100 times smaller than the hydraulic conductivity of two types of soil. The pitcher was buried in the center of a box of soil and water was taken from a Mariut tube to keep the water level inside the tube constant. The results showed that the rate of infiltration decreased linearly beyond capability even when the soil was dry from the beginning. The advancement of the front slowed down the moisture and was sometimes limited to a radius of depth not exceeding 30–40 cm, for the two types of soil being tested. The soil moisture around the pitcher was within the required range of plant growth. The different depths of pot placement in the soil create different access distances to the moisture front, but there was a significant difference in water availability. Careful placement of the depth of the pitcher in the soil is important to ensure effective soil moisture in the root zone and reduce evaporation. The appropriate location of the pitcher should be determined based on the hydraulic properties of the pitcher and the soil. In this study, a depth of 5 cm for pot planting was announced for the application of suitable pot irrigation.

Naik et al. (2008) examined saline profile in pot irrigation under saline conditions. The electrical conductivity of water varied between 5°C/m and 20°C/m. The results showed that with increasing salinity of irrigation water, the intensity of water leakage and the amount of moisture in the soil profile decreased.

12.3 Pot Irrigation Performance versus Other Systems

Subsurface irrigation of pots has many benefits, including more than ten times more water efficiency than surface irrigation from the aspects of water conservation, plant performance, irrigation efficiency and WUE (Table 12.5). In Mexico, where corn has been grown only once a year, it has been harvested twice using this irrigation method. Also, this method works with direct germination and good seedling and helps germination and seed establishment even in very hot and dry conditions (Masri, 2002, Daka, 2001). The average production of each pitcher in for some plants is 5.8 kg for tomatoes, 2.5 kg for cabbage, 4.8 kg for cabbage, 11.3 kg for watermelon and 3.5 kg for grapes (Keller, 2002; Stein, 1990).

Mehdizadeh (1977) has examined the possibility of using pots in creating landscapes in some Iranian cities such as Bandar Abbas, Saveh and Sabzevar by cultivating fruitful and fruitless trees at the Research Institute of Forests and Rangelands. In this study, the effect of pitcher use on seedling placement, the effect of pitcher application on different species and the effect of pitcher on increasing species growth have been studied. The results of his research indicate the success of pot irrigation in this regard. In moderate conditions in arid regions, with the available water, trees and plants can be planted about 68 times more with this system.

TABLE 12.5 Estimated Water Use Efficiency under Different Irrigation Systems (Bainbridge, 2001)

Irrigation Method	WUE (kg/m^3)
Closed furrow (basin)	0.7
Sprinkler	0.9
Trickle	1–2.5
Porous capsule (pressure)	1.9+
Porous capsule (no pressure)	2.5+
Buried clay pot	2.5–7

Micro-irrigation methods improve irrigation efficiency by reducing soil surface evaporation and deep percolation losses, as well as creating and maintaining soil moisture conditions that are suitable for crop growth.

Experiments with water balance in Zimbabwe have shown that more than 50% of the water used in traditionally irrigated orchards is lost in surface irrigation as evaporation from the soil surface (Batchelor et al., 1996).

Given that the centuries-old potted irrigation system in Iran's arid regions, including Yazd and Ardakan, has been used to produce vegetables and other crops, mechanization and standardization are an important step in improving irrigation water efficiency in Iran (Zarei, 2003). Bastani (1995) reported using pottery pipe in underground irrigation by consuming water in the underground irrigation method at the rate of 35% of groove irrigation.

Arab Fard et al. (2000) examined and determined how moisture distribution and efficiency evaluation in pot irrigation, and the quantity of water required from the equations, if the required amount of water in the pot irrigation has not been lowered. The drip irrigation method is calculated and this amount of water is given to the plant through the pitcher and by applying pressure, the efficiency of the pitcher is doubled. Finally, they introduced the rubber tube system made of permeable pipes as an alternative to the pitcher system.

Malejinejad (2003) compared the efficiency of water consumption and crop yield in watermelon and cucumber by two methods of pot and furrow irrigation in three replications in Yazd Province. The increase in pot irrigation was estimated about 10% for watermelon irrigation and about 25% for cucumber irrigation, while water consumption for potted irrigation for both plants was estimated to be about a quarter of that for furrow irrigation. In this study, 5 L pitchers were used. The author stated that the use of pitchers with a volume of 4.5–5 L was suitable for this irrigation method. Smaller pitchers reduced the distance between plants due to the formation of smaller moisture bulbs around them and could not provide the required moisture for the plant. Due to sufficient moisture around the pitchers, the time difference between irrigation intervals reached 10 days.

Zarei (2003) compared the efficiency of pot subsurface irrigation method in terms of water consumption, yield and WUE with other similar irrigation methods (drip and subsurface). The results of this 3-year study showed that using this method, in addition to maintaining or improving the performance and quality of products produced, it is possible to increase the area under irrigated cultivation of the studied products due to limited water resources.

Keikha and Azar (2001) evaluated the subsurface pottery irrigation system in soils with saline water for greenhouse cucumber production. They showed that although in the early stages of growth due to high soil EC (about 18 dS/m), a lot of damage was done to cucumber plants, but by setting the irrigation program and the special ability of this system in controlling and providing permanent moisture in the root development area of cucumbers, these damages were completely eliminated in the next stage of growth. The volume of water used in this study was estimated to be 200 m^3 in a 540 m^2 greenhouse. Part of the water used in terms of volume was related to the soil leaching using to control salinity.

Keikha and Akbari moghaddam (2003) surveyed subsurface pottery and drip irrigation systems in cucumber greenhouse cultivation in farmers' conditions. The results showed that the clay pottery irrigation system had the highest yield of cucumbers, producing 288.888 t/ha (15.6 tonnes per 540 m^2) while using only 2,407.41 m3/ha of water, as opposed to drip irrigation's performance of 207.41 t/ha (11.2 tonnes per 540 m^2) while using 3,277.77 m^3/ha (177 m^3 per 540 m^2). According to these results, pot irrigation system performance and water saving compared to drip irrigation system were 28.2% and 26.6%, respectively.

A study with the aim of water saving and irrigation efficiency improvement in the cultivation of a number of horticultural, agricultural and vegetable crops was carried out using clay permeable pipes (Majidi et al., 2009). For this purpose, trees of pistachios, pomegranates, and olives; and crops of corn, sunflower and cotton, as well as cucumber and tomato were studied. For pot irrigation of the trees, clay pottery was used with the length of 45 cm and an outer and inner diameter of 6 and 5 cm, respectively.

For field and greenhouse crops, pottery pieces of 25 cm long with an outer 6 cm diameter and 5 cm inter-diameter were used vertically near each plant and 30 cm long tubes with similar external and internal diameters were used to create the stripe moisture. The results showed water saving, increasing crop yield, improving WUE and reducing the period of vegetative growth. Using this irrigation method, depending on the type of crop, water consumption has been reduced by 33%–65% and the yield has been quadrupled, except cotton. The quality characteristics of the products have been improved, and the use of chemical fertilizers has been reduced by up to 50% and the lowest number of pesticides has been used. The WUE of the products under the clay pottery irrigation system, depending on the type of crops, was 1.6–25 kg/m^3, while in the conventional irrigation system this value was 0.5–4 kg/m^3. In addition, in this method, due to the suitable environmental conditions in the soil, the length of the vegetative growth period of the plants was reduced and the length of the reproductive period can be adjusted by adjusting the discharge. When trees are watered with pottery, the diameter of the trunk and branch end is greater than when they are watered with flood irrigation, and the amount of water utilized was only 25% of flood irrigation method.(Majidi et al., 2009).

In a study conducted by Bhople et al. (2014), the system used clay pipes buried under the ground and capillary water drains indirectly from the wall of clay pipes, which moistens it. Using this method, only the optimal amount of water was provided to crops. In this system, there was no water entry into the soil surface, so the problem of weed management and proper distribution of salt in the soil could be solved. They stated that this method is suitable for areas where the rainfall is less than 500 mm/y. The water saving rate of this method in plant products was 50%–70% (Bhople et al., 2014).

Woldu (2015) stated that this system could be more useful when located near small yards with small-scale water collection technologies (e.g. seasonal runoff ponds, roof water tanks, and the use of hose pumps attached to hoses).

12.4 Pottery Mechanism of Preparing and Firing

Usually, the number and size of pots needed depend on the type of crop, the distance between the plants and the time between the two pots filled by the farmer. In principle, storage capacity of 2–5 L is suitable, but pitchers larger than 10–20 L can be used for larger plants, or the interval between filling them can be extended. Pottery pitchers can be tested to see if they are porous enough for irrigation. The wall of the clay potteries can be sprayed or filled with water. If the surface of the pottery gets wet quickly, their porosity is appropriate.

If pitchers are only made for irrigation application, pitchers with different wall thicknesses and different types of clay should be made and tested. The firing temperature should be below 1,000°C. Because the copper metal melts at 1,083°C, a copper bead can be placed in the furnace to control the temperature according to its melting threshold. The edges of the pitcher can be painted with non-toxin white color to make the vapor less vaporizable. If the root of the plant is quite deep, it is better to paint the top of the pitcher so that less water is lost from the top of the pitcher. The pitcher also needs a lid. The lids should be tightened so that domestic animals do not get water in them and the mosquitoes and other animals inside the pitcher are prevented from living. Embedding a small hole in the cap allows rainwater to enter the pitcher (Daka, 2001).

Ghorbani-Vaghei et al. (2010) mixed 200, 400, 600 and 800 g of fine sand per kilogram of loam clay soil to prepare porous clay capsules with different watering regimes and then fired them in an oven at 900°C. The water flow rates of porous capsules were 200, 400, 800, 1,200 and 2,000 cm^3/h at a hydrostatic pressure of 2.5 m. The radius and wet depth values of this model were compared with the analysis of errors remaining on the farm. The mixture of clay, time and temperature of heating and selection of the type of clay should be done correctly to make sure that the pitcher is sufficiently porous for this method. Fortunately, it is easy to test the pitchers and prepare the right mixture and observe the time and amount of heating (Daka, 2001). The relationship between pottery firing temperature and its porosity has been determined and presented as Figure 12.10 (Bainbridge, 2001).

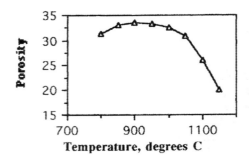

FIGURE 12.10 Porosity as a function of firing temperature.

TABLE 12.6 Cost Estimated for a Site in a Growing Season with 800 Plants

Irrigation Type	Cost ($)	Water Used	Survival
Deep pipe	3.25	Low	High
Clay pot, lid	4.50	Moderate	High
Porous capsule[a]	6	Low	High
Drip[b]	2.5	Moderate	Moderate
Basin	3	High	Very low

[a] It needs water tank, gravity pressure.
[b] It needs water tank, filters, pressure (pump or tower) and regular maintenance.

Bastani (1975) has studied subsurface irrigation with potted pipes to provide a suitable soil formula for the construction of high-strength, water-resistant pipes. The results showed that the pitcher tubes were able to deliver enough water to the root environment by leaking from the wall and had good resistance to the movement of agricultural machinery and plowing operations. In subsurface irrigation with pitcher tubes, the highest amount of crop was produced due to the prevention of surface evaporation with the least amount of water. This method is economically viable due to the low cost of production.

According to Stein (1990), in addition to the composition of materials, the method of production and the heating process also have a significant effect on the permeability of the pitcher. Also, the hydraulic conductivity of the saturation is not constant but changes with time when the pitchers are used continuously.

12.5 Economic, Social and Environmental Issues

Studies and experiences showed that clay pottery irrigation system is completely natural, applicable in terms of technical and economic aspects and durable in terms of useful life. The cost of installing an 8 L pitcher is about $0.25–$0.5. Usually, each pitcher is suitable for irrigating a cultivation area of 2–4 m². The average production of each pitcher in for some plants is 5.8 kg for tomatoes, 2.5 kg for cabbage, 4.8 kg for cabbage, 11.3 kg for watermelon and 3.5 kg for grapes. In addition, water quality up to 5.7 dS/m salinity does not affect tomato production. Also, water up to a maximum salinity of 9 dS/m does not affect the production of other products. Therefore, clay pottery irrigation system is a good solution for obtaining maximum production from very limited water resources and even from brackish water sources (Keller, 2002; Stein, 1990). However, this irrigation method is suitable for rural area and small-scale farmers where there is not electricity for water pumping (gravity-fed irrigation system). Table 12.6 compares the cost between different irrigation methods in very dry and controlled conditions.

12.6 Conclusions

Irrigation will be one of the most important factors in providing food resources to increase the world's population. The production of basic products must follow the same growing trend as in previous decades; therefore, their productivity must be increased. As a result, irrigation plays an important role, but must be adapted to the conditions of water scarcity and environmental sustainability. On the other hand, with the development of many regions of the world, urban activities have increased the demand for water and the competitive use of water resources has been limited. Hence, this increases the uncertainty as to whether the volume of water used in irrigation agriculture can be maintained. Thus, the need to use new technologies and more efficient irrigation methods is strongly felt. In this regard, farmers are advised to use irrigation methods that reduce water consumption and increase yield. Consequently, the prospects for SDI are promising, and stored irrigation water may be available for other uses. Pot irrigation as a subsurface irrigation method has the significant ability to minimize water loss by evaporation, runoff and deep percolation compared to other irrigation methods. It may also increase crop yield as it reduces fluctuations in soil water content and the root zone of the aeration plan.

References

Abu-Zreig, M.M., and Atoum, M.F. 2004. Hydraulic characteristics and seepage modelling of clay pitchers produced in Jordan. *Canadian Biosystems Engineering*, 46: 1.15–1.20.

Abu-Zreig, M., Khdair, A., and Alazba, A. 2009. Factors affecting water seepage rate of clay pitchers in arid lands. *University Sharjah Journal of Pure and Applied Science and Technology*, 6(1): 59–80.

Aghaii, A., Afrasiab, P., Keikha, M., and Keikha, G.H. 2015. Simulation and evaluation of soil moisture distribution pattern under pitcher underground irrigation system. *Iranian Journal of Irrigation and Drainage*, 2(9): 233–241.

Ajit, N., Virendra, K., Mahesh, K., Purohit, R.C., and Singhvi, B.S. 2007. Hydraulic study of earthen pitchers. *Journal of Agricultural Engineering*, 44 (2): 88–92.

Albaji, M., Eslamian, S., Naseri, A. and Eslamian, F. 2020. *Handbook of Irrigation System Selection for Semi-Arid Regions*, Taylor and Francis, CRC Group.Alemi, M.H. 1980. Distribution of water and salt in soil under trickle and pot irrigation regime. *Agricultural Water Management*, 3: 195–203.

Arabfard, M. 2007. Evaluation of efficiency and distribution of moisture in pot irrigation in comparison with several irrigation methods in one type of medicinal plant, Master Thesis, Shahid Chamran University of Ahvaz, Ahwaz Iran (in Persian).

Arabfard, M., Kashkouli, H.A., and Abedi Koupaei, J. 2000. Investigation and comparison of pot irrigation with furrow and drip irrigation. *Proceedings of the First National Conference on Strategies for Coping with Dehydration and Drought*, Kerman, Iran, March 30–21, 2000, 271 (in Persian).

Ashrafi, S., Gupta, A.D., Babe, M.S., Izumi, N., and Loof, R. 2002. Simulation of infiltration from porous clay pipe in subsurface irrigation. *Hydrological Sciences Journal*, 47(2): 253–268.

Bahrami, H.A., Ghorbani-Vaghei, H., Alizadeh, P., Nasiri, F., and Mahallati, Z. 2011. Fuzzy modeling of soil water distribution using porous clay capsule irrigation from a subsurface point source. *Sensor Letters*, 8(1): 75–80.

Bainbridge, D.A. 2001. Buried clay pot irrigation: a little known but very efficient traditional method of irrigation. *Agricultural Water Management*, 48: 79–88.

Bastani, S.H. 1975. Underground irrigation using clay pipe. Forests and Rangelands Research Institute. pp. 1–20.

Batchelor, C., Christopher, L., and Murata, M. 1996. Simple microirrigation techniques for improving irrigation efficiency on vegetable gardens. *Agricultural Water Management*, 32: 37–48.

Behnia, A.K. and Arab-Fard, M. 2005. Determination of discharge-pressure relation of pitchers using in pitcher irrigation. *Agricultural Science & Technology Journal*, 19(1): 1–12 (in Persian).

Bhople, B.S., Adhikary, K., Kumar, A., Singh, A., and Singh, G. 2014. Sub-surface method of irrigation-clay pipe irrigation system. *IOSR Journal of Agricultural and Veterinary Science (IOSR-JAVS)*, 7(11): 60–66.

Chandrasekaran, B., Annadurai, K., and Somasundaram, E. 2010. *A Textbook of Agronomy*. New Delhi: New Age International (p) Ltd.

Cote, C.M., Bristow, K.L., Charlesworth, P.B., Cook, F.G., and Thorburn, P.J. 2003. Analysis of soil wetting and solute transport in subsurface trickle irrigation. *Irrigation Science*, 22(3): 143–156.

Daka, A.E. 2001. Clay pot sub-surface irrigation as water-saving technology for small-farmer irrigation in development of a technological package for sustainable use of Dambos by small-scale farmers, PhD Thesis, University of Pretoria, South Africa.

Dehghanisanij, H. and Zarei, G. 2012. From clay pot irrigation to subsurface drip irrigation for sustainable agriculture in arid and semiarid region. *The International Conference on Traditional Knowledge for Water Resources Management*, Yazd, Iran.

Ghorbani-Vaghei, H., Bahrami, H.A., Alizadeh, P., Nasiri, F., and Mahallati, Z. 2010. Improving physical and hydraulic properties of porous clay capsules from a subsurface point source. *Twin International Conference on Geotechnical and Geo-Environmental Engineering CUM (7th) Ground Improvement Techniques*. June 23–25, Seoul, Korea.

Gupta, A.D., Babel, M.S., and Ashrafi, S. 2009. Effect of soil texture on the emission characteristics of porous clay pipe for subsurface irrigation. *Irrigation Science*, 27(3): 201–208.

Gupta, S.K. 1999. Some technologies to conserve irrigation water. *Proceedings of the 17th Congress on Irrigation and Drainage*, Vol. 1A, Granda, Spain, pp. 277–287.

Javaheri, P. 1976. Investigating the possibility of using clay pipes in irrigation. Institute of Soil Science and Soil Fertility, pp. 486 (in Persian).

Keikha, G.H. and Akbari Moghaddam, H. 2000. Investigation of the effect of planting density and underground clay irrigation system on the yield of greenhouse-cucumber cultivars. Sistan Agricultural and Natural Resources Research Center, Iran (in Persian).

Keikha, G.H. and Azar, E. 2003. Comparison and evaluation of groundwater and drip irrigation systems in cucumber greenhouse cultivation under farmer's conditions in Sistan region. *Monthy Journal of Moravej*, (44) (in Persian).

Keller, J. 2002. New Irrigation Technology for Smallholder: Revealed through an Innovative Contest. 2000 Irrigation Association's (IA) Annual Convention and Trade Show. Utah State University, and Ceo Keller, Bliesner Engineering, Logan, Utah, USA.

Kisi, O. and Cimen, M. 2009. Evapotranspiration modeling using support vector machines. *Hydrological Sciences Journal*, 54(5): 918–928.

Lazarovitch, N., Shani, U., Thompson, T.L., and Warrick, A.W. 2006. Soil hydraulic properties affecting discharge uniformity of gravity-fed subsurface drip irrigation. *Journal of Irrigation and Drainage Engineering*, 132(2): 531–536.

Majidi, A., Zarei, G.H., Keshavarz, A., and Hejazi, S.M. 2009. Evaluation of irrigation efficiency of various products (pistachio and pomegranate orchards and vegetable) using permeable clay pipes. Research Report No. 281/88 of Agricultural Research, Education and Extension Organization, Iran.

Malejinejad, H. 2003. Water use efficiency and crop yield under pot and furrow irrigation systems. *Journal of Agricultural Sciences and Natural Resources*, 10(1): 27–38 (in Persian).

Malekian, A., Mahdavi, M., Kholghi, M., Zehtabian, G.R., Saravi, M.M., and Rouhani, H. 2012. Optimal planning for water resources allocation (Case study: Hableh Roud Basin, Iran). *Desert*, 17(1): 1–8.

Masri, Z., 2002. *Time-tested pitcher irrigation helps green the slopes of Khanasser Valley*. ICARDA Caravan, ICARDA, Syria.

Mehdizadeh, P. 1977. Research on water saving and the use of earthenware jars to create green space and cultivate fruitful and non-fruitful trees in arid desert areas. Forests and Rangelands Research Institute, 21 (in Persian).

Mondal, R.C. 1974. Farming with pitcher: a technique of water conservation. *World Crops*, 26(2): 91–97.

Mondal, R.C. 1984. Pitcher Farming Technique for Use of Saline Waters. Annual Report 1984, Central Soil Salinity Research Institute, Karnal, India.

Mondal, R.C., Gupta, S.K., and Dubey, S.K. 1987. Pitcher irrigation. Leaftlet no. 11. Better farming in salt affected soil series. Central Soil Salinity Research Institute. Karnal, Haryana, India. Pp. 15.

Naik, B.S., Panda, R.K., Nayak, S.C., and Sharma, S.D. 2008. Hydraulics and salinity profile of pitcher irrigation in saline water condition. *Agricultural Water Management*, 95: 1129–1134.

Pacheco, F.A. 2020. Sustainable use of soils and water: the role of environmental land use conflicts. *Sustainability*, 12(3), 1163.

Saleh, E. and Setiawan, B.I. 2010. Numerical modeling of soil moisture profiles under pitcher irrigation application. *Agricultural Engineering International CIGR Journal*, 12(2): 14–20.

Sheng, H.S. 1974. *Fan Sheng-Chih Shu: An Agriculturist Book of China*, Written by Fan Sheng-Chih in the First Century BC. Science Books, Beijing, China, pp. 36Y37.

Siyal, A.A. 2009. Pitcher Irrigation: A Water Saving Technique. Pakissan.Com.

Siyal, A.A. and Skaggs, T.H. 2009. Measured and simulated soil wetting pattern under porous clay pipe subsurface irrigation. *Agricultural Water Management*, 96: 893–904.

Siyal, A.A., Van Genuchten, M.T., and Skaggs, T.H. 2013. Solute transport in a loamy soil under subsurface porous clay pipe irrigation. *Agricultural Water Management*, 121: 73–80.

Sohrabi, T. and Gazori, N. 1997. Subsurface irrigation with porous pipe. *Iranian Journal of Agriculture Science*, 28(3):145–156.

Stein, T.M. 1990. Development and Evaluation of Design Criteria for Pitcher Irrigation Systems. Ph. D. Dissertation. Silsoe College, Cranfield Institute of Technology, Cranfield, Bedfordshire Shrivenham, Oxfordshire, England.

Tanwar, B.S. 2003. Saline water management for irrigation. International Commission on Irrigation and Drainage. New Delhi, India, p.140.

Vasudevan, P., Thapliyal, A., Dastida, M.G., and Sen, P.K. 2007. Pitcher or clay pot irrigation for water conservation. *Proceedings of the International Conference on Mechanical Engineering 2007 (ICME2007)*, 29–31 December, Dhaka, Bangladesh.

Wang, D.Y., Tang, C.S., Cui, Y.J., Shi, B., and Li, J., 2016. Effects of wetting–drying cycles on soil strength profile of a silty clay in micro-penetrometer tests. *Engineering Geology*, 206: 60–70.

Wang, J., Long, H., Huang, Y., Wang, X., Cai, B., and Liu, W. 2019. Effects of different irrigation management parameters on cumulative water supply under negative pressure irrigation. *Agricultural Water Management*, 224: 105743.

Woldu, Z. 2015. Clay pot pitcher irrigation: a sustainable and socially inclusive option for homestead fruit production under dryland environments in Ethiopia (a partial review). *Journal of Biology, Agriculture and Healthcare*, 5(21): 157–167.

Yang, H., Du, T., Qiu, R., Chen, J., Wang, F., Li, Y., and Kang, S. 2017. Improved water use efficiency and fruit quality of greenhouse crops under regulated deficit irrigation in northwest China. *Agricultural Water Management*, 179: 193–204.

Zarei, G.H. 2003. Clay subsurface irrigation as an effective solution to adapt to drought. *Journal of Drought*, 8: 15–25 (in Persian).

Zarei, G.H., Keshavarz, A., Majidei, I., and Hejazei, M. 2003. Subsurface clay pipe irrigation for increasing water productivity. International Conference on Water-Saving Agriculture and Sustainable Use of Water and Land Resources. October 26–29, Yangling, China.

Zarei, G.H. and Shahpari, S.A., 2014. Hydraulic characteristics of porous clay capsules in a subsurface irrigation system at three soil textures. *Journal of Agricultural Engineering Research*, 14(4): 57–72.

V

Smart Irrigation

13
Automation and Smart Irrigation

13.1	Introduction 275
13.2	Developments in Agriculture 276
	Agricultural Revolutions • Irrigation • Internet of Things Inn an Agricultural Setting • Big Data within Agriculture • Blockchain in Agriculture
13.3	Artificial Intelligence 279
	Artificial Neural Networks • Fuzzy Logic • Expert Systems
13.4	Machine Learning 282
	Machine Learning Terminology and Definitions • Machine Learning Tasks • Analysis of Learning • Learning Models • Machine Learning in Irrigation • Water Management • Studies Using Different Learning Methods
13.5	Computer Vision 286
	Computer Vision Terminology and Definitions • Computer Vision in Today's World • Computer Vision in Agriculture • Computer Vision in Irrigation
13.6	Smart Manual Irrigation 288
13.7	Conclusions 288
References 290

Sajjad Roshandel
Xiamen University

Saeid Eslamian
Isfahan University of Technology

13.1 Introduction

With agriculture facing climatic and population demands – in 2019 the United Nations predicted that the global population will be 9.7 billion by 2050 (UNDP, 2019) – it is inevitable that producers need to use improved techniques to feed the population. Traditional farming methods are labour-intensive and time-consuming. When faced with a decreasing agricultural workforce, adopting smart technologies is a primary focus for food producers and stakeholders.

Agricultural Technology (also known as AgriTech, agrotech or agritech) and Digital Agriculture (also termed Precision Farming) have arisen as scientific fields that use data intense approaches to drive agricultural productivity whilst minimising the environmental impact of agriculture. The increasing use of technology to get from farm to plate is being referred to as the Fourth Agricultural Revolution (Lejon and Frankelius, 2015), with Carl-Albrecht Bartmer describing it as "Agriculture 4.0" in his opening speech at Agritechnica 2015.

Digital Agriculture gathers and disseminates data using a variety of different sensors and input such as drones or cameras. This data has created an improved understanding of the operational environment (i.e. an understanding of the dynamics between weather, soil, crops) and the operation itself (machinery, outliers), thus providing insight and knowledge to facilitate precision in the agricultural decision-making process.

In 1999, British pioneer, Kevin Ashton, coined the term 'Internet of Things' (IoT) (Smithsonian, 2015) to describe the way that the Internet is linked to our physical world – 'things' – using pervasive computing sensors and software. A 'thing to thing' communication tool with goals including cost saving, effective communication and automation.

The 21st century has seen technology develop further with the IoT evolving to have specific, focused domain applications including Multimedia Internet of Things (MIoT), Internet of Underground Things (IoUT) and Industrial Internet of Things. Without the inclusion of these notions, IoT systems cannot successfully attain the concept of pervasive computing. Such is the world today, where devices and their functionality have the ability to go beyond labour saving and rise to the challenge of decision making – machine learning (ML).

ML can be defined as the scientific field that gives machines the ability to learn without explicit programming. As technology develops, ML is applied to more scientific fields and everyday life: from climatology (Fang et al., 2017) to Fraud Detection (Raghavan and Gayar, 2019) and from Speech Recognition Tools (Delić et al., 2019) to Medicine (Sidey-Gibbons and Sidey-Gibbons, 2019).

Using the data gathered via sensors/drones/cameras, the IoT and IoUT, whereby communication devices and sensors are partially or completely buried underground for real-time soil sensing and monitoring, it is possible to 'train' machines in an agricultural setting. In this chapter, we look at the historical problems faced by agriculturalists when irrigating, explain ML, ML models and Computer Vision in greater detail and explore how ML and Computer Vision can be applied to agriculture, specifically irrigation. Water Management and Smart Manual Irrigation Systems will be mentioned before concluding on what can be expected for irrigation in the future. This study aims to provide a brief and useful guideline to the role of human and technological innovation in soil and water resource management, in particular, intelligent irrigation and managed food production for the coming years.

13.2 Developments in Agriculture

13.2.1 Agricultural Revolutions

The first Agricultural Revolution is generally recognised as the period humankind shifted from Hunter Gatherer to cultivator of the land. As populations grew and people settled, the need for a reliable source of food was established. Thought to have started in the 'Fertile Crescent' of the Middle East, the rest of the world began to practise agriculture and animal domestication soon afterwards.

The next recognised Agricultural Revolution was on the back of the European Industrial Revolution which saw the mechanisation of farming at the same time that transportation improved, crop rotation systems were developed and selective breeding introduced.

The mid-20th century saw the Third Agricultural Revolution, sometimes referred to as the 'Green Revolution'. Genetic engineering and hybridisation, coupled with the heightened use of fertilisers and pesticides whilst farming larger areas with heavy machinery, increased crop yields and lead to monoculture and the internationalisation of agricultural marketing.

The Fourth Agricultural Revolution is concerned with digital reinvention. Agriculture, as the rest of society and industry, is increasingly exposed to technology and the IoT. In order to survive, businesses have to evolve and adapt to this new age. Failure to do so will likely see that business collapse – think of Blockbusters declining the Netflix offer in 2000. Deere & Company – initially an agricultural machinery manufacturer – has been expanding its technological pedigree with investments in Artificial Intelligence (AI) and robotics, notably the acquisition of Blue River Technology in 2017. Much of Deere & Company's large machinery now comes with Connected Support Technology and some older large Ag Machines can have mobile communications and processing platform installed. Farmers have access to the data gathered by their machines through Deere & Company's own digital platform, which can then be used to improve the efficiency and operability of their farm.

13.2.2 Irrigation

Irrigation has been the answer to the worldwide problem of sustaining a regular and reliable water supply for intended crops. Inconsistent, or a simple lack of, water resources have been a challenge since Neolithic times. This challenge was resourcefully addressed by civilisations through the ages; ancient Egyptian engineers diverted Nile flood water to cultivate crops that wouldn't grow at other times of the year. Chinese and Peruvian societies developed terrace systems, whilst artificial reservoirs, that still exist today, were developed in Sri Lanka.

Although irrigation provides stability in food production which in turn gives food security and alleviates poverty, irrigation is not without its own issues. Soil erosion and water wastage are two problems traditionally encountered when irrigating. Water logging and consequent salinisation are another two. The use of AI and robotics – Smart Irrigation – addresses those concerns. Through analysis of soil moisture and climate conditions, Smart Irrigators can administer the appropriate volume of water accurately, reducing water wastage, increasing crop yields and preventing water logging.

13.2.3 Internet of Things Inn an Agricultural Setting

In a farm setting, IoT refers to devices used to measure data remotely. Devices gather information on moisture levels, livestock health and crop disease amongst other things. The initial sensors used in agriculture would log the data in an attached memory. Whilst it was valuable, it wasn't live/real-time data and developments using cloud computing have resulted in what is known as Smart Farming whereby management tasks are based on data as well as location. This need for real-time accurate information also gave rise to the IoUT where underground devices and sensors transmit data about the soil or plants to the cloud for real-time decision making. Figure 13.1 shows the cyber-physical management cycle of Smart Farming (Wolfert et al., 2017).

Soil sensors can read various properties about the soil such as the pH level, composition of the soil and the nutrients it contains. Soil moisture sensors are used in crop fields to measure the content of water held by the soil. This advancement has largely replaced manual soil moisture technologies which were arduous and proved time-consuming when taking readings in remote areas. With the development of wireless technologies transmitting the data, efficient water management decision making is made.

Weather sensors are used to sense weather and environmental conditions such as wind, precipitation, temperature and humidity. Accurate climate forecasting can predict crop needs and prevent over watering whilst also offering better labour optimisation and overall crop health. Deere & Company has incorporated weather sensors into their field and water management solutions (Deere & Company, 2017).

Drones – also known as Unmanned Ariel Vehicles – are increasingly used for image surveillance and have largely replaced satellite imagery. Drones can provide a soil moisture map and monitor crop growth. If linked to IoUT, drones can aid real-time decision making by giving aerial information as sensors emit data. Drones can also film or take images of crops which are then analysed for disease, readiness or yield predictions.

Robots, colloquially known as Agbots (McAllister et al., 2019), have been successfully trialled in coordinated weeding in simulated fields across a range of seed bank densities with different stages of weed growth. As weeds develop resistance to pesticides, this is a strategic development that can save a yield. In the United States, it has been estimated that herbicide-resistant weeds cost over half a billion USD a year in crop losses. Other Agbots are being developed for fruit harvesting (Livingston et al., 2016).

13.2.4 Big Data within Agriculture

Big Data has been described as the side effect of continuous data flow from a myriad of devices (Tzounis et al., 2017). Wolfert et al. (2017) define it as the "massive volumes of data of a wide variety that is captured, analysed and used for decision making". Either way, Big Data is the term for the large datasets generated every second of every day.

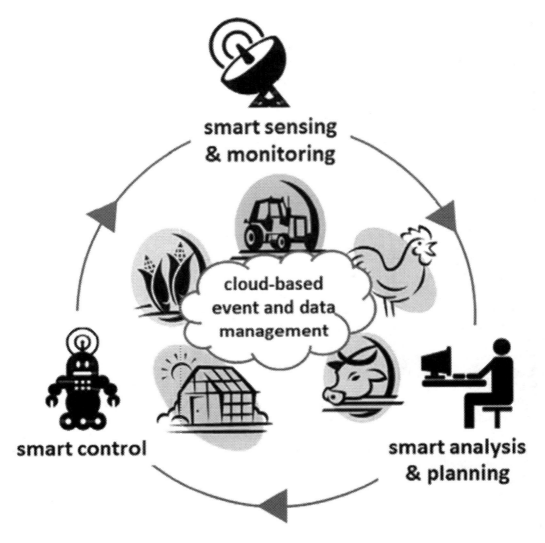

FIGURE 13.1 The cyber-physical management cycle of Smart Farming.

In an agricultural setting, with the limited processing abilities and energy considerations on farmland, it is not feasible to store the mass of data gathered in the field locally. Cloud seems to be able to withstand the volume of data whilst enabling real-time visualisation. Field data can be stored in public or private databases (Yan et al., 2017). Online marketplaces which store Big Data sets and agricultural apps can be utilised when analysing an area. Moreover, previous soil and contemporary meteorological data can be drawn from public services, combined with on-site data and utilised for precision decision making (Schönfeld et al., 2018). An automated system (Agile Actions) can control the farm equipment. For example, switch an irrigator on/off. A study in Ankara, Turkey, observed that the use of an Intelligent Irrigation System resulted in less temperature and moisture stress on the soil and efficient water consumption (Dursun and Ozden, 2011).

It is worth mentioning that the data gathered in the field cannot only be utilised by the farmers, but other stakeholders (AgBusiness and AgTech Organisations) can use the data for predictive insights and improve the efficiency of the supply chain whilst alleviating food security concerns.

Wolfert et al. (2017) noted that in order to make the Big Data applications work for Smart Farming, technical infrastructure where differing systems can effortlessly 'talk' to one another needs to be in place.

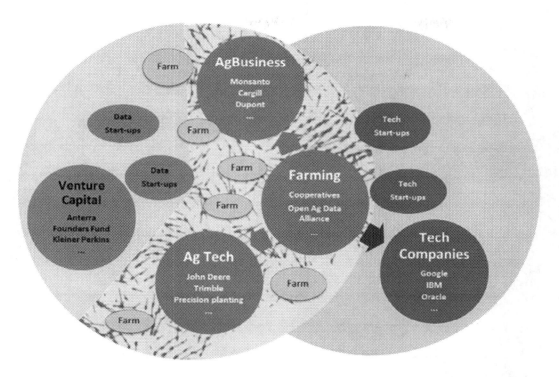

FIGURE 13.2 A landscape of the Big Data network for Smart Farming.

Wolfert et al. observe that applications from established AgTech and AgBusiness companies will be based on their existing infrastructures, whilst start-ups will likely be operating on open standards which can communicate with a variety of datasets. Figure 13.2 shows a landscape of the Big Data network.

13.2.5 Blockchain in Agriculture

The technology that started many of today's crypto currencies is known as Distributed Ledger Technology (DLT). DLT is a decentralised system which records, analyses, processes and validates transactions. Blockchain is an emerging extension of DLT that can be referred to as an Internet of Value (FAO, 2019). Within agriculture, blockchain can provide robust food tracing and validation of food sources as more consumers' demand reassurance that the food they purchase is of high quality and sourced responsibly. Blockchain provides a transparent, traceable system which can foster consumer confidence.

The blockchain is a unique dataset which stores data or transactions in a series of 'blocks', each of which has the hash value – or fingerprint ID – of the preceding block as well as a hash value of its own. Figures 13.3 and 13.4 show the examples of blockchain. Unlike a regular database, there are particular rules, or stages, in the method employed to add data to the blockchain (FAO and ITU, 2019, E-agriculture in Action). Ultimately, these stages ensure incontrovertible transaction records (Xu et al., 2020). Atlam et al. (2018) concluded that an integration of the blockchain with the IoT would be advantageous and improve any security issues within IoT.

13.3 Artificial Intelligence

AI refers to the ability that a machine has to perform activities that are normally thought to require intelligence. It also refers to the branch of computer science dedicated to developing this concept (Morris, 1970). Occasionally, AI is referred to as Synthetic Intelligence or Computational Rationality.

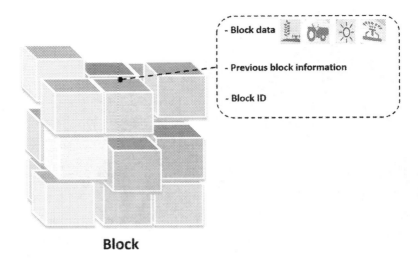

FIGURE 13.3 Example of one block in blockchain.

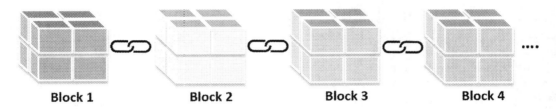

FIGURE 13.4 Example of connected blockchain.

The term Artificial Intelligence is attributed to John McCarthy in 1956, when he was Assistant Professor of Mathematics at Dartford College. McCarthy further defined AI as "the science and engineering of making intelligent machines". Nowadays, AI can be seen as the study and development of intelligent machines, or, the intelligence that a system demonstrates. Either way, studies and development of AI have continued since 1956 and has penetrated every aspect of 21st-century life. Without AI, there would be no Siri.

AI has assisted with the many challenges faced by agriculturalists daily, as described in Section 13.2.3. As Banerjee et al. (2018) state, agriculture is a dynamic domain. A key feature of AI is that it does not generalise a problem but provides bespoke solutions for idiosyncratic problems. AI requires a learning process of the machine, with ML a subdomain of AI (Sil et al., 2019).

There are three major AI techniques – Artificial Neural Networks (ANNs), Fuzzy Logic and Expert Systems.

13.3.1 Artificial Neural Networks

ANNs are supervised models designed to simulate the human brain, specifically cognition, learning and how the brain analyses and processes complex information (Chen, 2021).

Just as the brain is home to millions of neurons that communicate collectively, an ANN consists of interconnected processing units known as 'nodes' which are arranged in 'layers'. The nodes work together to process information and generate results using adaptive weight paths or weighted connections. Typically, a neural network will have an Input Layer where the ANN receives the raw data. The middle of the ANN is host to the Hidden Layers where the mathematical calculations and analysis occur. The Output Layer is where the results/decision/prediction is obtained. Here is an example of the three layers translated into everyday human terms;

You stroll by a pizzeria and your register the smell of pizza (Data Input – Input Layer), "I love pizza!" (Thought – Hidden Layer) "I think I will get some pizza" (Decision Making – Hidden Layer) "Ah, but I am supposed to be reducing the amount of junk food I eat." (Memory – Hidden Layer) "Perhaps just a slice won't hurt…" (Reasoning – Hidden Layer) "OK! I will just get one slice then." (Action – Output Layer) (Marr, 2020). ANNs are divided into two categories: 'Traditional ANNs' and 'Deep ANNs'. Both categories use a variety of learning algorithms.

Deep ANNs are also known as deep learning (DL) or deep neural networks (DNNs). DL allows multi-layered computational models to learn representations of complex data with numerous levels of abstraction (LeCun et al., 2015). A key advantage of DL is that they can identify sophisticated structures within large datasets through use of a backpropagation algorithm, a term popularised by Rumelhart, Hinton and Williams, in 1986. The backpropagation algorithm can be used to indicate how the machine should adjust its internal parameters to calculate the representation in each layer from the representation of the prior layer (LeCun et al., 2015). DL models have expanded the scope of ANN's application in many sectors and industries, including agriculture, as ANNs and DL are often used in classification and regression tasks which are devoted to agri-specific problem solving. DNNs can be supervised, partially supervised or even unsupervised.

13.3.2 Fuzzy Logic

Fuzzy Logic is a reasoning method not dissimilar to human reasoning to deal with imprecise information. Mathematically, it is a multi-valued logic form with a degree of truth values of 0.0 and 1.0, like probability theory. Fuzzy Logic was developed by Professor Lotfi A. Zadeh, a mathematician, in 1965 (Singh et al., 2013). Fuzzy Logic allows for vague human assessments within computing problems. Fuzzy Logic has been found to be useful in low-level machine control.

Fuzzy Control, a small set of initiative rules, was integrated with electronics, e.g. vacuum cleaners and elevators. Fuzzy Control means that the electronics could adapt to a slightly different, although similar, situation. In computing, it can adapt to an ambiguous dataset, providing a supplementary component for situations where conventional logic will fail to effectively solve a problem.

13.3.3 Expert Systems

Within computer science, Expert Systems mimic the decision-making ability of a human in order to solve complex problems. The Expert System simulates the knowledge that a human expert would have in a designated field. It uses reasoning to solve problems by matching the facts in the working memory with the knowledge base and gets new information.

Developed in 1965 by Edward Feigenbaum and Joshua Lederberg of Stanford University in California to analyse chemical compounds, Expert Systems successfully design, diagnose and monitor in a range of industries and everyday life such as accounting, gaming and medicine.

Forsyth (1984) states that an Expert System replaces the software tradition of

Data + Algorithm = System.

With a new architecture centred around the 'Knowledge Base' and an 'Inference Engine',

Knowledge + Inference = System.

This, although similar, has profound operational and rational differences.

13.4 Machine Learning

13.4.1 Machine Learning Terminology and Definitions

ML is a branch of AI whereby systems or applications learn from inputted data and experience to become more accurate. ML methodologies require a 'learning process'. The intention is that the training data are obtained as though learning from 'experience'. Building upon the sample data, algorithms are created to improve through experience and can, essentially, make independent decisions.

A variety of statistical and mathematical models are used depending on the desired outcome. Once the learning process has been completed, the trained model will be used to gather or predict new testing data. At the end of the learning process, the trained model can be used to classify, predict or cluster new examples (testing data) using the experience obtained during the training process. Figure 13.5 shows a typical ML approach (Liakos et al., 2018).

13.4.2 Machine Learning Tasks

Learning type and models dictate the category and classification of ML tasks. ML approaches, also known as 'tasks', typically have two categories depending on the learning type. They are supervised and unsupervised learning.

Supervised learning is an approach whereby the computer is offered inputs ('labelled examples') and desired outputs with the idea that a general rule, which will map the inputs and outputs, is learned. Essentially, the supervised ML algorithm will apply what it has learned to new data, thus predicting future events/outcomes. Within supervised learning, reinforcement learning can be used. Much like Pavlov's dogs (1906), reinforcement is given for correct answers. The computer interacts with a dynamic environment to learn correct responses.

Unsupervised learning is an approach whereby leading algorithms are not labelled but the computer has to find its own structure to the data presented. Basically, the system will study the unlabelled data provided and make conclusions to describe patterns.

A fourth method is occasionally referred to as semi-supervised ML; this falls between supervised and unsupervised learning. It uses both labelled and unlabelled data to classify problems but learns from that at the same time. Within these approaches there is further ordering such as Dimensionality Reduction, ANNs, Decision Trees or even the learning models employed to implement the selected task. Figure 13.6 shows ML algorithms that are used in some existing irrigation systems (Janani and Jebakumar, 2019).

13.4.3 Analysis of Learning

Dimensionality Reduction is a technique that is applied to both supervised and unsupervised learning types, with the aim of providing a more efficient, lower-dimensional representation of a dataset by reducing the number of variables, thus safeguarding much of the original data. Dimensionality

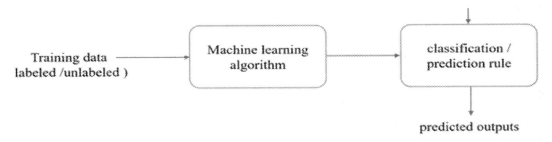

FIGURE 13.5 A typical machine learning approach.

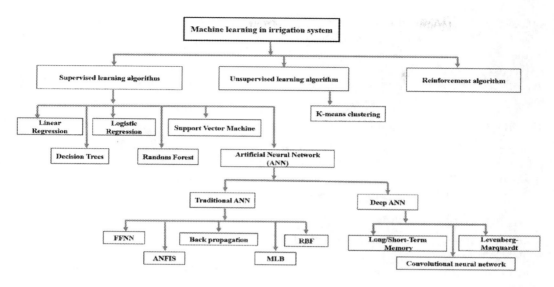

FIGURE 13.6 Some of machine learning algorithms used in irrigation system.

Reduction is often performed prior to applying a classification or regression model in order to avoid poor performance of algorithms which can occur with large numbers of input features. Dimensionality Reduction techniques are divided into linear and non-linear techniques (Cutura et al., 2018).

13.4.4 Learning Models

Here follow brief descriptions of some learning models and algorithms that are applied in ML.

13.4.4.1 Clustering

Clustering is an unsupervised learning method where the objective is to locate similar traits in the data presented. This is then 'clustered' in groups that share similarities but are dissimilar to other groups. There are different clustering algorithms including hierarchical clustering, K-means technique and fuzzy K-means.

13.4.4.2 Decision Tress

A Decision Tree is a type of supervised learning that uses a tree-like model of decisions and possible outcomes. Decision Trees are easy to use and interpret and so a popular predictive modelling algorithm (Kotu and Deshpande, 2014). In Decision Trees, the dataset is sorted into smaller, similar subsets as an inverted tree graph is generated. An issue with Decision Trees is that, with too little pruning, they can be poor predictors.

13.4.4.3 Regression

Regression Analysis is a supervised learning model used to predict a continuous value based on the known input variable. It is used for forecasting and analysing cause-and-effect relationships. The most known algorithms include linear regression and logistic regression which differ based on the number of variables (independent and dependent) and their relationship with one another.

13.4.4.4 Support Vector Machines

Another supervised learning model is the Support Vector Machine (SVM), which is based on statistical learning theory (Cortes and Vapnik, 1995), and is generally used for regression and classification, although SVM has been utilised for clustering (Finley and Joachims, 2005). SVM is seen as highly accurate whilst using little data and can deal with overfitting problems.

The purpose of the SVM is to locate a linear hyperplane to distinguish and classify data points in an n-dimension space. To enhance the capabilities of the SVM, a 'kernel trick' is used. A 'kernel trick' provides an efficient way to convert non-separable problems into separable problems, i.e. it can transform the original feature space into one of a higher dimension.

13.4.4.5 Instance-Based Models

Instance-Based Models – sometimes known as Memory-Based Models – are adaptable models that learn by comparing new instances with specific instances stored in the memory that were seen in training (Daelemans and Bosch, 1999). Hypotheses are formulated from the data available. The disadvantage of these models is that their hypotheses complexity can grow with the data.

13.4.4.6 Ensemble Learning

Ensemble Learning (EL) models combine ML techniques with the aim of producing one predictive model whilst decreasing variance (bagging) or bias (boosting).

EL models can be sub-divided into Sequential and Parallel ensemble methods which exploit dependence/independence between base learners, respectively, and therefore enable a fusion of hypotheses. Recently, Khalaf et al. (2020) noted that Ensemble Learning provided a dependable tool to predict the severity of flood levels.

13.4.4.7 Bayesian Models

Bayesian Models fall under supervised learning and are frequently applied to regression and classification questions. Bishop (2006) defined Bayesian approaches as statistical methods which can be used to derive probability distributions of sets of variables.

13.4.5 Machine Learning in Irrigation

Water scarcity is increasingly an issue with climatic changes lowering groundwater levels required for irrigation. Irrigation traditionally has a high labour, energy and resource use that is not often reflected in the crop yield. Inefficient and unreliable methods of irrigating see too much wastage of its precious raw material – water. In particularly arid regions, this can be critical to the livelihood as well as quality of life of the farmer.

Accurate detection and distribution of water can significantly cut the associated costs of crop production. IoT, and subsequent MIoT, is one way to battle this. ML devices are progressively more intelligent, smaller and financially accessible to farmers and growers worldwide. Smart Irrigation is paramount to securing and improving crop yields whilst becoming a low-cost solution. This section is intended to highlight the importance of ML in the overall Smart Irrigation solution by describing water management methods and detailing some studies that have analysed the success of various Smart Irrigation tools before describing Measured Irrigation (MI).

13.4.6 Water Management

Water Management, when applied to irrigation, is the practice of managing and controlling the distribution of water, ensuring the most efficient use of a limited resource for irrigation purposes. Water Management is important, not only as a planning tool, but to allow growers to manage irrigation expectations and forecast yields more accurately.

Two methods are used for 21st-century irrigation water management, computational and statistical. Computational methods were the earliest to be used and relied on a correlation between physical elements such as soil type and average rainfall. Statistical methods use prediction algorithms (as previously described in the Learning Models section) to automate the irrigation process (Krupakar et al., 2016).

So far, the most advanced ML-based applications are connected with estimation of daily, weekly or monthly evapotranspiration, which also allows predictions of daily dew point temperature.

Various studies continue to be conducted to assess the real-time and practical application of Smart Irrigation using IoT – sensors connected to a Wireless Support Network that will transmit real-time data relating to soil moisture, temperature, humidity and water level indicators, amongst other things and ML – to support predicted weather patterns against individual crop requirements.

13.4.7 Studies Using Different Learning Methods

Liakos et al. (2018) evaluated four studies focusing on the estimation of daily, weekly or monthly evapotranspiration. They noted the complexity of accurate evapotranspiration estimation against its importance as water resources and their availability decline. Of the studies they analysed, ANN and SVM models were used the most. Hinnell et al. (2010) conducted a study seeking to provide easily accessible, rapid computation and valid illustrations of spatial and temporal subsurface wetting patterns from drip emitters with the goal of maximising drip irrigation systems. Using ANN to predict the wetting patterns via an emitter placed on the surface of the land. The concern was that water was not penetrating to the lower level of the soil. Hinnell et al. (2010) developed the Neuro Drip, an Excel-based ANN that successfully provides rapid illustrations of soil wetting patterns.

Using ML and IoT sensing devices, Diedrichs et al. (2018) successfully predicted the occurrence of frost with the aid of Internet-enabled devices to gather data on water collection. The reference data were humidity and temperature sensors from five meteorological stations in Argentina over a period of 15 years.

In a study with a slightly different focus, Kwok and Sun (2018) suggested plant detection using DL. Once the software learned the plant type, the appropriate irrigation amount required was released upon recognition of the plant through the use of cameras.

Anneketh Vij et al. (2020) analysed various open source databases available online and ML algorithms (classification and regression) to provide a collective and detailed proposal for a solution to irrigation needs. They concluded "IoT and Machine Learning algorithms such as SVM (Support Vector Machine) and SVR (Support Vector Regression) with Radial basis function kernel helps in classification and quantitative predictions of soil type, crop type and amount of irrigation required by the crops" (Figure 13.7).

However, shortfalls were noted. Namely, climatic predictions cannot be 100% reliable and will continue to pose a serious threat; if the set-up of the Smart Farm Technology is not installed correctly, predictive accuracy is compromised; ML needs to be trained on larger data as opposed to just region-specific; nature can pose a threat to the hardware installed; the accuracy of the model depends on the data available, of which there was not sufficient; dedicated server storage or networks are required, which is not always feasible in some parts of the world.

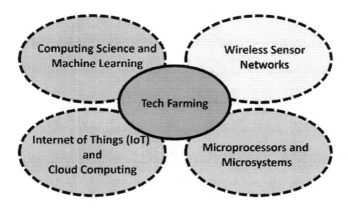

FIGURE 13.7 Some of the technology enablers that are used in a tech farming.

13.5 Computer Vision

13.5.1 Computer Vision Terminology and Definitions

The Former Director of Stanford University's Artificial Intelligence laboratory, Fei-Fei Li (2015), is credited with the statement "If we want machines to think, we need to teach them to see". Li is a leader in Computer Vision. Computer Vision is another branch of AI. Computer Vision is a field of research where computers are trained to recognise and interpret our visual world. Machines are being developed that correctly identify and comprehend what they 'see' before reacting to that image. Computer Vision uses embedded, 'learned' digital images and makes a decision on how to respond to the visual data.

13.5.2 Computer Vision in Today's World

Through a combination of ML and Computers, visual systems that can contextualise are being established. As 'seeing' is a multifaceted process, the technology is increasingly intricate. Sight is one of the five senses, which means that the eye is a sensor. When light reaches the retina photoreceptors (a special cell), it converts that light into signals which are sent to the brain via the optic nerve. The brain reads and converts those signals into images and their meaning. Stimuli for that meaning or interpretation include colours, shape and motion. Computer Vision is developed from the same principles whereby cameras capture images by letting in light. The images are then processed for their meaning through the use of ANNs – specifically Convolutional Neural Networks (CNNs).

CNN is a grid-based DL model used to process data (Yamashita et al., 2018). CNNs are robust enough to interpret the multiple components that go into interpretation of an image. Andrew Ng was amongst the first to use labelled data to generate mathematical models from images (Quoc V. Le, 2013). In 2012, a Google research team used 16,000 processors to create a neural network with 1 billion connections. Instead of hard coding the image of a cat, labelled images of cats in different positions, different colours and of different breeds were presented. The model developed its own idea of what a cat looks like in a replication of how humans learn to recognise things (Markoff, 2012).

Nearly 10 years later, Computer Vision has developed exponentially with Computer Vision appearing in everyday situations; facial recognition software used on social media platforms such as Facebook or in security footage is a facet Computer Vision. Technology has developed enough that Tesla are bringing out cars with autopilot features such as cruise control that can keep a safe distance from other vehicles. Tesla is predicting full self-driving features in the future (Tesla, 2021).

13.5.3 Computer Vision in Agriculture

As mentioned throughout this chapter, agriculture is facing increasing demands to produce higher yields in an economic and environmentally friendly way for a burgeoning population amongst global warming and climate concerns. Computer Vision tools have been becoming increasingly crucial to the successful management and maintenance of crops. Agricultural production management is increasingly relying on Computer Vision-based agricultural automation technology to aid efficiency and productivity.

Traditional farming techniques are notoriously labour-intensive with heavy reliance on environmental elements for successful and voluminous yields. Human judgement and historical knowledge affect decision making, yet poor climate or soil nutrition can destroy months of hard work and effective assessments.

It has been noted that herbicide-resistant weeds are estimated to cost over half a billion USD a year in crop losses (Livingston et al., 2016). In 2017, Deere & Company acquired Blue River Technology. In June 2021, See & Spray Select will be commercially available (John Deere, 2021). See & Spray Select uses Computer Vision and ML technology to reduce up to 77% herbicide usage. See & Spray Select is installed on a John Deere sprayer and uses two rows of the cameras to make on-the-spot decisions. The front row of cameras makes the decision as to whether a plant is a weed or not using colour differentiation as the rear cameras calibrate the system. The machine is capable of processing up to 20 images a second. Once a weed is identified, the robotic nozzles target and spray the weed. If required, machine operators can switch to broadcast spraying. John Deere claims that the targeted weed kill is as effective as broadcast spraying.

This effective use of Leverage Vision and ML will not only provide cost savings in the targeted application of herbicides, but farmers can utilise more expensive yet effective mixes and still achieve a lower cost over all. Additionally, the ability to switch from spot spraying to broadcast spraying is estimated to save time. Finally, the targeted use of spray means less wastage and time taken to refill.

Culman et al. (2017) conducted a study using *PalmHAND* to identify nutrient deficiencies in oil palm crops, in real time. The identification of nutrient deficiencies can contribute to control fertilisers. *PalmHAND* is an application that uses visual indicators found in the oil palm foliage to identify nutrient deficiency. This costly, time-consuming task is traditionally performed manually and not without error. The application uses digital image processing and pattern recognition algorithms within a smart device. One advantage in using a smart device is that they are readily available and not as costly as other specialised equipment. Additionally, knowledge from experts is encoded into in software which can be deployed to areas where there is a scarcity of those said professionals. Finally, the use of a smart device allows for real-time monitoring of large-scale plantations. Predictions of nutrient deficiency in oil palms are geo-located and uploaded to a cloud computing service that uses a Linux and Microsoft Windows operating systems. Furthermore, predictions are presented using Power BI reports. *PalmHAND* has been demonstrated to be capable of handling multiple devices simultaneously using infrastructure developed by the IoT. Culman et al. (2017) noted the "effective, low cost and scalable solution to precise nutrient management policies" that can be provided by the *PalmHand*.

13.5.4 Computer Vision in Irrigation

Computer Vision is also extended to irrigation. Studies and practices are being undertaken to confirm the advantages these systems can offer. This next section aims to examine some of those studies further.

Many traditional irrigation methods rely on indiscriminate spraying which is not always effectual. Studies have and are being conducted to find bespoke answers to unique irrigation issues. Chang and Kin (2018) proposed a scheme using Computer Vision and multi-tasking processes to develop a small-scale agricultural machine that would automatically weed whilst performing variable rate irrigation. In their study, image processing was used to determine the position of plants and weeds. Additionally, data relating to the wet distribution area of the surface soil was presented to a Fuzzy Logic controller which in turn would initiate irrigation pumps for variable rate irrigation as required. The use of Fuzzy Logic as a control method ensured that the amount of water distributed to crops was based on the data received from the soil moisture sensors placed around the field at root depth. This ensured that crops were not over-watered and soil maintained the appropriate moisture content.

Lam et al. (2006) developed a ground-based remote-sensing feedback control system to monitor the rate of advancement of water down a furrow. Furrows are widely used to irrigate row crops in California. Furrows were historically found to be inefficient and have a large amount of water runoff. The development of the control system used a camera to capture images of water flow down a funnel. These images were analysed using a Computer Vision system. A signal was sent from the machine vision system to an automated gate valve which would adjust the rate of water into the furrow.

Gutiérrez Jagüey et al. (2015) used an automated irrigation sensor within an Android smartphone to capture and process images of soil by the root of a crop. The sensor, buried at the root of the crop with controlled illumination, would optically estimate the water contents. An Android app was developed to autonomously control the digital camera and Wi-Fi network. Once the app initialises the phone to prompt the camera to take an image through the anti-reflective glass window where upon and RGB to grey process is used to estimate the ration between wet and dry area of the image. Using the Wi-Fi, the ratio is transmitted via a router node to a gateway which controls the irrigation pump.

It was concluded that app offered a simple and practical implementation, enhanced by the increasing availability of smartphone devices. When compared with traditional sensors, the installation in the field required more effort and time. However, no other significant additional labour was noted. It was felt that the irrigation sensor held an advantage over other soil moisture sensors.

13.6 Smart Manual Irrigation

Not all Smart Technology is mechanised. MI is an economic manual irrigation scheduling technique that responds to prevailing environmental factors. The term Measured Irrigation is associated with the work of Bernie Omodei who is acknowledged to have invented MI in late 2010. According to Omodei, MI is an irrigation method that satisfies the following two conditions:

1. Variations in the water usage throughout the year are controlled by the prevailing net evaporation rate (evaporation minus rainfall).
2. The volume of water discharged by each emitter during an irrigation event is controlled directly without the need to control the flow rate or the duration of the irrigation event (Figure 13.8).

MI is an unpressurised gravity-fed irrigation system that controls the amount of water issued without adjusting the flow rate. Omodei found that MI nozzles exhibited accuracy of approximately 95% and uniformity between 90% and 95% when tested (Omodei, 2020).

In developing countries where the Smart Irrigation systems requiring costly electronics and ready access to cloud-based digital platforms, the solar-powered MI unit has proved to be a cheap and efficient technology for smallholders that still results in higher yields and water preservation (Omodei, 2015).

13.7 Conclusions

In this chapter, we have explored the changing face of agriculture with particular focus on the digital reinvention needed to be a successful part of the Fourth Agricultural Revolution. Agriculture, as an industry that has a large amount of water dedicated to it, must focus its efforts to conserve water and use it wisely. Water wastage is costly, and as water resource is becoming scarce amid increasing concerns about global warming, better water management is paramount. The awareness of sustainable consumption among many governments, businesses and citizens will also contribute to the conservation of water (Al-Omran et al., 2021). Smart Irrigation developments in the 21st century have many positive impacts. Primarily, food security can be assured and growers can move from a labour-, energy- and resource-intensive approach to a tactical management approach. The use of robotics and sensors can resolve water wastage issues, maximise food production and provide a solution to the decreasing agricultural workforce problem. In countries like India, where approximately one-third of the country's capital is derived from farming, Smart Irrigation can be seen as an answer to the development of the nation.

From the point of view of the grower, Smart Irrigation can save time. Without the time spent travelling to remote areas or analysing weather predictions, growers can focus their efforts elsewhere and become more tactical in their business plans. The predicted world population of 9.7 billion by 2050 (UNPD, 2019) will not only need to be fed but space will become an issue. Smart Irrigation not just produces increased yields whilst using less water, but it saves space. From a financial point of view, Smart Irrigation will increase revenues whilst cutting costs overall. As prices for electronics comes down, growers find that the IoT systems become more available to them. Simultaneously, progressive ecological effects such as the transformation of arid plains into oxygen-producing green areas and a range of habitats where wildlife can thrive have also been developed. Use of Computer Vision technology which targets the application of pesticides and herbicides will reduce spray drift and residuals entering the water board. Jakku et al. (2016) ran a study on perceptions of, and experiences with, Big Data and digital technologies in

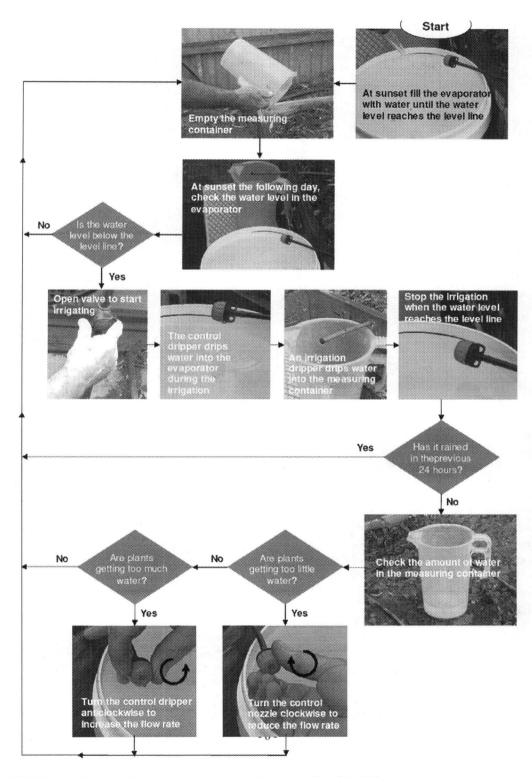

FIGURE 13.8 Flowchart for using manual Measured Irrigation (Omodei, 2020).

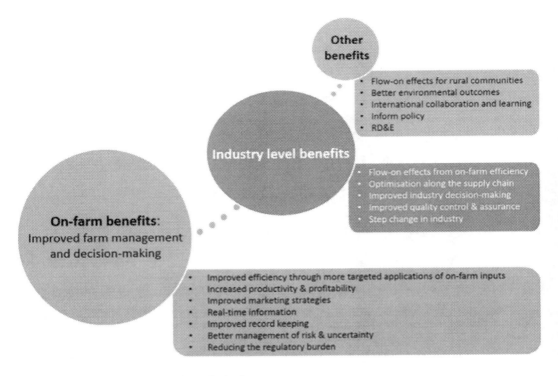

FIGURE 13.9 Interesting range of benefits by the interviewees.

agriculture. Twenty-six grains industry stakeholders from the state of Victoria in Australia were interviewed with particular attention laid on how Big Data may affect relationships, including trust, between members of the supply chain. Figure 13.9 shows the interesting range of benefits felt by the interviewees (Jakku et al., 2016). The Fourth Agricultural Revolution has seen that when ML is coupled with the networked capabilities of MIoT, cooperation and facilitation are fostered between diverse devices that can provide further economic and social benefits such as flood control, hydropower and rural development. With continued big investment in IoT and Smart Irrigation, we will see hardware developed that can withstand the elements and unpredictable environmental conditions such as wild animals. Ownership of data may be clarified and a Universal Standard could be implemented that will reduce competition and encourage collaboration. Interoperability is a challenge for wireless networks, so research and development into robust and accurate networking systems is likely to be a key focus.

References

Al-Omran, A. M., Al-Khasha, A., & Eslamian, S. (2021). Irrigation Water Conservation in Saudi Arabia, Ch 25 in *Handbook of Water Harvesting and Conservation*, Vol. 2: Case Studies and Application Examples, Ed. By Eslamian, S. & Eslamian, F., John Wiley & Sons, Inc., New Jersey, USA, 373–384.

Atlam, H., Alenezi, A., Alassafi, M., & Wills, G. (2018). Blockchain with Internet of Things: Benefits, Challenges and Future Directions. *International Journal of Intelligent Systems and Applications*. doi: 10.10.5815/ijisa.2018.06.05

Banerjee, G., Sarkar, U., Das, S., Ghosh, I. (2018). Artificial Intelligence in Agriculture: A Literature Survey. *International Journal of Scientific Research in Computer Science Applications and management Studies*, 7(3), 1–6.

Bishop, C. (2006). *Pattern Recognition and Machine Learning*. Springer, New York. doi: 10.1117/1.2819119

Chang, C.-L. & Lin, K.-M. (2018). Smart Agricultural Machine with a Computer Vision-Based Weeding and Variable-Rate Irrigation Scheme. *Robotics*, 7(3), 38. doi: 10.3390/robotics7030038

Chen, J. (2021). Neural Network. Received from https://www.investopedia.com/terms/a/artificial-neural-networks-ann.asp#:~:text=An%20artificial%20neural%20network%20(ANN)%20is%20the%20piece%20of%20a, by%20human%20or%20statistical%20standards

Cortes, C. & Vapnik, V. (1995). Support-Vector Networks. *Machine Learning*, 20, 273–297. doi: 10.1007/BF00994018

Culman, M., Escobar, J., Portocarrero, J., Quiroz, L., Tobon, L., Aranda, J., Garreta, L., & Bayona, C. (2017). A Novel Application for Identification of Nutrient Deficiencies in Oil Palm Using the Internet of Things. In 5th IEEE International Conference on Mobile Cloud Computing, Services, and Engineering (MobileCloud), Pages 169–172. doi: 10.1109/MobileCloud.2017.32

Cutura, R., Holzer, S., Aupetit, M., & Sedlmair, M. (2018). VisCoDeR - A Tool for Visually Comparing Dimensionality Reduction Algorithms. University of Vienna. European Symposium on Artificial Neural Networks, Computational Intelligence and Machine Learning (ESANN), Austria.

Daelemans, W. & Van den Bosch, A. (1999). *Memory-Based Language Processing*. Cambridge University Press, Cambridge. doi: 10.1017/CBO9780511486579

Deere & Company. (2017). Annual Report. Page 6. United States. https://www.annualreports.com/HostedData/AnnualReportArchive/d/NYSE_DE_2017.pdf https://www.deere.com/en/technology-products/precision-ag-technology/field-and-water-management

Delić, V., Perić, Z., Sečujski, M., Jakovljević, N., Nikolić, J., Mišković, D., Simić, N., Suzić, S. & Delić, T. (2019). Speech Technology Progress Based on New Machine Learning Paradigm. *Computational Intelligence and Neuroscience*. doi: 10.1155/2019/4368036. PMID: 31341467; PMCID: PMC6614991.

Diedrichs, A. L., Bromberg, F., Dujovne, D., Brun-Laguna, K., & Watteyne, T. (2018). Prediction of Frost Events Using Machine Learning and IoT Sensing Devices. *IEEE Internet of Things Journal*, 5(6), 4589–4597. doi: 10.1109/JIOT.2018.2867333

Dursun, M. & Ozden, S. (2011). A Wireless Application of Drip Irrigation Automation Supported by Soil Moisture Sensors. *Scientific Research and Essays*, 6, 1573–1582.

E-Agriculture in Action: Blockchain for Agriculture Opportunities and Challenges. (2019). Edited by Sylvester, G. Food and Agricultural Organisation of the United Nations (FAO). http://www.fao.org/3/CA2906EN/ca2906en.pdf

Fang, K., Shen, C., Kifer, D., & Yang, X. (2017). Prolongation of SMAP to Spatiotemporally Seamless Coverage of Continental U.S. Using a Deep Learning Neural Network. *Geophysical Research Letters*, 44, 11030–11039.

Fei-Fei, L. (2015). Stanford Institute for Human-Centered Artificial Intelligence (HAI), https://www.wired.com/brandlab/2015/04/fei-fei-li-want-machines-think-need-teach-see/

Finley, T. & Joachims, T. (2005). Supervised Clustering with Support Vector Machines. In Proceedings of the 22nd International Conference on Machine Learning. Germany. Pages 217–224. doi: 10.1145/1102351.1102379

Forsyth, R. S. (1984). The Architecture of Expert Systems. In: Forsyth, R.S. (ed.) *Expert Systems: Principles & Case Studies*. Chapman & Hall Ltd., London, UK, Pages 9–17.

Gutiérrez Jagüey, J., Villa-Medina, J. F., López-Guzmán, A., & Porta-Gándara, M. Á. (2015). Smartphone Irrigation Sensor. *IEEE Sensors Journal*, 15(9), 5122–5127. doi: 10.1109/JSEN.2015.2435516

Hinnell, A., Lazarovitch, N., Furman, A., Poulton, M., & Warrick, A. (2010). Neuro-Drip: Estimation of Subsurface Wetting Patterns for Drip Irrigation Using Neural Networks. *Irrigation Science*, 28(6), 535–544. doi: 10.1007/s00271-010-0214-8

Internet of Things. Smithsonian. (2015). https://www.smithsonianmag.com/innovation/kevin-ashton-describes-the-internet-of-things-180953749/

Jakku, E., Taylor, B., Fleming, A., Mason, C., & Thorburn, P. (2016). Big Data, Trust and Collaboration: Exploring the socio-technical enabling conditions for big data in the grains industry. Report number: EP164134. doi: 10.13140/RG.2.2.26854.22089

Janani, M. & Jebakumar, R. (2019). A Study on Smart Irrigation Using Machine Learning. *Cell & Cellular Life Sciences Journal*. doi: 10.23880/cclsj-16000141

John Deere Company. (2021). https://www.deere.com/en/our-company/news-and-announcements/news-releases/2021/agriculture/2021mar02-john-deere-launches-see-and-spray-select

Khalaf, M., Haya, A., Hussain, A., Baker, T., Maamar, Z., Buyya, R., Liatsis, P., Khan, W., Tawfik, H., & Al-Jumeily Obe, D. (2020). IoT-Enabled Flood Severity Prediction via Ensemble Machine Learning Models. *IEEE Access*, 8, 70375–70386. doi: 10.1109/ACCESS.2020.2986090

Kotu, V. & Deshpande, B. (2014). *Predictive Analytics and Data Mining: Concepts and Practice with RapidMiner. Predictive Analytics and Data Mining: Concepts and Practice with RapidMiner*. Morgan Kaufmann, Amsterdam, 1–425.

Krupakar, H., Jayakumar, A., & Ganesh, D. (2016). A Review of Intelligent Practices for Irrigation Prediction. In National Conference on Computational Intelligence and High-Performance Computing (NCCIHPC), India.

Kwok, J. & Sun, Y. (2018). A Smart IoT-Based Irrigation System with Automated Plant Recognition using Deep Learning. In ICCMS 2018: Proceedings of the 10th International Conference on Computer Modeling and Simulation, January, Pages 87–91. doi: 10.1145/3177457.3177506

Lam, Y., Slaughter, D.C., Wallender, W.W., & Upadhyaya S.K. (2006). Computer Vision System for Automatic Control of Precision Furrow Irrigation System Paper number 062078. American Society of Agricultural and Biological Engineers ASAE Annual Meeting. doi: 10.13031/2013.20693

Le, Q.V. (2013). Building high-level features using large scale unsupervised learning. In IEEE International Conference on Acoustics, Speech and Signal Processing, 2013, pp. 8595–8598, doi: 10.1109/ICASSP.2013.6639343

LeCun, Y., Bengio, Y., & Hinton, G. (2015). Deep Learning. *Nature*, 521, 436–444. doi: 10.1038/nature14539

Lejon, E. & Frankelius, P. (2015). *Sweden Innovation Power—Agritechnica 2015*. Elmia, Jönköping, Sweden.

Liakos, K.G., Busato, P., Moshou, D., Pearson, S., & Bochtis, D. (2018). Machine Learning in Agriculture: A Review. *Sensors*, 18(8), 2674. doi: 10.3390/s18082674

Livingston, M., Fernandez-Cornejo, J., & Frisvold, G. (2016). Economic Returns to Herbicide Resistance Management in the Short and Long Run: The Role of Neighbour Effects. *Weed Science*, 64(S1), 595–608. doi: 10.1614/WS-D-15-00047.1

Markoff, J. (2012). How Many Computers to Identify a Cat? 16,000, *New York Times*, https://www.nytimes.com/2012/06/26/technology/in-a-big-network-of-computers-evidence-of-machine-learning.html?_r=1&pagewanted=all

Marr, B. (2020). Deep Learning Vs. Neural Networks – What's the Difference? https://bernardmarr.com/deep-learning-vs-neural-networks-whats-the-difference/

McAllister, W., Osipychev, D., Davis, A., & Chowdhary, G. (2019). Agbots: Weeding a Field with a Team of Autonomous Robots. *Computers and Electronics in Agriculture*, 163, 104827. doi: 10.1016/j.compag.2019.05.036

Morris, W. (1970). The American Heritage dictionary of the English language. New York: American Heritage, Fifth Edition. Retrieved January 19, 2021. https://www.thefreedictionary.com/Artificial+intellegence

Omodei, B. (2015). Accuracy and Uniformity of a Gravity-Feed Method of Irrigation. *Irrigation Science*, 33. doi: 10.1007/s00271-014-0452-2

Omodei, B. (2020). Unpowered Measured Irrigation Training Manual for Smallholders. www.measuredirrigation.com.au.

Pavlov, I.P. (1906). The Scientific Investigation of the Psychical Faculties or Processes in the Higher Animals. *Science*, 24, 613–619.

Raghavan, P. & Gayar, N. (2019). Fraud Detection Using Machine Learning and Deep Learning. In 2019 International Conference on Computational Intelligence and Knowledge Economy (ICCIKE), pp. 334–339. doi: 10.1109/ICCIKE47802.2019.9004231

Rumelhart, D.E., Hinton, G.E., & Williams, R.J. (1986). Learning Representations by Back Propagating Errors. *Nature*, 323(6088), 533–536.

Schönfeld, M., Heil, R., & Bittner, L. (2018). Big Data on a Farm—Smart Farming. In: Hoeren, T. & Kolany-Raiser, B. (eds) *Big Data in Context. Springer Briefs in Law*. Springer, Cham. Switzerland. doi: 10.1007/978-3-319-62461-7_12

Sidey-Gibbons, J. & Sidey-Gibbons, C. (2019). Machine Learning in Medicine: A Practical Introduction. *BMC Medical Research Methodology*, 19, 64. doi: 10.1186/s12874-019-0681-4

Sil, A., Roy, R., Bhushan, B., & Mazumdar, A.K. (2019). Artificial Intelligence and Machine Learning based Legal Application: The State-of-the-Art and Future Research Trends. In International Conference on Computing, Communication, and Intelligent Systems (ICCCIS), India, pp. 57–62, doi: 10.1109/ICCCIS48478.2019.8974479

Singh, H., Gupta, M., Meitzler, T., Hou, Z-G., Garg, K., Solo, A., & Zadeh, L. (2013). Real-Life Applications of Fuzzy Logic. *Advances in Fuzzy Systems, Advances in Fuzzy Systems*. doi: 10.1155/2013/581879

Tesla. (2021). Cars come to improve functionality over time. https://www.tesla.com/autopilot

Tzounis, A., Katsoulas, N., Bartzanas, T., & Kittas, C. (2017). Internet of Things in Agriculture, Recent Advances and Future Challenges. *Biosystems Engineering*, 164, 31–48. doi: 10.1016/j.biosystemseng.2017.09.007

United Nations Development Programme (UNDP). (2019). Population/world-population-prospects. Machine Learning. https://www.un.org/development/desa/en/news/population/world-population-prospects-2019. Machine Learning.

Vij, A., Singh, V., Jain, A., Bajaj, S., Bassi, A., & Sharma, A. (2020). IoT and Machine Learning Approaches for Automation of Farm Irrigation System. *Procedia Computer Science*, 167, 1250–1257. doi: 10.1016/j.procs.2020.03.440

Wolfert, S., Ge, L., Verdouw, C., & Bogaardt, M.J. (2017). Big Data in Smart Farming – A Review. *Agricultural Systems*, 153, 69–80.

Xu, J., Guo, S., Xie, D, & Yan, Y. (2020). Blockchain: A New Safeguard for Agri-Foods. *Artificial Intelligence in Agriculture*, 4. doi: 10.1016/j.aiia.2020.08.002

Yamashita, R., Nishio, M., Do, R., & Tagashi, K. (2018). Convolutional Neural Networks: An Overview and Application in Radiology. *Insights Imaging*, 9, 611–629.

Yan, Q., Yang, H., Vuran, M., & Irmak, S. (2017). SPRIDE: Scalable and Private Continual Geo-Distance Evaluation for Precision Agriculture. In IEEE Conference on Communications and Network Security (CNS), 2017, pp. 1–9. doi: 10.1109/CNS.2017.8228620

14
Intelligent Irrigation and Automation

Hamideh Faridi
University of Manitoba

Babak Ghoreishi
Forest, Range and Watershed Management Organization

Hamidreza Faridi
University of Tehran

14.1 Introduction 295
  Water Use Efficiency (WUE) • Factors Affecting the WUE Process • Importance of Using Intelligent Irrigation Systems
14.2 Advanced Irrigation Management Equipment and Techniques...298
  Remote Sensing (RS) • Sensor and Communication Networks • Artificial Intelligence
14.3 Intelligent Irrigation Systems...303
  Climate-Based Intelligent Irrigation System • Intelligent Irrigation System Based on the Soil Moisture Sensors
14.4 Applications of Intelligent Methods in Irrigation 312
  Application of Renewable Energy in IIS
14.5 Conclusions.. 316
Notes 317
References...317

14.1 Introduction

Irrigation is one of the oldest human activities with the main purpose to increase the production of agricultural crops. This shows the importance and role of irrigation in agriculture, especially in arid and semi-arid regions. Thus, it can be said that irrigation systems are the basic infrastructure for preparing food in the world, which is the best result in agriculture by choosing the appropriate irrigation method.

The optimal use of water and its efficient management is the most important challenge in the agricultural sector in many countries, today. According to estimates, 25% of the world's farms (399 million hectares) utilize around 70% of the freshwater available to irrigate them, meeting 45% of the world's food demands (Thenkabail et al., 2011). About 20% and 10%, respectively, of the total amount of water consumed worldwide are utilized for industrial and household purposes. In general, it seems that the irrigation process in agriculture will face a water shortage crisis in near future. More than half of the water used in conventional irrigation systems is wasted as a result of their inefficiency. The implementation of intelligent irrigation systems (IISs) is the greatest remedy for this issue. This chapter's goal is to go over how to use water resources more effectively.

14.1.1 Water Use Efficiency (WUE)

Water productivity (WP) and efficiency are the main characteristics of the optimal water consumption. The ratio of product production to water use, measured in kilograms per cubic meter (kg/m^3), is known as water productivity. The term water use efficiency (WUE), which is the same as the ratio of production to evaporation and transpiration, was first proposed by Weitz in 1966 (Stanhill, 1986). Since then, the term has been used to determine the ratio of performance (in terms of photosynthesis performance, biological performance, or economic performance) per unit volume of water consumption (as water volume, evaporation and transpiration, or volume of water used). WUE, or simply irrigation efficiency, is the technical term used to describe the need to use water resources more effectively. From an engineering standpoint, the WUE is frequently described as the portion of irrigation water that is utilized economically or effectively by the plant (Eqs. 14.1 and 14.2) using a volumetric or hydrological technique. The primary use of this term is in farm size water management and irrigation. However, it should be emphasized that WUE may also be evaluated on a catchment or basin scale (Paydar and Qureshi, 2012).

The two most common performance indicators of irrigation systems are as follows:

i. Application Efficiency (AE), and
ii. Requirement Efficiency (RE),

which can be defined as Eqs. (14.1) and (14.2), respectively.

$$AE = \frac{\text{Volume of water stored in the root zone}}{\text{Total volume of water applied}} \quad (14.1)$$

$$RE = \frac{\text{Volume of water stored in the root zone}}{\text{Water deficit prior to irrigation}} \quad (14.2)$$

There are new and creative options to significantly enhance WUE in agricultural irrigation thanks to technological advancements, long-term investments, and agricultural research and development. Use of distant data (from satellites or unmanned aerial vehicles, or UAVs), communication networks, and the availability of inexpensive sensors are a few examples of these. In particular situations when water is a production constraint, improving irrigation WUE may result in water savings and the ability to irrigate more area with a given amount of water.

14.1.2 Factors Affecting the WUE Process

With regard to the above discussion, it is clear that the process of WUE in irrigation is influenced by a wide range of factors widely classified as (1) engineering and technology, (2) environmental factors, (3) socio-economic factors, and (4) advancements in plant and pasture science. The development of irrigation systems on farms, the expansion of water distribution networks, irrigation planning, real-time control and optimization, remote sensing, and sensor and communication networks are all examples of engineering and technology-related elements. By lowering water losses, these elements enhance irrigation. A variety of hardware and software technologies have recently become commercially accessible and are being utilized to improve WUE irrigation. Advances in plant genetics have led to higher yielding, and disease-resistant species with higher WUEs. Environmental awareness (Figure 14.1) has led some governments around the world to offer water-saving schemes, realizing that stored water is released as an environmental flow. WUE is influenced significantly by social and economic issues as well. This section focuses on how users of irrigation water accept new technologies and make decisions.

When farmers have limited access to water, they have to make a challenging decision on how to make the most of it. Farmers are frequently observed irrigating a portion of their land while leaving the remainder to be supplied by rain in places like Australia where water is mostly a productivity limiting issue but land is nearly limitless (Koech and Langet, 2018). According to a research conducted in southern Spain, the majority of farmers had significant irrigation difficulties when cultivating olive trees and relied on dehydration to increase water value. It should be highlighted, nevertheless, that

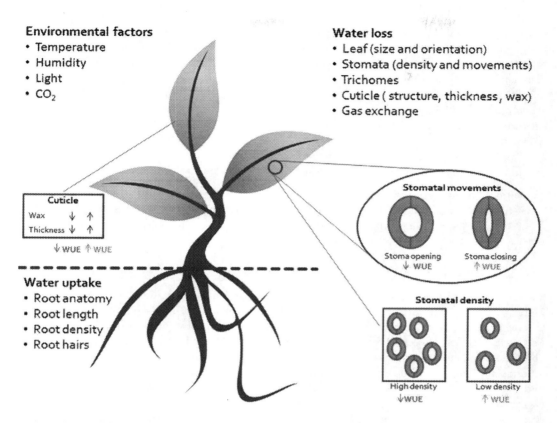

FIGURE 14.1 Environmental factors influencing water use efficiency (Ruggiero et al., 2017).

technological adoption is a complicated sociological process, and it depends heavily on consumer willingness to embrace new values. Consumers of irrigation water are more inclined to stick with farming practices that they are already more comfortable without fear of the unknown than are other members of the public.

14.1.3 Importance of Using Intelligent Irrigation Systems

Farmers used to use traditional irrigation methods, but today there have been improvements in irrigation systems. Due to the importance of water in agriculture, new and modern methods have been adopted for irrigating greenhouses and agricultural lands. Under these conditions, agricultural lands and greenhouses can be irrigated at the lowest cost and manpower. Smart irrigation systems can be the best option currently because of the concerns that most people have. Due to the fact that most of the agricultural lands and greenhouses are outside the city and it takes a long time for traveling to irrigate and take care of them, this can be done easily using the automatic irrigation method (Mostafazadeh-Fard et al., 2006). The number of research studies looking at ways to cut back on water use in irrigation systems has increased as a result of these factors. Some of these studies recommend implementing social and economic strategies as well as technology advancements to enhance water management. In recent decades, the phrases "irrigation rehabilitation" and "automation" have proliferated in the literature on irrigation. Renovation is the process of replacing outdated irrigation infrastructure and procedures, which were frequently created with more contemporary, "modern," equipment and technology. Water conservation, better water distribution, and lower operational and labor expenses are all goals of modernizing irrigation, which will increase farmers' lives and promote sustainable agricultural productivity

(Plusquellec, 2009). The use of machinery that enables irrigation to take place with little to no human involvement, aside from normal maintenance and periodic inspections, is known as automation of irrigation systems. Automation in agriculture can be used with components such as Arduino, Roseberry-Pi, and many more efficient components (Burt, 2005). Currently, there is lack of enough place in database to store data, so cloud technology has been introduced to solve this problems, as some similar problems are solved by this cloud technology. Agriculture can be done using model technology, which is digital agriculture, and this data analysis is big data analysis (Schaible and Aillery, 2012).

Developed countries and some developing countries usually have favorable conditions for automation and modernization of irrigation systems, in particular on a broad scale (Plusquellec, 2009). Regarding the automation and modernization of irrigation systems, the following aspects can be mentioned:

- Procedure to organize water for users
- Those irrigation systems that are typically kept up with require less money to update
- Strong and environmental policies and oversight as well as the willingness and capacity to enforce water use laws
- Access to technical knowledge and resources
- Roads and other well-developed infrastructures.

There are numerous research studies, which have been done on the modernization and automation of irrigation in different countries (González-Cebollada, 2015; Lecina et al., 2010; Burt, 2005). For example, important reforms and renovations of the Spanish government, similar to Australia, began in 2002 to manage water demand. These projects were funded mainly by the tax budget, and aimed at rehabilitating the irrigation and water conservation sector. These initiatives included enhancing the irrigation water distribution infrastructure and supporting water-saving strategies like the drip system. Similar to how drip structures were modified, return systems were used, measurement and control were improved, and SCADA systems were used, contemporary irrigation systems in the United States were intended to improve the water delivery system (Burt, 2005). However, according to a US Department of Agriculture report from a year ago, at least half of US farms continue to use low-efficient irrigation techniques (Schaible and Aillery, 2012). Due to the improvement in low-cost sensors for agricultural and water monitoring, new low-cost sensors have also been surveyed in research. These comprise a leaf water-stress tracking sensor (Daskalakis et al., 2018), a multi-level soil moisture sensor (SMS) constructed of copper rings adjacent to a PVC pipe (Guruprasadh et al., 2017), a water salinity monitoring sensor made of copper coils (Parra et al., 2013), or a turbulent water sensor using for color and infrared transmitters and receivers (Sendra et al., 2013).

14.2 Advanced Irrigation Management Equipment and Techniques

In general, investment in research projects and development and advances in technology in recent decades have created new opportunities for increasing WUE in irrigation. This takes the shape of cutting-edge tools and methods as well as less-expensive and more precise alternatives. Utilizing modern technology, such as satellite-based information gathering and processing, software, and support systems, is crucial for planning and managing the limited water and soil resources. Satellite remote sensing data are utilized due to the features like wide and integrated vision, using various wavelengths of the electromagnetic spectrum to record phenomena' characteristics, repetitive coverage, speedy transmission, a wide range of data formats, and the ability to use both hardware and software. The use of computers specifically has been warmly welcomed throughout the world, and they have shown to be an effective instrument for evaluating, exploring, monitoring, controlling, and managing water and soil resources, forests and pastures, agriculture, and the environment. In the last decades, the increase in the amount of information available and the need to combine this information have led to the formation of another technique called Geographic Information Systems (GISs). GISs are the software that provide data entry, management and analysis, and output.

14.2.1 Remote Sensing (RS)

In irrigation, WUE is achieved by optimizing the time and the amount of irrigation water. Described planning methods, whether plants, soil, or meteorology (evaporation and transpiration), are commonly used on the field. These procedures often cost money, take time, and cannot be carried out automatically (Jones, 2004). Additionally, they are best suited for small spaces and should not be used in expansive spaces.

The benefits of regular measurements in a specific area and time, the ability to cover large areas, an active area of irrigation water management research, and the ability to integrate into models and GIS have made remote sensing, which is not at all a new concept in agriculture, the focus of much attention in recent years (Saadi et al., 2018). The application of remote sensing technology in various studies related to irrigation and drainage has three main categories as follows:

A. Processing of irrigation and drainage activities on a large scale
B. Identify the type and extent of crop yields and soil salinity hazards
C. Preparation of farm maps, farm boundaries, and legal aspects.

Duplicate images are required for category A. The second category is based on multi-spectrum images that are intended to monitor terrestrial resources, and high-precision images must be provided to identify small landforms in order to produce items in category C.

Each of the mentioned categories needs to be taken separately or together for applications such as land use maps, updating existing maps, agricultural studies, determining plant characteristics, determining soil texture, estimating soil organic matter content, reviewing soil salinity, measurement of soil moisture and identification of static water layers, determination of subsurface area of irrigated lands, geological studies, surface water resources studies, identification of groundwater resources, study of water pollution, study of water outflows, creation of reservoir operation model, farm management, management of large-scale irrigation and drainage projects, improving traditional irrigation systems, control soil erosion and desertification, improve irrigation systems, planning water resources management, determining the location of small dams, and snow removal.

In intelligent irrigation, there are mainly two components: the unit of measurement and the main station. The unit of measurement may be installed on the farm and the main station may be far from the unit of measurement. Data are transferred between the two blocks (Juang et al., 2007). In any irrigation method, efficient use of water and labor is very important.

Advances in communication science and computational technology, as well as the rapid growth and pervasiveness of Internet use in societies, have created an increasing demand for improved intelligent and remote control of performance systems. This has also had a significant impact on data collection systems in industry, agriculture, meteorology, and so on. Advances in speeding up data exchange, computing, and data storage have led to the ability to transfer and store measured data with higher accuracy. These advances have also made it possible to measure in different parts of the world out of reach. In the Internet of things (IOT) -based model, each sensor alone or with a hardware interface has the ability to connect to the Internet and exchange data over the Internet. Also, each displayers can only connect to the Internet and receive data from each of them. Smart greenhouses have tools and systems that have been used to improve the quantity and quality of the product and to minimize human activities. The main function of this equipment can put the climate inside the greenhouse (temperature, relative humidity, light, CO_2, etc.) in proper conditions. Creation of this climate controlled are performed by common facilities and equipment in greenhouses such as heating/cooling systems, humidification/dehumidification systems, ventilation valves, fans, additional lighting systems, covers, and carbon dioxide injection systems. As a result, by creating a suitable climate conditions for cultivation, it is provided in all seasons of the year, without any weather conditions or environmental stresses.

Wireless communication involves the transmission of information between two or more points while no electrical interface is used to make this connection. This allows data and communication to be transferred without any hassle. A wireless communication network has numerous advantages, especially the ability to move network equipment freely. It is also possible to easily add a communication device to the network or remove another communication device from the network without disrupting the rest of the system. Aside from the initial cost of setting up a wireless communication network, operating and maintenance costs are minimal. Basically, using wireless technologies in greenhouse projects, the values related to environmental parameters can be collected from all over the greenhouse and sent to a central control system. In this case, the wires that occupy the space inside the greenhouse will no longer be needed. Also, since the nature of the connection between the sensors and the control system is wireless, the central control system can be located outside the greenhouse and within the range of the wireless board. On the other hand, if this data are sent to the Internet, the distance limit will be completely eliminated for sending information from sensors to control centers or servers. Satellite imagery is used in many agricultural applications, such as operations and disease monitoring. Several algorithms have been developed in recent years to extract vegetation indices from satellite photos and combine them with terrestrial observations to predict evapotranspiration (ET) across vast regions (Rajalakshmi and Mahalakshmi, 2016; Prathibha et al., 2017).

Using Landsat's Infrared Thermal Images[1] (TIR) to obtain information on the ET spatial diversity in farm scale and consistency of water use and uniformity of water consumption to improve WUE is a step toward improving irrigation water management (Roselin and Jawahar, 2017; Agarwal and Agarwal 2017). Remote sensing is not employed in monitoring, assessing, and regulating the use of irrigation water, among other things, due to issues with temporal and geographic segregation, the quality of data, once-in-a-lifetime syndrome, and location. However, the current Landsat-8 satellite series has a spatial resolution of 30 m and can be used to accurately assess the evaporation and transpiration of a crop and to use agricultural water on a farm scale. There are several commercial satellites in operation at present such as Sentinel-2[2] and Planet[3] that may be used for agricultural purposes.

The combination of drone-collected multi-spectrum data with thermal imaging is another form of remote sensing to assess crop condition. According to research, canopy temperature may be utilized to control irrigation water since it correlates with plants' water status (Cozzolino, 2017). Reflections in the near-infrared and infrared regions of the electromagnetic spectrum have been used in applications to gauge the water condition of grains, fruit trees, vines, and pastures (Cozzolino, 2017).

With conventional techniques like soil probes or plant-based plant methodologies, it is impossible to assess the status of the aquifer in relation to geographical scales. This is where remote sensing has a major benefit. It is also anticipated that as UAV technology advances, their costs will go down, making them more affordable for many farms. To achieve economies of scale, further work must be done to integrate irrigation and remote sensors.

The collection of very high-resolution data (<10 m) using extremely sensitive sensors, such as those found on the current IKONOS and Quickbird commercial satellites, quick access to data from numerous sensors with a wide range of space, radiometric features, and multi-data synthesis are some of the key opportunities and improvements.

14.2.1.1 Application of GIS in Irrigation and Drainage

The data and parameters available in irrigation and drainage networks are very extensive due to the existence of various technical, managerial, economic, social, and environmental aspects, so that the volume of paper information related to a network from the study stage to the operation can occupy a lot of useful space from office environments. Under these circumstances, network operators are faced with a lot of cluttered and unclassified information, and sometimes this situation is distorted (lost, stolen, damaged, events) or forgotten a lot of useful information from irrigation networks. One of the most effective ways to protect, organize, access, and reduce the size of information on a CD is to use GIS software (Elbeih, 2015; Yang et al., 2011).

The GIS is able to integrate all data and spatial and descriptive information of irrigation networks in a computer environment and make it accessible to users. In GIS software environments, it is possible

Intelligent Irrigation and Automation

to create different layers of information, including images, numbers, and text, and after entering the information, part of the required information can be called as needed. These facilities can be used in irrigation and drainage networks as follows. The images section mainly includes schematic maps and locations, including items such as; Canal and Drainage Map, Farm Schematic Map, Schematic Map of Cultivated Products (Cultivation Pattern), Land Classification Map, Well Location Map, Structural Location Map, Underground Water Table Depth Map, Soil Profile, Road Network Map, Geological Map, and so on. Much of network information includes statistics, each of which can provide a layer of information in the GIS software environment, including:

- Channel information (including channel length, design flow, flow in terms of time, restoration operation, dredging volume, design specifications, etc.)
- Drainage (including discharge flow at different time periods, salinity, and other chemical parameters and design specifications)
- Wells' information, water needs, evaporation and transpiration, crop yields in different years (on each farm or the entire network)
- Irrigation efficiency (farms, canals, and networks)
- Condition of reservoirs and valves (identification details, amount of water delivered, repair date)
- Water price (including the amount of water collected in each canal, farm, agricultural block and network) and dozens of other parameters.

The last section contains detailed written information on irrigation and drainage networks, which can also be placed in different layers of information in GIS software. These layers can be a detailed description of events, actions taken, decisions made, how to solve problems, and recommendations for leading a section of the channel. The user can use or review the information of one layer or information of different layers simultaneously according to their needs.

14.2.2 Sensor and Communication Networks

For better agricultural management, sensors are tools that monitor and gather a variety of data, such as soil moisture and climate. SMSs and weather stations are typical examples of sensors used in irrigation water management. By connecting cables, equipment was traditionally used to monitor a product's condition or the amount of water in the soil. This method frequently required reading the user manual and data used to schedule following irrigations. In addition to being inaccurate, such manual techniques are time-consuming and frequently expensive when used to manage water in the future. In order to increase WUE in irrigated agriculture, wireless sensor technologies are increasingly being used.

In order to monitor numerous factors in this sector, such as soil moisture and weather information, a number of wireless sensors are often employed. This is particularly true now that technology has advanced recently and that there are more inexpensive sensors accessible because of competition. A wireless sensor network is made up of several different sensors (sensor nodes), a communication system, and a sink node or ball node for processing and receiving data from sensor nodes (Rehman et al., 2014). Actuators in sensor networks may also be utilized to automate the irrigation system. In various literature, including Rahman et al. (2014), wireless communication technologies utilized for agricultural objectives, including water management, have been examined. ZigBee, Bluetooth, WiFi, General Packet Radio Service (GPRS)/3G/4G, LoRa Radio, and SigFox are the current communication technologies utilized in agriculture. Due to its extensive range, low cost, energy economy, and dependability, ZigBee technology is typically utilized in irrigation water management (Jawad et al., 2017; Rehman et al., 2014; Ojha et al., 2015).

14.2.2.1 ZigBee Communication Protocol

One of the most important wireless technologies that can be used in communication projects is the ZigBee communication protocol, which was developed by the Institute of Electrical and Electronics

Engineers (IEEE) in 2003 according to the IEEE 802.15.4 standard (Ponce et al., 2015). Depending on the power consumption and environmental characteristics of this method, the range of information exchange varies between 10 and 100 m. Devices equipped with the ZigBee communication protocol can also transmit data to distant devices by passing through graded networks and intermediaries. ZigBee, in its simplest form, is a wireless technology created as a free (open) worldwide standard to address the particular requirements of low-cost, low-cost M2M wireless networks. The IEEE 802.15.4 Physical Radio Specification of the ZigBee standard allows it to operate in unlicensed bands like 2.4 GHz, 900 MHz, and 868 MHz. The radio package used in low-power and battery-powered devices is specified by this standard. The protocol enables devices to interact over a broad range of diverse network topologies, with battery life that can last for many years.

14.2.2.2 XBee Communication Protocol

XBee is named as Digi International for families of ZigBee-compatible radio modules (Ponce et al., 2015). The first XBee radio-wave protocol was introduced in 2005 under the brand name MaxStream, which is based on the 802.15.4.2003 standard and was used in point-to-point and star communications with data transfer speeds in excess of 250 kbit/s. Initially, the XBee's 1-MW model was offered at a low price, and the 100-MW XBee-PRO with a long range. After their unveiling, a number of new XBee radio devices have also been introduced to the market, all of which are now known and marketed under the Digi brand. All XBee radios have a minimum number of connections, a working voltage of 3.3V, a direct connection to ground, a data input and output (Universal asynchronous receiver-transmitter (UART)), a sleep mode to reduce battery power and restart. In addition, most XBee families have other varied flow controls, including I/O, A/D, and internal indicator lines. A programmable version of XBee has an additional onboard processor for the user code.

14.2.3 Artificial Intelligence

Science and technology have made a significant leap in human life over the years. The development of technology in the modern digital world has led to a branch of computer science has received attention with the main goal of producing smart machines with the ability to perform tasks that require human intelligence in various fields such as medicine, education, agriculture, industry and many other sciences. AI is actually a simulation of human intelligence for a computer, and it is actually a machine that is programmed to think like a human being and has the ability to mimic human behavior.

The term AI was coined by John McCarthy in 1956. The implementation involves the process of machine learning, which is driven by the "machine learning" domain. Machine learning serves the only aim of providing a device with historical data and statistical information to enable it to carry out its given work and resolve a particular problem. There are several applications available now that analyze historical data and experience recognize voice and faces, forecast the weather, and diagnose medical conditions. With the development of AI scientists, there is a critical need to find a system that can be utilized to train a device. There is no method to solve large issues with the aid of simpler algorithms prior to mastering the machine, except from the most fundamental problem-solving strategies. The discipline of big data and data science has advanced significantly as a result of machine learning. In order to create intelligent machines, machine learning is a mathematical strategy. Crop contamination, improper storage management, the misuse of pesticides, weed management, and a lack of irrigation and drainage systems are just a few of the sectors where AI and machine learning have invaded.

In conventional irrigation methods, the main problems are water wastage and water shortage. Much research has been done to address the irrigation process' issues. IISs based on sensors have been developed by several firms. These technologies were created to monitor water contamination, use water efficiently, and address certain other major issues. Without the involvement of farmers, sensors measuring soil moisture and temperature communicate directly with agricultural equipment to distribute

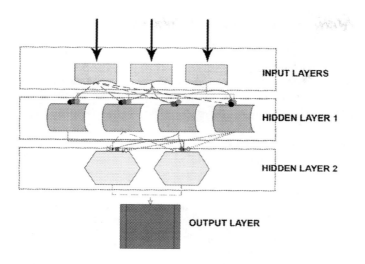

FIGURE 14.2 Layers of artificial neural networks (Jha et al., 2019).

the necessary water to crops. Smart irrigation or any other popular technique must use high-quality water to provide fields. Separately, Bannerjee et al. (2018) have suggested improvements in AI categorization and a quick overview of several AI methodologies. From 1983 onward, computers and technology started to permeate the industry. Since then, a variety of ideas and techniques have been put up to enhance agriculture, from the database to the decision-making procedure. AI has led to the development of new techniques that simplify the process of addressing problems. The following approaches are listed by Jha et al. (2019):

1. Fuzzy logic
2. Artificial neural network (ANN)
3. Fuzzy-neural logic
4. Special systems.

ANN is the technique that is most frequently and consistently utilized for research. Artificial neurons are a collection of interconnected units or nodes that function similarly to biological neurons in an animal's brain. Each neuron-to-neuron link (synapse) has the ability to convey a signal. The signal(s) can be processed first by the signaling neurons coupled to the receiving neuron (synaptic post). Three layers make up the ANN architecture: input, hidden, and output layers (Figure 14.2). Neurons are often arranged in layers. Different layers may convert their input in different ways. Signals go through many layers as they move from the first layer (input) to the last layer (output).

14.3 Intelligent Irrigation Systems

Smart irrigation systems regulate irrigation planning without human intervention, in accordance with the climate and soil moisture. These programs and controllers significantly improve the efficiency of water use in irrigation, both outdoors and indoors, such as greenhouses. Intelligent irrigation controllers automatically modify the watering program to the real circumstances of the agriculture, in contrast to traditional irrigation controllers, which work according to a standard and predefined timetable. For example, by increasing the temperature in the field or reducing rainfall, intelligent irrigation controllers adjust the sprinkler schedule according to their specific variables, such as soil type, sprinkler function, humidity of soil, air relative humidity, and so on. IISs are often split into two categories: soil-based IISs and IISs based on climate.

14.3.1 Climate-Based Intelligent Irrigation System

Climate-based controllers actually adjust irrigation programs based on local climatic conditions. These systems collect weather information and even some of them in a specific area (types of plants, soils, slopes, etc.) to adjust the timing of irrigation so that the area is always in the right amount of the water. There are various products in this group with different weather input options and special adjustment factors. Water saving can be significant, and the convenience of these self-regulating controllers is another advantage.

Evapotranspiration controllers are another name for climate-based controls (ETs). Evaporation and transpiration is the sum of soil surface evaporation and plant transpiration. These controllers gather meteorological data and modify when to irrigate the plants so that they receive an adequate amount of water. Meteorological data uses four climatic parameters including temperature, wind, sunlight, and humidity. This is the most accurate way to calculate the amount of agricultural land water needs. The process of evaporation and transpiration is essential for keeping the hydrological cycle stable, using sustainable irrigation techniques, and managing water. Area altitude; mean, maximum, and lowest daily temperatures; wind speed; relative humidity; sunny hours; daylight hours; and latitude are effective variables on plant evaporation and transpiration. There are more than 20 different ways to calculate ET, and they all depend on different factors.

14.3.1.1 Irrigation Planning

The balance of water flowing into and out of the root zone, also known as soil water balance, can be used to estimate a plant's need for water (Figure 14.3). The root zone receives input from irrigation and rainfall. The plants' requirement for water can also be met by the shallow-water table via capillary action. Runoff, deep percolation, evaporation, and transpiration are all considered outputs from the soil water balance since they allow water to leave the soil and plant system. Water is lost by transpiration and evaporation from the soil's surface into the atmosphere, respectively (Allen et al., 1998). Evaporation

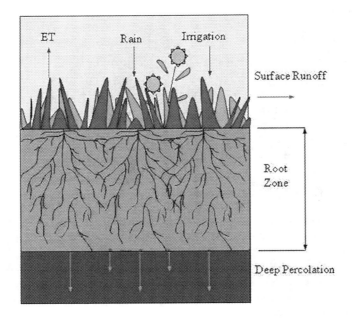

FIGURE 14.3 Water-based elements emerge in a plant's root zone under well-drained circumstances with no shallow-water table (Dukes et al., 2009a).

and transpiration are integrated into a single quantity known as evapotranspiration for determining the soil water balance (ET).

Calculating the change in soil moisture in the root zone requires Eq. (14.3) (Anonymous, 2005). In general, in common sandy soils, it can be assumed that surface runoff is negligible, unless the soil is very compacted or has additional characteristics that prevent water penetration. Also, irrigation is planned in such a way that, ideally, there is little damage. Irrigation events that do not exceed the soil's water-holding capacity limit the basin's infiltration, whereas surface runoff from irrigation events is insufficient to permeate the soil but instead runs off (i.e. cycles/soaks). In sandy soils, changes in storage between irrigation intervals are often insignificant. These presumptions add Eq. (14.3) to the equation that determines the necessary irrigation depth as follows:

$$\Delta S = R - ET_C + I - D - RO, \tag{14.3}$$

where:
ΔS = Variation of soil water storage (in)
R = Rainfall (in)
ET_C = Crop evapotranspiration (in)
I = Net irrigation (in)
D = Deep percolation (in)
RO = Surface runoff (in).

The soil water balance equation, often known as Eq. (14.3), is used to balance the change in soil water storage in a plant's root zone. The simplified version of Eq. (14.3), which is used to determine the necessary net irrigation depth under the assumption of minimal drainage, runoff, and storage change, is Eq. (14.4).

$$I = ET_C - R_E, \tag{14.4}$$

where
R_E = the effective rainfall within the root zone (in)

Effective precipitation is how much falls in the root zone (Anonymous, 2005). Soil water storage capacity is defined as soil water capacity (Anonymous, 2005). Rainfall that surpasses the storage capacity of soil water in the root zone is thought to be exhausted or destroyed and is no longer useful to the plant. Therefore, how much water can soil store across all soil types?

14.3.1.2 Calculating the Water Content of Soil

The portion of soil that is above ground in a plant's root zone that it may use for physiological functions is called the root zone. The texture of the soil affects how much water can be retained in the root zone. The level of water in the root zone that permanently wilts plants without allowing them to recover is known as the permanent wilting point (PWP) of soil (Anonymous, 2005; Figure 14.4). Field capacity (FC) is also described as the water level when the pace of downward movement in the root zone is greatly decreased as a result of post-saturation gravity (Anonymous, 2005). Irrigation should, in theory, be completed before PWP and FC fills. PWP and FC differ according to the soil's texture, where more sand makes the soil less water-retentive while more moisture makes the soil more water-retentive.

The amount of water that is readily accessible to plants for usage should fall between FC and PWP, and is referred to as accessible water (AW) in accordance with the aforementioned parameters (Anonymous, 2005; Eq. 14.5). To avoid stressing plants, AW shouldn't go to PWP before watering planning. When the AW level reaches the maximum permitted discharge, irrigation should be used (maximum allowable depletion; Anonymous, 2005). If there is no specific information available, 50% is usually a general rule. RAW refers to the volume of water that may be utilized for irrigation prior to other uses and is readily accessible (Anonymous, 2005; Eq. 14.6). Through the ET_C, water is gradually drained from the root

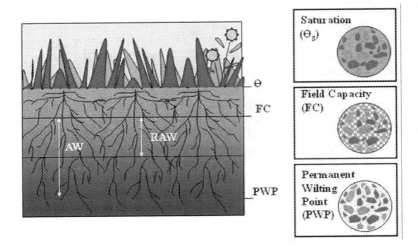

FIGURE 14.4 Water content diagram in the root zone, showing (Dukes et al., 2009a) that FC is the amount of water that remains in soil pores after gravity drainage has stopped. Permanent wilting point is reached when there are only thin films of water left that are unavailable to plants. Saturation is the point at which all soil pore space is filled with water. The entire quantity of water stored by the soil for plant usage is known as accessible water (AW), and the amount of water that is immediately usable by plants (RAW) prior to stress is known as rapidly available water (RAW).

canal. Effective rainfall is taken into account while calculating the daily ET_C numbers as long as it is not equal to or higher than RAW. Irrigation should be carried out once RAW is finished in order to fill the soil reservoir and maximize farm productivity.

$$AW = \frac{(FC - PWP) \times RZ}{100} \tag{14.5}$$

$$RAW = MAD \times AW \tag{14.6}$$

where:

RAW = Rapidly available water (in)
MAD = Maximum allowable depletion.

14.3.1.3 Calculation of Evaporation–Transpiration

Grass at a height of 0.12 m that is appropriately watered, actively growing, completely covers the soil, and has a consistent surface resistance that is used to calculate reference evaporation (ET_o), which is characterized as ET from the reference surface (Allen et al., 2005). When calculating ET, the ASCE reference standard evapotranspiration equation is typically utilized, and Eq. (14.7)—the ASCE standardized reference evapotranspiration equation—is frequently used to compute ET_os (Allen et al., 2005). This equation, which is dependent on wind speed, temperature, relative humidity, and solar radiation, is used to determine the daily ET_o.

$$ET_o = \frac{0.408\Delta(R_n - G) + \gamma \frac{C_n}{T + 273}(e_s - e_a)u_2}{\Delta + \gamma(1 + C_d u_2)} \tag{14.7}$$

where:

ET_o = Reference ET (mm/day)
R_n = Net radiation (MJ/m²/day)
G = Heat flux (MJ/m²/day)
u_2 = Wind speed (m/s)
T = Temperature (°C)
Δ = Vapor pressure (kPa/°C)
γ = Psychometric constant (kPa/°C)
e_s = Saturation vapor pressure (kPa)
e_a = Actual vapor pressure (kPa)
C_n = Constant (900)
C_d = Constant (0.34).

With the use of a crop coefficient, Eq. (14.8) may be used to determine the ET loss for a particular crop or plant based on reference ET.

$$ET_C = K_C \times ET_o \qquad (14.8)$$

where:

ET_C = Crop evapotranspiration (in/day)
K_C = Crop coefficient.

The variables used in the ASCE standardized reference evapotranspiration equation (Allen et al., 2005). Crop ET (ET_C) is defined as ET usable for a particular plant other than the reference product. ET_C can be calculated using the crop coefficient (K_C) for a particular plant material. Depending on the particular crop type, horticulture techniques, and geographic area, crop coefficients are available in a variety of resources.

14.3.1.4 Evapotranspiration (ET) Controllers

ET controllers are irrigation planning tools that schedule irrigation based on the principles of soil water balance. Evapotranspiration controllers (ET) are another name for climate-based controls (Figure 14.5).

There are three main categories of ET controllers:

1. *Signal-based controllers*: In order to gather meteorological data, either the Meteorological Station Network or the general public must reach an agreement. The ET value for the hypothetical grass level is calculated for that site. The ET data are then sent to nearby controllers via a wireless connection. In some cases, the ET values are set for controllers that are not close to the meteorological

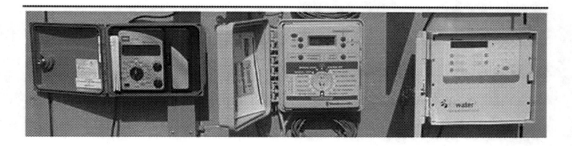

FIGURE 14.5 Three ET controller types are being assessed at the University of Florida's Department of Agricultural and Biological Engineering.

data collection point. Throughout the year, the ET controller modifies the number of irrigation days or hours based on the weather.

2. *Standalone controllers*: This method employs a pre-programmed water use curve for various locations using ET controllers. A sensor that detects the local meteorological conditions, such as a temperature or sun radiation sensor, may change the curve.
3. *Climate measurement on site*: The continuous ET is calculated using observed air data in the controller, and the watering schedule is modified in accordance with the weather.

14.3.1.5 Signal-Based Controllers

The ET_o data are received by these controllers by wired (telephone) or wireless (mobile or pagination) transmission. Local public or private weather stations that are close to the controller's location are used to acquire climate data. Some producers gather weather data from meteorological stations, compute an ET_o daily amount, and then feed the daily computed amount straight to the controller each day. The ET_o controller may be computed by other manufacturers after distributing weather information from meteorological stations. Depending on the type of plant used, ET_C is determined by using the product coefficients and ET_o. These controllers have the benefit of being configured in response to actual weather conditions. The drawback of this strategy is that the weather at the data source for the meteorological station could not be indicative of the circumstances at the control point. Particularly, because rainfall may supply the majority of the water required for the plant, special rainfall in various places is particularly significant. For computations of the soil water balance to be correct, the signal must be appropriate. Prior to the controller receiving the signal again, certain ET controllers make use of past data. Every day that the controller is silent, others utilize the most recent ET_O transmission. Signal-based controllers often permit the use of an external antenna when the device's antenna is insufficient.

14.3.1.6 Standalone Controllers

These controllers gather data from on-site sensors that measure the weather to determine the ET_O in real time. A daily ET_O is derived from the data collected by the sensors at intervals of 1 second to 15 minutes. Temperature, sun radiation, and even the whole range of weather station characteristics are examples of on-site sensors (Riley, 2005). However, it is neither practicable nor cost-effective to build meteorological stations wherever. Thus, straightforward ET estimate techniques are frequently employed. The Weathermatic Smartline controller, for instance, substitutes the Hargreaves equation for the traditional ET_O ASCE equation. The sensor can only measure temperature since the Hargreaves equation depends on temperature (Jensen et al., 1990). The benefit of this approach is that the ET is measured on the spot and no further signal expense is needed. Simple approaches are inaccurate in a wide range of climatic circumstances, which is a drawback of this strategy (Jensen et al., 1990).

14.3.1.7 Add-On ET Controllers

Some ET controllers, like Rain Bird ET Manager, are just capable of adding devices to the automated timer; they do not have the capacity to determine the execution time. Instead, they gauge whether an irrigation event will take place using the soil water balance. Rain Bird ET Manager is a software program to help development of the right irrigation program according to the specific conditions of the place. The software adds the timer and inserts the ET manager's water depth for each irrigation event that was estimated by the program. When deciding when to cease irrigation, the ET management considers depth input together with rainfall and daily ET.

14.3.2 Intelligent Irrigation System Based on the Soil Moisture Sensors

A suitable technique is employed by IISs associated with SMSs (Figure 14.6). When buried in the root zone of plants, grasses, trees, or shrubs, sensors precisely monitor the soil's moisture level and

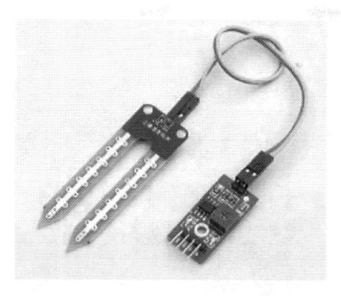

FIGURE 14.6 Soil moisture sensor.

subsequently send this data to the controllers. With the SMS controller, two different types of control techniques are employed: "bypass" and "on-demand."

14.3.2.1 Bypass Irrigation System

Bypass configuration is usually used for small areas. The soil moisture threshold is often changed by a bypass SMS controller from "dry" to "wet". The point at which the irrigation system permits water to be customized to the unique demands of the plant, soil, and dust may be lowered or rose using this threshold. If this current soil moisture level is more than the adjustable threshold, this type of controller will pass the irrigation events on time. A rain sensor and the bypass mode are extremely similar. Most of these SMS controllers are added to the watch (Figure 14.7). Many of these systems have just one SMS, which must be buried in the driest irrigation region and the timing of other areas altered to minimize surplus water. Multi-sensor controllers allow the sensor to be installed in any irrigation area. In this method, the controllers are set up like the traditional method, which has a start and end time for irrigation. The advantage of this system is that if there is enough moisture in the soil, it stops the next planned irrigation.

14.3.2.2 On-Demand SMS Irrigation Controller

An on-demand SMS irrigation controller is defined as a low and high threshold for controllers, which can be started if there is not enough moisture in the soil and below the required minimum level. In fact, it starts irrigation on the low-moisture threshold of pre-planned soil and ends the irrigation on the upper threshold. This kind of controller is frequently utilized in situations requiring a high degree of customization or control, such as commercial sites or other sorts of sites with sizable irrigation fields. The bypass controller only permits irrigation events (i.e., day of the week, time of day, and time of execution) to be selected from 1 hour of time, whereas the controller starts and stops irrigation events. Therefore, over time, careful program planning is crucial.

14.3.2.3 Soil Humidity Sensors Function

The majority of SMSs are made to calculate the bulk water content of the soil using the bulk permittivity of the soil. The capacity of soil to carry electricity may be thought of as its dielectric constant. The soil's

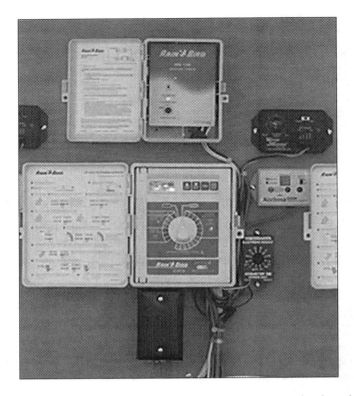

FIGURE 14.7 Four distinct types of controllers for soil moisture sensors equipped with an electronic timer.

dielectric constant rises as soil moisture content does. This reaction results from the fact that water has a substantially higher dielectric constant than other soil constituents, such as air. Therefore, a constant dielectric measurement predicts water content. Soil moisture irrigation controllers use a sensor type after water information and allow or not to irrigate based on specified time periods (Figures 14.8 and 14.9). A user-adjustable threshold setting on the SMS controller prevents the program from running if the soil water content is higher than that value. The user determines the threshold for soil water content. With SMS devices, there is also an "on-demand" control approach in which the controller initiates irrigation at the lower threshold and terminates it at the upper threshold.

The basic connection between an SMS and an autonomous watering system is depicted in a straightforward schematic in Figure 14.8. A hot wire and a common wire are used to link the irrigation timer to an electric valve. The SMS system is spliced into the common cable. In this system, a controller serves as a switch, and a sensor that determines the amount of water in the soil is buried in the root zone. The SMS controller utilizes the data from the SMS system's reading of the quantity of water in the soil to turn on or off the switch. The controller shuts off the switch when the soil water content falls below the user-set threshold, enabling the timer to get to the irrigation valve and begin watering. In this illustration, the controller turns on the irrigation when rain causes the soil near the SMS to get wet. Due to the dry conditions in the soil near the SMS in Figure 14.9, the controller shuts the switch that activates the irrigation system.

14.3.2.4 Sensor Installation

A unique sensor can be employed to regulate irrigation in several regions if the irrigation area is established by a solenoid valve, or multiple sensors can be used to control irrigation in various areas individually. When employing a sensor in many locations, it must be placed in the region that is typically the

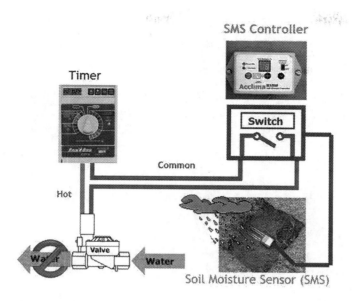

FIGURE 14.8 A simple diagram of how to connect a soil moisture sensor (SMS) to an automatic irrigation system (Dukes et al., 2009b).

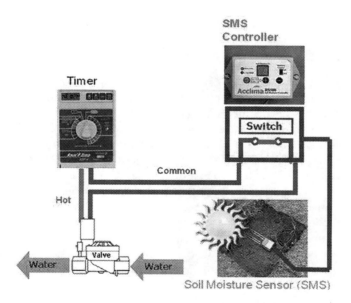

FIGURE 14.9 The operation of the switch controller (Dukes et al., 2009b).

driest or where irrigation requires the most water. Every section has a specific location for the sensor to guarantee optimum watering. The following are a few general guidelines for burying SMSs:

- The irrigation area's soil should be represented by the soil in the burial area.
- For irrigation, sensors should be buried in the root zone of plants since this is where the plants draw their water from. Burial in the root zone contributes to the quality of a lawn or landscaping. The sensor should typically be placed three inches deep in turf grass.

- The sensors need to have solid soil contact after burial. The air around the sensor must be impenetrable. The sensor shouldn't be too far from the dirt, but it should be firmly hammered.
- To make sure that all areas are properly irrigated, a sensor that will regulate the complete irrigation system must first be buried in the necessary water area. This region will often be the sunniest or have the greatest sunlight.
- Sensors should be positioned 3 ft from a planted bed area and at least 5 ft away from the dwelling, property border, or any other impervious surface (such as a driveway).
- The sensors must also be placed in the center of the irrigation area, at least 5 ft away from the irrigation heads.
- The sensors shouldn't be buried in busy locations to prevent excessive soil compaction around the sensor.

The sensor has to be calibrated or the soil water content threshold must be chosen once the sensor is buried and the SMS controller is linked to the irrigation system. To calibrate or regulate the SMS's threshold, take the following actions:

Step 1: Water should be distributed to the spot where the sensor is installed. In any event, set the irrigation area to consume at least 1 inch of water, or apply straight to the buried sensor using a 5 gallon bucket.

Step 2: Suspend irrigation for a day. After 24 hours of rain, the procedure should be repeated. The sensor threshold that determines whether to accept or disallow scheduled watering activities is now the water content after 24 hours. The landscape will still need to be closely watched to make sure that enough irrigation is being provided, even if this threshold is somewhat lowered (by around 20%) to allow for additional storage for rainfall.

14.4 Applications of Intelligent Methods in Irrigation

The number of studies targeted at reducing water use during irrigation has grown over time. Farmers with limited financial resources cannot deploy this sort of system since conventional commercial sensors are prohibitively costly for agricultural irrigation systems. The ability to link inexpensive sensors to nodes, however, has enabled manufacturers to create low-cost irrigation management and agricultural monitoring systems. Summary of research on IISs in this section is provided, given the recent advances in various intelligent technologies and remote control, such as IOT and Wireless sensor networks (WSN), as well as AI that can be used in the development of these systems.

Using neural networks, Robinson and Mort (1997) created a prediction model. This model received raw data for the first time, including humidity, temperature, precipitation, cloud cover, and wind direction. The data were collected between 1980 and 1983. Over a 3-year period, two special systems were created to increase cotton production. The first one was the COMAX (COtton MAnagement eXpert) system, developed in 1986 for continuous year-round activity on cotton farms based on three parameters, including irrigation planning, maintaining nitrogen levels on the farm, and growing cotton. The second special system was COTFLEX, which was introduced by Stone and Toman (1989) for cotton products. The system is built on the P Pyramid 90 computer that uses UNIX as the operating system. The system uses a farm database to provide important information about cotton products to the farmer so there is a possibility to make important and tactical decisions for farmers. Maier and Dandy (2000) used neural networks to predict water resource variables.

An in-depth study was conducted in the Dehrodun Valley of India to assess the importance of adding ANN by several ET estimation methods. Evaporation has a crucial role in irrigation and water management; therefore, this study showed that if the ANN structure is implemented properly, it will show appropriate predictions (Nema et al., 2017). Hinnell et al. (2010) presented studies on neuro-drip irrigation systems in which ANNs were applied to predict the distribution of subsurface spatial water distribution. For the proper functioning of the drip irrigation method, the distribution of water at low soil levels is of great importance. ANNs make a prediction that is useful to the user, which in turn leads to a quick decision-making process. Neural network models provide wetting patterns after water penetrates the soil.

In one study in Ankara, Turkey, IIS found achievements such as lower humidity and soil temperature stress, efficient water consumption, and elimination of human intervention in the event of flood irrigation. The developed system works in three units. The base unit (BU), the valve unit (VU), and the sensor unit (SU) of this system is powered by solar panels. After successfully installing each unit, BU sends the address to which the data are to be sent to SU. The SU sensors sense moisture and send the identified data to a specific address in the BU. If necessary, the BU sends back a signal to the VU to calibrate the valve position to supply soil water. Although the special use of the automatic irrigation system was introduced in the early 21st century; this method was very successful because it reduces the cost, feasibility, and complexity of the developed system. In addition, it is possible to create a unit that carries pesticides and herbicides on the farm using the same method. For this purpose, new sensors must be calibrated to transmit accurate information (Dursun and Ozden, 2011).

Research has been conducted to test ET-based technology, Information and communications technology (ICT)-based technology, and Internet Information Services (IIS) -based technology. In Riyadh, research was conducted on wheat and tomatoes, in which two methods of sprinkler and drip irrigation were tested using ICT and IIS. The water depth diagram was plotted against the product growth period (weekly) for each method. It is very desirable, based on the results obtained by IIS on water consumption compared to ICT and ET-based system (Al-Ghobari and Mohammad, 2011).

An automated irrigation system was introduced by Gutiérrez et al. (2014) that used the GPRS module as a means of communication. The system is programmed in a microprocessor-based gateway that controls the amount of water. Water storage has been shown to be 90% higher than in conventional irrigation systems (Gutiérrez et al., 2014). Kim et al. (2008) used a wireless network distributed to measure and control the irrigation process from a remote location (Kim et al., 2008).

In a study done by Gondchawar and Kawitkar (2016), For duties like weeding, spraying, moisture monitoring, frightening birds and animals, keeping awareness, and intelligent irrigation with intelligent control were offered based on real-time field data, a GPS-based remote control robot was made available. Any smart device or remote computer with Internet access may control all of these activities. The operations are carried out via the sensor interface, Wi-Fi or ZigBee modules, the camera, and the stimulus with the microcontroller and Raspberry Pi (Gondchawar and Kawitkar, 2016).

Thermal imaging is a non-invasive, uncontrolled method that examines farm surface temperature and gives the farmer insightful feedback. Cloud-based thermal imaging was employed in a study by Roopaei et al. (2017) to assist with irrigation and identify the farm region that need the most water.

The use of cloud-based assistance and decision-making in agriculture is expanding currently. Farmers may manage all applications through their online portals with the assistance of the Decision Support and Automation System (DSAS). The multi-step DSAS technique may link several devices at once and provide the farmer with real-time data. The farmer has a crucial role since they may utilize software to operate the entire gadget and examine real-time data. A certain amount of pesticides are sprayed on the farm via systems like spray controllers. Similar to how the fertilizer controller manages the fertilizer, the irrigation controller aids in managing irrigation. Data gathered by numerous sensors, including nitrogen and SMSs, are used by DSAS to operate (Tan, 2016).

Kumar (2014) created a drip irrigation system based on a wireless sensor and used the fertilizer and pH data to determine the proportion of soil elements to be eliminated. The IC 89c52 microcontroller was utilized by Ingale and Kasat (2012) to create an IIS. The prototype uses some water and only produces when the air humidity falls below a certain preset value.

A semi-automatic irrigation system for okra (*Abelmoschus esculentus*) was developed and tested. This system uses four humidity sensors and a PIC16F877A processor. The valves in the system are activated only when the voltage drop across both sensors is below a constant value and stays on, until the value reaches the set threshold value (Soorya et al., 2013). Using an IIS, water consumption can be reduced by up to 20% (Gupta et al., 2016).

Wall and King (2004) used a smart system to control sprinkler valves with the help of temperature and humidity sensors located on the ground. However, this system does not consider the problem of water pollution. Miranda et al. (2003) presented a distributed irrigation system for measuring soil water. M2M (machine-to-machine) technology allows devices to interact independently and store data directly on a cloud-based server online. This M2M technology is in its infancy and is developing rapidly. Shekhar et al. (2017) developed a technology that allows machines to communicate on their own. Yang et al. (2007) also developed a fully self-organized sensor-based intensive irrigation method. The system is made up of bottom and top layers. Pawar et al. (2018) provided an example of a small-scale IIS. Savitha and UmaMaheshwari (2018) considered automation and IOT for an IIS.

Nhamo et al. (2020) investigated the use of groundwater in agricultural irrigation using multi-dimensional remote sensing in the Vanda-Guzancolo area of Limpopo Province in South Africa using evaporation and transpiration of plants and the irrigation area resulting from the Normalized Difference Vegetation Index (NDVI) vegetation index. Evaporation and transpiration data from WP were obtained through open access of remotely sensed actual evapotranspiration and interception (WaPOR) dataset (250 m resolution) and irrigated areas using the Landscape NDVI data from Landsat 8. The results of this study can be used as a guide to support groundwater management in terms of planning and licensing of groundwater.

Thakur et al. (2020) used the Internet to evaluate intelligent irrigation and disturbance detection objects. In this study, they provided a cost-effective and reliable field for irrigating farms when water was needed and provided information on how to detect any disturbances in agricultural fields. This information was sent to farmers using a cloud program. The performance of this system was measured based on the diagnosis of soil disturbance and erosion for irrigation. The diagram of this system can be seen in Figure 14.10.

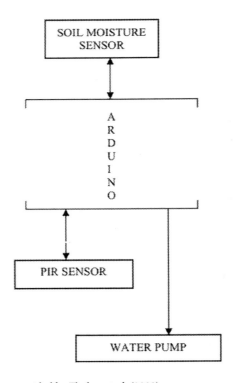

FIGURE 14.10 System diagram provided by Thakur et al. (2020).

14.4.1 Application of Renewable Energy in IIS

The study, presented by Sudharshan et al. (2019), includes an idea of an IIS based on renewable energy. The irrigation system uses three sensors: temperature, air humidity, and soil moisture, and fuzzy logic is used to operate the solenoid valve. The data collected by the sensors are then sent to the cloud using adafruit.io, and the farmer can observe the level of humidity, relative humidity, and temperature recorded by the sensors (Figure 14.11). All operations are handled by the Arduino, and the Arduino's power supply is powered by a solar panel that uses light-dependent resistance and turns it into an automatic tracking system. Crop production was cultivated in two tubs and the growth of the plant was compared with automatic irrigation and normal irrigation (Figure 14.12). The autonomous IIS has been successfully built, and it can assess the soil conditions and determine whether or not to water the farm using fuzzy logic and a variety of sensors, including the Digital Humidity and Temperature (DHT) sensor and SMSs. Solar energy was utilized to power the Arduino and the sensors, so the system could be installed without the usage of electricity. This system's design takes into account the problems with conventional irrigation systems.

Joaquin Gutiérrez et al. (2014) proposed an irrigation system that uses a photovoltaic solar panel for the electrical system due to the high cost of electricity. A program has been developed to use water efficiently and optimally, an algorithm with temperature threshold value and soil moisture that is programmed to a microcontroller gate (Gutiérrez et al., 2014).

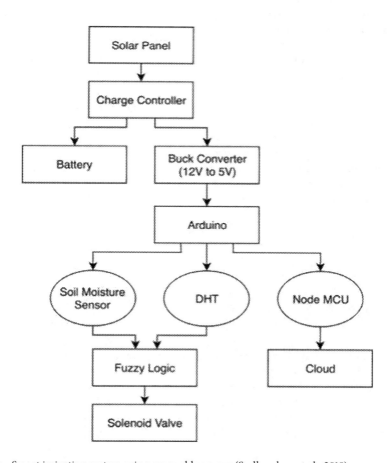

FIGURE 14.11 Smart irrigation system using renewable energy (Sudharshan et al., 2019).

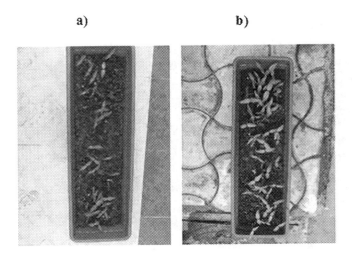

FIGURE 14.12 Crops cultivated using (a) conventional and (b) smart irrigation systems (Sudharshan et al., 2019).

A technique that enables farmers to water farms effectively using an automated irrigation system based on soil moisture has been presented by Parameswaran and Sivaprasath (2016). A humidity sensor measures soil moisture, and a microprocessor controls the solenoid valve based on that information. Using a computer, irrigation status is added to the server or local (Parameswaran and Sivaprasath, 2016).

A method for monitoring and managing soil moisture and temperature within the greenhouse was created by Akubattin et al. (2016). The Raspberry Pi-powered device determines whether to water the plant after detecting soil moisture.

A system based on a microcontroller, wireless radio-frequency (RF) module, and data transmission unit (DTU) was created by Zhang et al. (2015). The expansion capability is increased when an RF module is used in the acquisition terminal. In this system, the relay station puts together the soil moisture data, which is subsequently sent to the monitoring center by the DTU through the GPRS network.

In order to regulate water delivery in low-water locations, Kumar et al. (2014) introduced a smart system that makes use of a SMS. The theory behind how this sensor operates is to alter the moisture-dependent resistance between two sites in the soil.

14.5 Conclusions

Automation in agriculture is essential and it can be done in many ways. Automation is the most important and essential thing for optimal water use. In order to irrigate the field when the moisture level falls below a threshold, the soil humidity sensor assists in determining the soil moisture levels. A portable device that monitors agricultural water levels without involving humans was made possible by embedded systems and the Internet of Things. There are various techniques that can be implemented through various forms such as the use of machine learning, AI, deep learning, neural network, fuzzy logic, and automation. Each of these main techniques is intended to minimize the need for human interaction and effort. Although each of these approaches has pros and cons of its own, how they are used sets them apart from one another.

GIS and RS have penetrated almost all branches of irrigation and drainage and are widely used around the world. These two tools have created great features in terms of timely data collection and storage and processing. Harvesting terrestrial data and conducting field studies are very costly and time-consuming. So, sometimes, due to the difficulty of the work or the lack of the necessary financial support, such a time gap is created between conducting studies and implementing projects that other projects do not have economic justification or even changes to the design require the data to be retrieved. On the other hand, the information obtained is usually scattered and not available at once. Therefore, efficient tools are needed to store this information and process it, both statistically and in terms of answering the questions.

Developing the application of GIS and RS in irrigation and drainage and using satellite imagery will reduce both the time interval between field studies and project implementation, and the information will be integrated in one place. Also, access to information will be easier due to the ability of GIS to answer conditional questions.

In short, a smart controller is a "smart" soil moisture meter because of the feedback received from the irrigation system. The watering schedule is then modified to meet the demands of the plants using this feedback. A "dumb" irrigation clock, on the other hand, only administers water at the predetermined time and date. There hasn't been much information accessible until lately, despite the idea of intelligent controllers appearing to have the ability to cut wasteful water use while retaining the quality of the environment.

The possibility for increased water storage has also been demonstrated through sophisticated real-time irrigation planning, management, and optimization. The irrigation business is beginning to deploy remote sensing networks and communication sensors, which is anticipated to increase WUE. But developing the WUE presents difficulties. These include a lack of public support, particularly when the techniques employed are inefficient and don't usually encourage the adoption of new technology.

This research has highlighted that adopting water-saving techniques has optimized water use at the farm size. However, it's not always possible to reduce clean water usage on a basin scale.

Notes

1 https://lta.cr.usgs.gov/L8
2 https://sentinel.esa.int/web/sentinel/missions/sentinel-2
3 https://www.planet.com/markets/monitoring- There are for-precision-agriculture

References

Agarwal S., Agarwal N. 2017. An algorithm based low cost automated system for irrigation with soil moisture sensor. In International Conference on Computer Communication and Informatics, Coimbatore, India (ICCCI), pp. 1–5. IEEE.

Akubattin V.L., Bansode A.P., Ambre T., Kachroo A., SaiPrasad P. 2016. Smart irrigation system. *International Journal of Scientific Research in Science and Technology*, 2(5): 343–345.

Al-Ghobari H.M., Mohammad F.S. 2011. Intelligent irrigation performance: evaluation and quantifying its ability for conserving water in arid region. *Applied Water Science*, 1: 73–83.

Allen R.G., Pereira L.S., Raes D., Smith M. 1998. Crop evapotranspiration – Guidelines for computing crop water requirements. FAO Irrigation and Drainage Paper 56, Italy.

Allen R.G., Walter I.A., Elliot R., Howell T., Itenfisu D., Jensen M. 2005. The ASCE standardized reference evapotranspiration equation. American Society of Civil Engineers Environmental and Water Resource Institute (ASCE-EWRI). 59.

Anonymous. 2005. Irrigation Association [IA], Landscape Irrigation Scheduling and Water Management. Available at: http://www.irrigation.org/gov/pdf/IA_LISWM_MARCH_2005.pdf (accessed January 12, 2007).

Bannerjee G., Sarkar U., Das S., Ghosh I. 2018. Artificial intelligence in agriculture: A literature survey. *International Journal of Scientific Research in Computer Science Applications and Management Studies*, 7(3): 1–6.

Burt C. 2005. Irrigation district modernization in the US and worldwide. In *Irrigation Systems Performance*. Options Méditerranéennes: Série B; Lamaddalena, N., Lebdi, F., Todorovic, M., Bogliotti, C., Eds., CIHEAM, Paris, France, 52: 23–29.

Cozzolino D. 2017. The role of near-infrared sensors to measure water relationships in crops and plants. *Applied Spectroscopy Reviews*, 52: 837–849.

Daskalakis S.N., Goussetis G., Assimonis S.D., Tenzeris M.M., Georgiadis A. A. 2018. backscatter-morse_leaf sensor for low-power agricultural wireless sensor networks. *IEEE Sensors Journal*, 18: 7889–7898.

Dukes, Michael D., Mary Shedd, and Bernard Cardenas-Lailhacar, 2009b "Smart Irrigation Controllers: How Do Soil Moisture Sensor (SMS) Irrigation Controllers Work?:AE437/AE437." University of Florida, Institute of Food and Agricultural Sciences, FL, USA: 1–5.

Dukes, M. D., Shedd, M. L., & Davis, S. L. (2009a). Smart Irrigation Controllers: Operation of Evapotranspiration-Based Controllers: AE446/AE446, University of Florida, Institute of Food and Agricultural Sciences, FL, USA: 1–5.

Dursun M., Ozden S. 2011. A wireless application of drip irrigation automation supported by soil moisture sensors. *Scientific Research and Essays*, 6(7): 1573–1582.

Elbeih S.F. 2015. An overview of integrated remote sensing and GIS for groundwater mapping in Egypt. *Ain Shams Engineering Journal*, 6: 1–15.

Gondchawar N., Kawitkar R.S. 2016. IoT based smart agriculture. *International Journal of Advanced Research in Computer and Communication Engineering*, 5(6): 838–842.

González-Cebollada C. 2015. Water and energy consumption after the modernization of irrigation in Spain. WIT Trans. *Built Environment*, 168: 457–465.

Gupta A., Mishra S., Bokde N., Kulat K. 2016. Need of smart water systems in India. *International Journal of Applied Engineering Research*, 11(4): 2216–2223.

Guruprasadh J.P., Harshananda A., Keerthana I.K., Krishnan K.Y., Rangarajan M., Sathyadevan S. 2017. Intelligent Soil Quality Monitoring System for Judicious Irrigation. In Proceedings of the 2017 International Conference on Advances in Computing, Communications and Informatics (ICACCI), Udupi, India, 13–16 September.

Gutiérrez J., Medina J.F.V., Garibay A.N., Gándara M.A.P. 2014. Automated irrigation system using a wireless sensor network and GPRS module. *IEEE Transactions on Instrumentation and Measurement*, 63(1): 1–11.

Hinnell A.C., Lazarovitch N., Furman A., Poulton M., Warrick A.W. 2010. Neuro-drip: estimation of subsurface wetting patterns for drip irrigation using neural networks. *Irrigation Science*, 28: 535–544.

Ingale H.T., Kasat N.N. 2012. Automated irrigation system. *International Journal of Engineering Research Development*, 4(11): 51–54.

Jawad H.M., Nordin R., Gharghan S.K., Jawad A.M., Ismail M. 2017. Energy-efficient wireless sensor networks for precision agriculture: A review. *Sensors*, 17: 1781.

Jensen M.E., Burman R.D., Allen R.G. 1990. Evapotranspiration and Irrigation Water Requirements. ASCE Manuals and Reports on Engineering Practices No. 70. American Society of Civil Engineers. New York.

Jha K., Doshi A., Patel P., Shah, M. 2019. A comprehensive review on automation in agriculture using artificial intelligence. *Artificial Intelligence in Agriculture*, 2: 1–12.

Jones, H.G. 2004. Irrigation scheduling: Advantages and pitfalls of plant-based methods. *Journal of Experimental Botany*, 55: 2427–2436.

Juang J., Ekong D., Carlson U., Longsdorf C.W., Miller M. 2007. A computer-based soil moisture monitoring and hydrating system. In: 2007 Thirty-Ninth Southeastern Symposium on System Theory, Macon, GA, 2007, pp. 142–144.

Kim Y.J., Evans R.G., Iversen W.M. 2008. Remote sensing and control of an irrigation system using a distributed wireless sensor network. *IEEE Transactions on Instrumentation and Measurement*, 57(7): 1379–1387.

Koech, R., and Langat, P. 2018. Improving irrigation water use efficiency: A review of advances, challenges and opportunities in the Australian context. *Water*, 10(12): 1771.

Kumar, G. 2014. Research paper on water irrigation by using wireless sensor network. *International Journal of Scientific Engineering and Technology*, pp. 123–125, IEERT Conference Paper. San Jose, CA, USA.

Kumar A., Kamal K., Arshad M., Mathavan S., Vadamala T. 2014. Smart irrigation using low-cost moisture sensors and XBee-based communication. In IEEE Global Humanitarian Technology Conference (GHTC), pp. 333–337.

Lecina S., Isidorob D., Playánc E., Aragüésb R. 2010. Irrigation modernization and water conservation in Spain: The case of Riegos del Alto Aragón. *Agricultural Water Management*, 97: 1663–1675.

Maier H.R., Dandy G.C. 2000. Neural networks for the prediction and forecasting of water resources variables: A review of modeling issues and applications. *Environmental Modeling & Software*, 15(1), 101–124.

Mayer P.W., DeOreo W.B., Opitz E.M., Kiefer J.C., Davis W.Y., Dziegielewski B., Nelson J.O. 1999. Residential End Uses of Water. AWWA Research Foundation and American Water Works Association. Denver, Colorado.

Miranda F.R., Yoder R., Wilkerson J.B. 2003. A Site-specific Irrigation Control System. Presented at the ASAE Annu. Int. Meeting, Las Vegas, NV, July 27–30.

Mostafazadeh-Fard B., Osroosh Y., Eslamian E. 2006. Development and evaluation of an automatic surge flow irrigation system. *Journal of Agriculture and Social Sciences*, 2(3): 129–132.

Nema M.K., Khare D., Chandniha S.K. 2017. Application of artificial intelligence to estimate the reference evapotranspiration in sub-humid Doon valley. *Applied Water Science*, 7, 3903–3910.

Nhamo L., Ebrahim G.Y., Mabhaudhi T., Mpandeli S., Magombeyi M., Chitakira M., Magidi J., Sibanda M. 2020. An assessment of groundwater use in irrigated agriculture using multi-spectral remote sensing. *Physics and Chemistry of the Earth, Parts A/B/C*, 115: 102810.

Ojha T., Misra S., Raghuwanshi N.S. **2015**. Wireless sensor networks for agriculture: The state-of-the-art in practice and future challenges. *Computers and Electronics in Agriculture*, 118: 66–84.

Parameswaran G., Sivaprasath K. 2016. Arduino based smart drip irrigation system using Internet of Things. *International Journal of Engineering Science and Computing*, 6(5): 5518.

Parra L., Ortuño V., Sendra S., Lloret J. 2013. Low-Cost Conductivity Sensor based on Two Coils. In Proceedings of the First International Conference on Computational Science and Engineering, Valencia, Spain, 6–8 August.

Pawar S.B., Rajput P., Shaikh A. 2018. Smart irrigation system using IOT and raspberrypi. *International Research Journal of Engineering and Technology*, 5(8): 1163–1166.

Paydar, Z., and Qureshi, M. E. 2012. Irrigation water management in uncertain conditions—Application of Modern Portfolio Theory. *Agricultural Water Management*, 115: 47–54.

Plusquellec H. 2009. Modernization of large-scale irrigation systems: Is it an achievable objective or a lost cause. *Irrigation and Drainage*, 58: 104–120

Ponce P., Molina A., Cepeda P., Lugo E., MacCleery B. 2015. *Greenhouse design and control*. CRC Press. 378 pp.

Prathibha S.R., Hongal A., Jyothi M.P. 2017. IOT based monitoring system in smart agriculture. In International Conference on Recent Advances in Electronics and Communication Technology (ICRAECT), pp. 81–84. Bangalore, India.

Rajalakshmi P., Mahalakshmi S.D. 2016. IOT based crop-field monitoring and irrigation automation. In 10th International Conference on Intelligent Systems and Control (ISCO), pp. 1–6. Coimbatore, India.

Rehman A.U., Abbasi A., Islam N., Shaikh Z.A. 2014. A review of wireless sensors and networks' applications in agriculture. *Computer Standards & Interfaces* 36: 263–270.

Riley M. 2005. The cutting edge of residential smart irrigation technology. *California Landscaping*. July/August, pp. 19–26.

Robinson C., Mort N. 1997. A neural network system for the protection of citrus crops from frost damage. *Computers and Electronics in Agriculture*, 16(3): 177–187.

Roopaei M., Rad P., Choo K.K.R. 2017. Cloud of things in smart agriculture: Intelligent irrigation monitoring by thermal imaging. *IEEE Computer Society*, 10–15.

Roselin, A.R., Jawahar A. 2017. Smart agro system using wireless sensor networks. In International Conference on Intelligent Computing and Control Systems (ICICCS) (pp. 400–403). Madurai, India.

Ruggiero A., Punzo P., Landi S., Costa A., Van Oosten M.J., Grillo S. 2017. Improving plant water use efficiency through molecular genetics. *Horticulturae*, 3(2): 31.

Saadi S., Boulet G., Bahir M., Brut A., Delogu E., Fanise P., Mougenot B., Simonneaux V., Chabaane Z.L. 2018. Assessment of actual evapotranspiration over a semiarid heterogeneous land surface by means of coupled low-resolution remote sensing data with an energy balance model: Comparison to extra-large aperture scintillometer measurements. *Hydrology and Earth System Sciences*, 22: 2187.

Savitha M., UmaMaheshwari O.P. 2018. Smart crop field irrigation in IOT architecture using sensors. *International Journal of Advanced Research in Computer Science*, 9(1): 302–306.

Schaible G., Aillery M. 2012. Water Conservation in Irrigated Agriculture: Trends and Challenges in the Face of Emerging Demands; EIB-99; U.S. Department of Agriculture, Economic Research Service, September, Available online: https://www.ers.usda.gov/publications/pub-details/?pubid=44699 (accessed on 1 December 2018).

Sendra S., Parra L., Ortuño V., Lloret L. 2013. A Low Cost Turbidity Sensor Development. In Proceedings of the Seventh International Conference on Sensor Technologies and Applications, Barcelona, Spain, 25–31 August.

Shekhar Y., Dagur E., Mishra S., Tom R.J., Veeramanikandan M., Sankaranarayanan S. 2017. Intelligent IoT based automated irrigation system. *International Journal of Applied Engineering Research*, 12(18): 7306–7320.

Soorya E., Tejashree M., Suganya P. 2013. Smart drip irrigation system using sensor networks. *International Journal of Scientific & Engineering Research*, 4(5): 2039–2042.

Stanhill, G. 1986. Water use efficiency. *Advances in agronomy*, 39: 53–85.

Stone N.D., Toman T.W. 1989. A dynamically linked expert-database system for decision support in Texas cotton production. *Computers and Electronics in Agriculture*, 4(2): 139–148.

Sudharshan, N., Karthik, A. K., Kiran, J. S., and Geetha, S. 2019. Renewable energy based smart irrigation system. *Procedia Computer Science*, 165: 615–623.

Tan L. 2016. Cloud-based decision support and automation for precision agriculture in orchards. *IFAC-Papers OnLine* 49(16): 330–335

Thakur D., Kumar Y., Vijendra S. 2020. Smart irrigation and intrusions detection in agricultural fields using IoT. *Procedia Computer Science*, 167: 154–162.

Thenkabail P.S., Hanjra M.A., Dheeravath V., Gumma M. 2011. Global Croplands and Their Water Use from Remote Sensing and Non remote Sensing Perspectives. In *Advances in Environmental Remote Sensing-Sensors, Algorithms, and Applications*. Weng, Q., Ed.; CRC Press, Boca Raton, FL.

Wall R.W., King B.A. 2004. Incorporating Plug and Play Technology into Measurement and Control Systems for Irrigation. Management, 2004, Ottawa, Canada, August 1–4.

Yang X., Smith P.L., Yu T., Gao H. 2011. Estimating evapotranspiration from terrestrial groundwater-dependent ecosystems using Landsat images. *International Journal of Digital Earth*, 4: 154–170.

Yang W., Liusheng H., Junmin W., Hongli X. 2007. Wireless Sensor Networks for Intensive Irrigated Agriculture, Consumer Communications and Networking Conference, 2007 (CCNC 2007). 4th IEEE, pp. 197–201, Las Vegas, Nevada.

Zhang D.N., Zhou Z.N., Zhang M. 2015. Water-saving irrigation system based on wireless communication. *Chemical Engineering Transactions*, 46: 1075–1080.

Zhou Y., Yang X., Wang L., Ying Y. 2009. A wireless design of low-cost irrigation system using ZigBee technology. In International Conference on Networks Security, Wireless Communications and Trusted Computing. Vol. 1, pp. 572–575, IEEE, Wuhan, China.

15
Smart Irrigation in Urban Development Using Treated Wastewater: Irrigation Systems and Management

Leonor Rodríguez-Sinobas, Freddy Canales-Ide, and Sergio Zubelzu
Universidad Politécnica de Madrid

15.1	Introduction	321
	Treated Wastewater for Urban Irrigation • Quality of Wastewater for Urban Irrigation	
15.2	Irrigation Systems in Urban Parks	323
	Elements • Control and Automatization Systems	
15.3	Irrigation Scheduling	328
	Landscape Water Requirements • Determination of Irrigation Doses and Its Frequency	
15.4	Conclusions	335
Notes		335
References		335

15.1 Introduction

It is foreseen that about 68% of the world's population will live in urban areas (United Nations Department of Economic and Social Affairs, 2018) in 2050, and it will require a continuous and safe water supply. However, the climate change scenarios predict a decrease in the availability of fresh, pollution-free water in many areas of the planet, especially in cities or urban centers located in arid and semi-arid climate zones (Boretti and Rosa, 2019). Nowadays, most of the world's regions face supply restrictions on water resources and drinking water, which is expected to continue in the future; moreover, vulnerability to water scarcity and low-quality water will also increase if no efficient policies on the integrated management of water resources (conventional and non-conventional) are established, as well as on the revaluation of waste or contaminated water. Within this framework, the integrated and proper management of water resources in urban areas is a key factor to achieve the sustainability of water supply (Makarigakis and Jimenez-Cisneros, 2019). This will have to ensure the operation and distribution of water to its multiple uses such as food production, domestic consumption, water treatment, industry, sanitation, irrigation and energy production. Moreover, the water supply for population is a cross-cutting challenge for the decision makers (Madonsela et al., 2019).

In our vision for the cities in the future, the perception of modernity has changed from large mirrored buildings and concrete esplanades toward sustainable buildings with green roofs and rooftops. Also, it has changed the perception of large parks covered with grass toward areas with ornamental species adapted to local climate providing a natural urban landscape in contrast to the traditional English- or French-style parks which have higher water requirements. However, the feeling of greenery, which benefits the physical and mental health of the inhabitants, is often confused as a type of sustainability, but it is not. Although green roofs, trees on sidewalks and urban parks decrease the temperature across a limited area outside or inside buildings (reducing the needs for air conditioning during the summer), this decrease is mainly caused by the constant water evaporation in ornamental plants increasing water consumption. Moreover, this water demand must be often met with an unsustainable use of water resources or high energy consumption, which increases greenhouse gas emissions.

Considering the above, the reuse of wastewater will aid in the integrated management of water resources at the macro-basin level of urban developments. The main user of wastewater is irrigation or parks and green areas, and its source can be either from water treatment plants or from industries. The sections below illustrate that the reused water quality is dependent on the way the wastewater is applied in the field, meaning that reclaimed water should be used in an "intelligent", localized way with a strict control of its effect on plants.

In the cities of the 21st century, the stakeholders must carry out a centralized irrigation control, with clear criteria for efficient water use; this will be achieved when both irrigation systems and their management are properly designed/operated. A good irrigation network will deliver water with high efficiency to the active plants' root zone, and, in its operation and management, will develop an adequate irrigation scheduling fitting the actual water requirements of plants. Therefore, in smart cities, there is an opportunity for the control and management of urban irrigation systems. Sensors and drivers will be used although, as it is shown below, not only proper technical elements are needed but also an adequate irrigation design and scheduling. Thus, it could result that the expensive cost on irrigation material and technical elements will not return to social, environmental or economic benefits as devised.

This chapter addresses the use of treated wastewater to irrigate parks and gardens in urban developments and the management of the irrigation.

15.1.1 Treated Wastewater for Urban Irrigation

There is a major difference between the quality of wastewater and reclaimed water; the latter comes from wastewater that has undergone physical-chemical treatment to reach a standard quality (depending on the local/national standards) specified for its use in irrigation.

The current treatment methods allow the use of wastewater for irrigation of agricultural products, even edible green leafy vegetables, where the quality requirements to the tolerance for the presence of viruses and bacteria, with a constant control of levels of fecal coliforms, are very low; moreover, the pathogenic thresholds depend on local legislation. Likewise, not only microorganisms can affect the reclaimed water quality, but also many contaminants are not erased through the treatments such as hormone traces, toxins, chemical compounds such as endocrine-disrupting compounds, and pharmaceuticals. These are considered even more dangerous when absorbed by edible plants; thus, this issue must be considered in cities whose gardens/parks could be irrigated with reclaimed water.

On the one hand, it is also highlighted that contamination with metal (oids) such as Cd, Pb, Mn, Zn and Cu can generate phytotoxicity problems if they accumulate in the soil or substrates, after multiple irrigation seasons. On the other hand, nitrogenous compounds, phosphorus and potassium are included in the plants' fertilizers dissolved in the water to be applied according to the needs, although it must be considered that the electrical conductivity in the reclaimed water can be very high, so it is important to make an adequate control of the amount of salts, especially sodium.

Plants exposed to high levels of heavy metals can suffer physiological problems such as decreased absorption of water and nutrients, reduced photosynthesis and micronutrient imbalance; these will result in the development of leaf chlorosis, drying or browning, slow growth, root damage and in severe cases, plant senescence and subsequent death (Yadav, 2010). Likewise, research studies have shown that some ornamental plants are tolerant to the presence of high levels of heavy metals in the soil and even others, and they are also considered hyperaccumulators of these pollutants (Suman et al., 2018). Therefore, water quality and the soil capacity to adsorb pollutants should be considered in the design or management of irrigation systems for reclaimed water use.

Depending on the level of wastewater treatment, treated water has multiple uses, such as industrial (processes and cooling), toilet flushing, car and street washing, dust control, artificial lagoons, fountains and pools, water contributions to both surface and subway water courses and bodies (aquifer recharge), agricultural irrigation and urban irrigation, such as parks, sports platforms (golf courses, stadiums), medians and slope containment in avenues and highways.

15.1.2 Quality of Wastewater for Urban Irrigation

Green areas and public spaces should be designed considering that citizens can make full and complete use of them. The United Satres Environmental Protection Agency, 2012; in the Guidelines of Water Reuse regulations, proposes two different water qualities for urban use. One applies to areas restricted to citizens, and the other to non-restricted areas where the population can move freely. Lastly, people can be in contact with the treated wastewater; thus, the request on quality parameters will be stricter than in the other (US Environmental Protection Agency, 2012; Eslamian et al., 2015) as: pH=6.0–9.0; biochemical oxygen demand ≤10 mg/L; Nephelometric Turbidity Units=≤2; no detectable fecal coliform/100 mL; and 1 mg/L residual Cl_2

The Spanish government (Royal Decree 1620/2007) also set out these quality parameters regarding reclaimed water for residential use, although their values are smaller for irrigation in green areas and sports platforms.

In 2020, the European Union has announced the development of a regulation bringing together the requested characteristics for the wastewater used in irrigation. To date, many studies have reported the effects of this water on urban grounds; however, the standards do not include the quality loss in urban grounds. The main factor affecting soil properties negatively is the high salt concentrations, e.g. high levels of sodium in parks with heavy soils, where sodium can affect the stability of soil aggregates, reducing the hydraulic conductivity (Chen et al., 2013; Zalacaín et al., 2019).

The urban irrigation networks are composed of large irrigation pipes running throughout cities and urbanizations whose length can extend to several kilometers as in the case of Madrid city (Figure 15.1). Although it is advisable to disinfect the water with techniques such as the UV radiation and ozonation, the chlorination of water is still widely applied since its residual effect allows long residence times of water in the pressurized networks (Du et al., 2017). Several regulations recommend a minimum concentration of Cl_2 of 1 mg/L to ensure at least a concentration of 0.5 mg/L at the emission points of irrigation; it will avoid the proliferation of microorganisms, especially algae, that can change the hydraulic operation of water distribution and irrigation infrastructures. Likewise, ornamental species sensitive to chlorine levels described above should be avoided.

15.2 Irrigation Systems in Urban Parks

15.2.1 Elements

The reclaimed water is generally conveyed from the treatment plants to the pumping stations of irrigation systems, which will supply it to irrigation network across the green areas. Since the water requirements for the area to be irrigated can be higher than the water supply capacity, water reservoirs must be designed to store the water which is needed to provide an instantaneous discharge to operate the irrigation network with the frequency established in the irrigation scheduling. This will be determined by the balance between the plants' evapotranspiration in the park and the water retention of the soil/substrate,

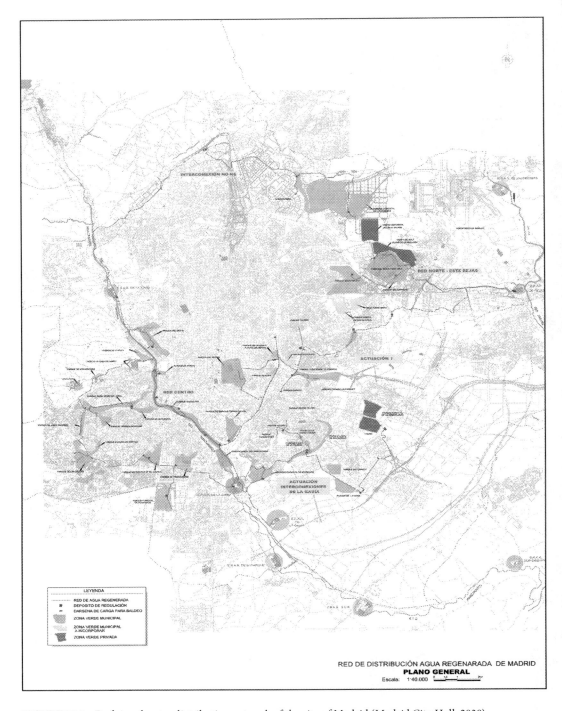

FIGURE 15.1 Reclaimed water distribution network of the city of Madrid (Madrid City Hall, 2020).

FIGURE 15.2 Water reservoir for reclaimed water.

FIGURE 15.3 Installation of drip irrigation laterals for reclaimed water.

where they grow, during the maximum demand month. Figure 15.2 shows a 5,000 m³ water reservoir, to irrigate an urban park' of 20 ha, with a nominal flow of 45 m³/h. The reservoir for reclaimed water storage must be closed to avoid aerosolization of pollutants and pathogens.

The pumping stations must include mesh or disks filters, located at the reservoir's outlet, whose high filtering capacity enables an efficient screening of particles that may clog irrigation emitters and pipes. The reservoirs can store water for several hours, days and even weeks during the months of lower water demands. Thus, the storage time should be considered to assess the application of chemical or physical treatments to prevent the proliferation of pathogenic microorganisms or algae, avoiding further clogging and health problems.

Following international regulations, the pipes conveying reused water are painted in purple as shown in Figure 15.3. It is recommended to advertise that the parks, gardens and sports facilities are irrigated with treated wastewater, in order to warn the public and pedestrians of health risks. This figure also shows a typical irrigation unit which is composed of laterals (where emitters are evenly spaced) and tertiary pipes (where the laterals are connected). The parks/green areas are recommended to be divided into hydrozones; thus, one or more units (depending on the area) will irrigate them. A hydrozone is composed of plant's species with similar water requirements that would need a similar flow rate. In cases where no simultaneous irrigation to all hydrozones can be performed, it is advisable to split the irrigated area into several irrigation sectors so that water would be supplied as needed.

FIGURE 15.4 Irrigation zone valve cabinet.

FIGURE 15.5 Algae growth into an irrigation valve box.

The basic elements for the operation of an irrigation sector are (see Figure 15.4): electric valves (if it is automated), a relief valve, which removes air within the hydraulic elements, and an additional gate valve to close the sector for repair. If reclaimed water is used, an extra filter is recommended that should be frequently self-cleaning, in order to avoid emitter clogging and not to increase pressure losses.

The emitters are the devices which controls water discharge through one/two orifices after moving across a narrow labyrinth (between 0.8 and 1.2 mm). Each emitter model has their own configuration and is more prone to clog among all the elements from the irrigation unit.

The current regulations on reclaimed water for irrigation propose a routine control of water quality parameters as the one in ISO 20760-1:2018. Water quality must be monitored and, if necessary, chemical/physical procedures must be applied, to control the development of microorganisms' colonies that could block irrigation emitters (drippers, micro-sprinklers and sprinkler nozzles). Emitter clogging decreases the water application uniformity and the energy efficiency of the irrigation networks. Figures 15.5 and 15.6 present the proliferation of algae in a valve leakage of an irrigation chest, and the clogging of a dripper, respectively.

Inside the emitters is often observed the presence of particles, with diameters greater than the passage section of filters, as well as deposits of non-organic and/or organic material (biofilm develop in wastewater). They may be caused by a poor selection or maintenance of the filter (s) and for the suction of soil particles produced when the valves close at the end of irrigation.

Smart Irrigation in Urban Development Using Treated Wastewater

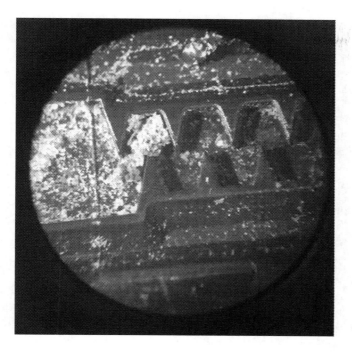

FIGURE 15.6 Inside view of a clogged emitter.

The clogging affects not only the emitters, but also the pipes whose water-conveying capacity decreases as biofilm and/or iron oxide incrustations accumulates in their walls, affecting both energy and water application efficiency (Figure 15.6).

15.2.2 Control and Automatization Systems

The concept of "smart irrigation" in cities is still developing, although it highlights the level of automation without emphasizing, properly, the efficiency in water management and distribution. It deals with precision irrigation whose main objective is to achieve good efficiency in the management of irrigation and water application, which is possible if the following conditions are satisfied:

- The irrigation system must be efficient both in the distribution and application of water.
- The irrigation method must be localized and water the active root zone of the plants.
- The frequency of watering must be such that it provides an ornamental and functional quality so that the plants last over time.
- The evapotranspirative demand must be calculated every day.
- The irrigation systems must be autonomous and modify the irrigation dose according to the climatic variables or, if available, to the water content of the soil.

In the smart cities, the control and automation of the irrigation are a key factor in the management of the urban water. Thus, the project and the scheduling of irrigation in parks and gardens must be done taking into account the water needs of the different plant species in the green area. For this purpose, it is necessary to determine the instantaneous flow of each sector and the proper irrigation time to fulfill the plants' water requirement.

The garden coefficient should not be selected or fixed arbitrarily since it could cause an over–irrigation of some plants. Consequently, it is important to define hydrozones (groups of plants) with similar water needs, and consider them as a basic unit to design the irrigation sectors operated by an electromechanically operated valve or "solenoid" valve. When the garden area is small, the number of sectors will also be small, and can be easily automated; however, when it is large, an irrigation control center will be required to control the opening and closing of the valves in all sectors.

The irrigation sectors, depending on the controlled area by the irrigation central, can be connected to irrigation programmers, by means of wiring and decoders, or be activated/programmed with technologies such as radio frequency, bluetooth or telecommunication networks (GPRS, 3G). Thus, in order to achieve an efficient irrigation control, control elements are necessary for visualizing any failure and then, to manage the reparation. The most frequent failures are due to lack of pressure in sectors and water leaks that can be monitored through water meters and remote manometers connected to the irrigation central.

15.3 Irrigation Scheduling

15.3.1 Landscape Water Requirements

Investments in the control and automation of irrigation systems are not useful if the estimation or calculation of plant water needs is not done correctly, to which the quality of the water must be added. In the case of wastewater, it must be taken into account whether it may contain contaminating compounds which can reach the groundwater if irrigation doses are applied in excess.

There are many studies dealing with the efficiency of irrigation management in green areas that highlight that water is applied in excess. Although the current trend goes to a naturalistic landscaping, or xeriscape, in the case of temperate or arid climates, large areas of turf or plants with high water requirements still predominate worldwide. In lawns, the determination of its water needs and the irrigation management are easy; lawns generally comprise different kinds of plants and cold-weather grasses, which have a crop coefficient close to 1, are watered frequently. However, in the green areas composed of trees and ornamental plants that are irrigated with this criterion, overirrigation is an issue and can cause phytosanitary problems and affect ornamental quality (Figure 15.7). This will result in extra cost not only for the loss of plants but also for water (which is scarce), the energy for pumping and the increase of drainage.

In established parks or green areas, the main difficulty in irrigation management is the determination of the water requirements of the ornamental plants. Unlike agricultural crops, most parks and gardens are designed with a heterogeneous spatial distribution of plant species, with different water requirements. Among the methods proposed to determine the water needs in gardens, the most used is the Water Use Classifications of Landscape Species (WUCOLS).[1]

FIGURE 15.7 Remains of corky roots from overwatered dead tree.

15.3.1.1 WUCOLS Method

Costello proposed the WUCOLS, which follows a methodology similar to the one by FAO[2] for the calculation of water requirements of agricultural crops (FAO, 2006). In WUCOLS, the crop coefficient K_c has been replaced by a garden coefficient K_L for ornamental species as shown in Eq. (15.1) (Costello, 1994; Costello and Jones, 2014).

$$ET_L = ET_0 \times K_L, \tag{15.1}$$

where:

ET_L = Landscape evapotranspiration [mm/d]
ET_0 = Reference evapotranspiration [mm/d]
K_L = Landscape factor [–].

ET_0 is estimated from the Penman–Monteith method as follows:

$$ET_0 = \frac{0.408 \times \Delta \times (R_n - G) + \gamma \times \dfrac{900}{T+273} \times v_2 \times (e_s - e_a)}{\Delta + \gamma \times (1 + 0.34 \times v_2)}, \tag{15.2}$$

where:

R_n = Net radiation on the plant surface [MJ/m²/d]
G = Ground heat flow density [MJ/m²/d]
T = Average daily air temperature at 2 m altitude [°C]
v_2 [[Tab]] = Wind speed at 2 m height [m/s]
e_s = Vapor saturation pressure [kPa]
e_a = Current vapor pressure r [kPa]
$e_s - e_a$ = Saturation pressure deficit [kPa]
Δ = Slope variation of the vapor pressure curve [kPa/°C]
γ = Psychrometric constant [kPa/°C].

While the garden coefficient is expressed as follows:

$$K_L = K_s \times K_{mc} \times K_d, \tag{15.3}$$

where:

K_L = Landscape Coefficient
K_s = Specie factor
K_{mc} = Microclimate factor
K_d = Density factor.

The K_s is estimated by an expert committee that has classified the plant species into various water demand categories shown in Table 15.1.

The K_{mc} coefficient refers to the specific microclimate in the garden locations and depends on the shading degree by buildings, wind speed, radiation intensity and humidity. Its value varies between 0.5

TABLE 15.1 Specie Coefficient K_s for the WUCOLS Categories

Category	K_s Range
High	0.7–0.9
Moderate	0.4–0.6
Low	0.1–0.3
Very low	<0.1

and 1.4 (see Table 15.2): the lowest value corresponds to very shady areas and the highest value to gardens located in paved areas or that receive light reflections from glass roofs.

K_d refers to water use considering the vegetation density in the garden or leaf area indices. It varies between 0.5 and 1.3 (see Table 15.3). The lowest value corresponds to areas with low planting density, gardens in formation or those composed of not very robust species, and 1.4 for dense gardens composed of shrubs, groundcovers and dense grasses.

Figure 15.8 presents the variability of K_L coefficient across the parks in the Valdebebas urbanization located in Madrid.

TABLE 15.2 Microclimate Coefficient K_{mc}

Category	K_{mc} Range
High	1.1–1.4
Moderate	1.0
Low	0.5–0.9

TABLE 15.3 Density Coefficient K_d for Different Categories

Category	K_d Range
High	1.1–1.3
Moderate	1.0
Low	0.5–0.9

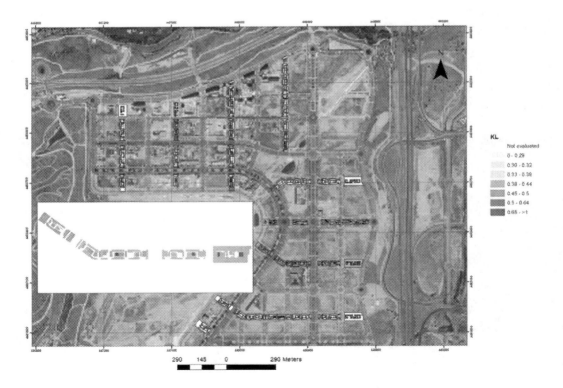

FIGURE 15.8 Garden coefficient K_L variation across the parks in the smart city of Valdebebas (Madrid).

15.3.1.2 Deficit Irrigation Factor Method

Another method for calculating water needs has been developed in Australia, although its use is not very widespread. However, it would be suitable for irrigation management in areas with scarce water availability for irrigation and in greenhouses/areas which have been developed mainly for lawn species (Nouri et al., 2013; Shojaei et al., 2018). The method proposes a crop stress factor K_{st} to determine water needs after applying deficit irrigation; this will be able to maintain an adequate ornamental quality in plants without fulfilling the soil field capacity (FC) in each irrigation event.

$$ET_L = ET_0 \times K_{st} \tag{15.4}$$

It is recommended to check the plants' resistance to stress before applying the method; in ornamental species, it will be higher in arid climates than in humid and cold ones. The method is still underdeveloped, and it could save water if applied properly.

15.3.1.3 Measurement of Water Content in the Soil

One of the procedures to determine the irrigation schedule with more precision, adapted to each case, is by measuring soil water content. This procedure is considered interesting not only for improving irrigation efficiency but, especially in the irrigation with regenerated water, to avoid the leaching of nutrients and pollutants.

The measurement of soil water content is done through moisture sensors. Although in the market the variety of sensors is the most demanded ones with application in landscaping, are those of electromagnetic linear transmission. Its operation is based on the good linear adjustment between the dielectric permittivity of the soil and the volumetric water content (θ_V) (Davis and Dukes, 2015). Thus, commercial models use this relationship to estimate θV from different operating modes such as Time-Domain Reflectometry, Time-Domain Trasmissometry and Transmission Line Oscillation, impedance and capacitance.

The accuracy of sensors may vary with the chemical and physical properties of the soil, so it is advised that instead of using the manufacturer's calibration, sensors should be recalibrated for the soils in which they will be used (Blonquist et al., 2005; Cobos, 2009). As an example, Figure 15.9 shows the equipment required for the calibration process. The calibration usually is performed in the laboratory on samples of dry soil or substrate where the sensors will be installed. The calibration process measures different moisture contents: saturation, field capacity FC and intermediate points until the wilting point is reached. The measurements will be used to fit a calibration curve.

FIGURE 15.9 Detail of the equipment required for the calibration of soil water sensors.

There are sensors that can be connected to irrigation programmers to manage irrigation based on the availability of soil water content; nevertheless, a high number of sensors could be required to obtain an average characterization of plants' behavior.

15.3.2 Determination of Irrigation Doses and Its Frequency

Depending on the degree of automation of the irrigation system, the irrigation scheduling is based on the evapotranspiration of the garden's plants or the monitoring of θV. This considers that the water to be applied will increase the water content up to a maximum value fixed in advance (usually corresponding to soil FC). The manufacturers proposed several approaches for determining the available θV such as the use of tensiometers or probes, but their application in smart cities is less practical since a continuous and remote monitoring of available water variations will be expected.

The weather station is the basic element to estimate the reference evapotranspiration; then, when the garden coefficient is known, it will be used to calculate the plant water needs and, irrigation doses will be change accordingly. If no equipment is available to carry out the water balance in the park/garden, it can be estimated using the previous information of precipitation, temperature and evapotranspiration collected from the irrigated area.

The evapotranspiration is a measure to estimate the a water balance of plants in the parks and gardens that is useful to determine irrigation doses and frequencies. The soil water balance is expressed (Martin et al., 1990; Allen et al., 1998) as shown in Eq. (15.5):

$$\Delta\theta = P - E + I - D_p - ET_L, \tag{15.5}$$

where:

θ = Water content [mm]
P = Effective rain
E = Runoff [mm]
I = Irrigation depth [mm]
D_p = Depth infiltration [mm].

In drip irrigation systems, and especially in subsurface systems, the water fraction considered as runoff loss can be omitted since emitters' discharge is very small. In the case of irrigation with treated or regenerated wastewater, deep infiltration is a loss that must be avoided by a precise calculation of the evapotranspiration needs.

In each irrigation event, the effective rainfall is the one above 5 mm; if the values are smaller than 5 mm, water remains on the canopy or is evaporated directly from the soil. A threshold for optimal irrigation is considered, and it is known as easily exploitable water (EA) (Eq. 15.6):

$$EA = f \times TAW, \tag{15.6}$$

where:

f = Fraction of water that is depleted before water stress [%]
TAW = Total available water [mm].

The f factor for high-yielding crops is commonly set at 0.5, i.e. irrigation should be programmed when half of the water stored in the soil is depleted. Although there are no recommendations regarding the irrigation threshold for ornamental species, three-fifths (0.75) of the TAW is adequate. This defines the water fraction depleted without affecting the ornamental and functional quality of plants. The available water depth should be calculated considering changes in soil water content (Eq. 15.7):

$$TAW = 1.000 \left(\theta_{CC} - \theta_{WP}\right) \times Z_r \times (1 - P), \tag{15.7}$$

where:

θ_{CC} = Moisture content at FC [m³/m³]
θ_{WP} = Moisture content at the point of permanent wilt [m³/m³]
Z_r = Root depth [m]
$(1-P)$ = Percentage of soil diameter less than 2 mm [%].

Both the water content at FC and the wilting point are determined in a pressure chamber. In large urban parks, several soil samples must be taken randomly and sent to specialized laboratories for the determination of both parameters.

The recommended root depths for estimating water availability are presented in Table 15.4 (Martín Rodriguez et al., 2003).

The Figures 15.10–15.12 present the irrigation scheduling for 2017 in urban parks encompassing four plants typology: trees, shrubs, ornamental flowers and grasses. The parks are located at the smart city, fully automated, of Valdebebas (Madrid) which have a Mediterranean climate. In this case, the irrigation scheduling was programmed with the daily climatic variables provided by the weather station in the urbanization. Likewise, the soil water content was measured as depicted in the figures.

A high level of automation is not always synonymous of high efficiency in urban irrigation management (Canales-Ide et al., 2019) as shown in Figure 15.10. In this case, the irrigation of trees is applied separately to specific hydrozones at a fixed frequency; the dose for each irrigation event maintains the soil water level above the easily exploitation water (EA) and below the FC thresholds, which is calculated with the weather variables, garden coefficient and the water balance equation. However, it is observed that there are some irrigation events far beyond FC, and also there is moisture deficit (soil water content below EA) between the first week of July and the end of the month (Figure 15.10).

TABLE 15.4 Common Root Depth of Plant Typologies

Plant Typology	Z_r (m)
Lawn	0.25
Perennials and bulb flower	0.40
Shrubs	0.65
Trees	1.00

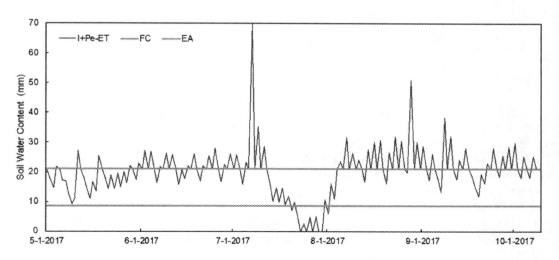

FIGURE 15.10 Irrigation scheduling, performed by water balance, for trees in the irrigation season 2017.

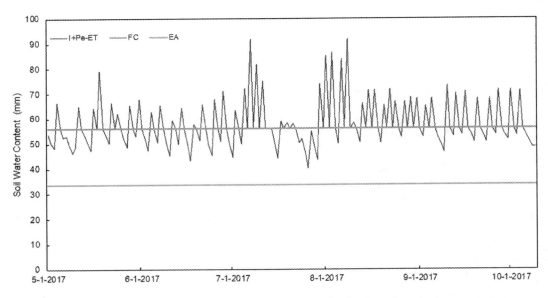

FIGURE 15.11 Irrigation scheduling, performed by water balance, for shrubs in the irrigation season 2017.

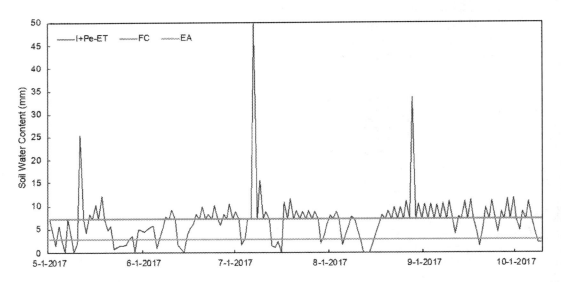

FIGURE 15.12 Irrigation scheduling, performed by water balance, for lawn in the irrigation season 2017.

In Valdebebas, the irrigation management and scheduling for the ornamental plants, mainly flowers, ornamental grasses and shrubs, follow the same pattern as for other trees, considering the same roots' depth and specific species coefficient. When water depths and garden coefficients are fitted to the most water-demanding species, water balance among these categories is different. In shrubs (see Figure 15.11), the available water drops from the two-fifths threshold (12.6 mm) and before the most significant rainfall event it drops to 0 mm. As in the trees, the controller considers all rainfall as available water, causing subsequent water stress.

Finally, turf areas are frequently irrigated and have a good adjustment between plant water needs and applied irrigation in the case presented, as shown in Figure 15.12. It is observed that the available moisture was 0 mm at several times but no problems are anticipated since the prairie is composed of Bermuda grass with high tolerance to water stress (Zhao et al., 2011).

15.4 Conclusions

On the one hand, the climate change scenarios foresee a reduction in the availability of water resources, and on the other hand, the use of drinkable water for irrigation increases. Both these issues will lead to the worldwide increase of wastewater reuse for activities requiring low water quality.

The irrigation in smart cities needs further development. Intelligent irrigation management systems are available to apply irrigation in green areas such as Internet of things. However, their decision-making algorithm must take into account that plants are complex living organisms whose water requirements are affected by several factors such as garden/park location in the city, soil/substrate where they grow, roots and physiological state.

A so-called intelligent water application must include localized irrigation. In irrigation with treated wastewater, the emitters are more prone to clogging than when irrigated with better-quality water. Thus, special care must be taken in selecting proper filter equipment and its maintenance; in addition, it is advisable to apply acid solutions to dissolve biofilm and salt accumulations in emitter, pipes and valves.

Notes

1. WUCOLS: Water Use Classifications of Landscape Species (Costello, 1994).
2. FAO: Food and Agriculture Organization.

References

Allen, R. et al. (1998) Crop Evapotranspiration. Guidelines for Computing Crop Requirements. Irrigation and Drainage Paper No56. Edited by FAO. Rome, Italy.

Blonquist, J. M., Jones, S. B. and Robinson, D. A. (2005) "Standardizing characterization of electromagnetic water content sensors," *Vadose Zone Journal*. doi: 10.2136/vzj2004.0141

Boretti, A. and Rosa, L. (2019) "Reassessing the projections of the world water development report," *npj Clean Water*. doi: 10.1038/s41545-019-0039-9

Canales-Ide, F., Zubelzu, S. and Rodríguez-Sinobas, L. (2019) "Irrigation systems in smart cities coping with water scarcity: the case of Valdebebas, Madrid (Spain)," *Journal of Environmental Management*, 247. doi: 10.1016/j.jenvman.2019.06.062

Chen, W. et al. (2013) "Impacts of long-term reclaimed water irrigation on soil salinity accumulation in urban green land in Beijing," *Water Resources Research*. doi: 10.1002/wrcr.20550

Cobos, D. R. (2009) "Calibrating ECH2O soil moisture sensors," Application Note, Decagon Devices Inc., Pullman, WA, USA.

Costello, L. R. (1994) WUCOLS Water Use Classification of Landscape Species, University of California, Cooperative Extension. Available at: http://ucanr.edu/sites/oc/files/132534.pdf (Accessed: December 12, 2018).

Costello, L. R. and Jones, K. S. (2014) WUCOLS IV, Water Use Classification of Landscape Species. Davis, California, USA.

Davis, S. L. and Dukes, M. D. (2015) "Methodologies for successful implementation of smart irrigation controllers," *Journal of Irrigation and Drainage Engineering. American Society of Civil Engineers*, 141(3), p. 4014055. doi: 10.1061/(ASCE)IR.1943-4774.0000804

Du, Y. et al. (2017) "Formation and control of disinfection byproducts and toxicity during reclaimed water chlorination: a review," *Journal of Environmental Sciences*. doi: 10.1016/j.jes.2017.01.013

Eslamian, F., Eslamian, S., and Eslamian, A. (2015) "Water Reuse Guidelines for Agriculture," in Eslamian, S. (ed) *Urban Water Reuse Handbook*. Taylor and Francis, CRC Group, Boca Raton, FL, pp. 177–186.

FAO (2006) Evapotranspiración del cultivo. Guías para la determinación de los requerimientos de agua de los cultivos. Rome, Italy.

Madonsela, B. et al. (2019) "Evaluation of water governance processes required to transition towards water sensitive urban design—an indicator assessment approach for the City of Cape Town," *Water. Multidisciplinary Digital Publishing Institute*, 11(2), p. 292. doi: 10.3390/w11020292

Makarigakis, A. K. and Jimenez-Cisneros, B. E. (2019) "UNESCO's contribution to face global water challenges," *Water (Switzerland)*. doi: 10.3390/w11020388

Martin, D., Stegman, E. and Ferreres, E. (1990) "Irrigation Scheduling Principles," in Hoffman, G., Howell, T., and Solomon, K. (eds.) *Management of Farm Irrigation System*. ASAE, American Society of Agricultural Engineers, St. Joseph, MI, pp. 115–203.

Martín Rodriguez, A., Avila Alvarces, L., Yruela Morillo, M.C., Plaza Zarza, F., Quesada Navas, A. and Fernandez Gomez, F. (2003) *Manual de riego de jardines*. Edited by Junta de Andalucía. Sévilla: Conserjería de Agricultura y Pesca.

Nouri, H. et al. (2013) "Water requirements of urban landscape plants: a comparison of three factor-based approaches," *Ecological Engineering*, 57, pp. 276–284. doi: 10.1016/J.ECOLENG.2013.04.025

Shojaei, P. et al. (2018) "Water requirements of urban landscape plants in an arid environment: the example of a botanic garden and a forest park," *Ecological Engineering*, 123, pp. 43–53. doi: 10.1016/j.ecoleng.2018.08.021

Suman, J. et al. (2018) "Phytoextraction of heavy metals: a promising tool for clean-up of polluted environment?," *Frontiers in Plant Science*. doi: 10.3389/fpls.2018.01476

United Nations Department of Economic and Social Affairs (2018) World Urbanization Prospects 2018, Webpage.

US Environmental Protection Agency (2012) "Guidelines for Water Reuse," Development, USEPA, USA.

Yadav, S. K. (2010) "Heavy metals toxicity in plants: an overview on the role of glutathione and phytochelatins in heavy metal stress tolerance of plants," *South African Journal of Botany*. doi: 10.1016/j.sajb.2009.10.007

Zalacáin, D. et al. (2019) "Influence of reclaimed water irrigation in soil physical properties of urban parks: a case study in Madrid (Spain)," *Catena*. doi: 10.1016/j.catena.2019.05.012

Zhao, Y., Du, H., Wang, Z., Huang, B. (2011) "Identification of proteins associated with water-deficit tolerance in C4 perennial grass species, Cynodon dactylon×Cynodon transvaalensis and Cynodon dactylon," *Physiol Plantarum*, 141, pp. 40–55. DOI: 10.1111/j.1399-3054.2010.01419.x

VI

Water Pumps for Irrigation

16
Water Pumps for Irrigation: An Introduction

Saeid Eslamian
Isfahan University of Technology

Mousa Maleki
Illinois Institute of Technology

Asal Soltani
The University of Toledo

16.1 Introduction 339
16.2 Types of Irrigation Systems 341
   Surface Irrigation • Localized Irrigation • Drip Irrigation • Sprinkler Irrigation • Center Pivot Irrigation • Lateral Move Irrigation • Subsurface Irrigation
16.3 Types of Water Pumps for Irrigation Systems...................................343
   Displacement Pump • Centrifugal Pumps • Deep-Well Turbine Pump • Submersible Pumps • Submersible Turbines • Propeller Pumps
16.4 Conclusions..351
References..352

16.1 Introduction

Today, pumps are one of the most important consumables in agriculture and industry. Proper use and operation can improve pump performance. The most important use of pumps in agriculture is related to their use in conveying water and in circulating water in pressurized irrigation systems. The use of pressurized irrigation networks results in increased water conservation under cultivation and increases crop yield per unit area. This chapter examines the types of pumps used in irrigation systems as well as their construction and related theories.

Water pumping systems driven by renewable energies are more environmentally sound and, at times, less-expensive alternatives to electric- or diesel-based ones. Of all pumps, hydro-powered pumps have additional advantages. Nevertheless, these seem to be largely ignored nowadays. More than 800 scientific and nonscientific documents contributed to assemble their fragmented storylines. A total of 30 pressure-based hydro-powered pumping (HPP) technologies worldwide have been classified and plotted in space and time. Although these do not present identifiable patterns, some noticeable clusters appear in regions such as Europe, South–Southeast Asia, and Eastern Africa, and in timeframes around 1960–1990, respectively. Some technologies have had a global impact and interest from their beginning until contemporary times; others have been crucial for the development of specific countries, and a few barely had almost imperceptible lives. All of them, nonetheless, have demonstrated to be a sound alternative to conventional pumping technologies, which can be unaffordable or inaccessible, particularly in remote and off-the-grid areas (Intriago Zambrano et al., 2019).

Currently, HPP technologies face a regained momentum, hence, a potentially promising future. However, researchers, manufacturers, and users need to be aware of the importance that management systems, as well as business models, pose for these technologies beyond their mere performance (Intriago Zambrano et al., 2019).

Given the considerable number of smallholders farms worldwide (Lowder et al., 2016), intensification of their crop farming is key for local and global food security (Tscharntke et al., 2012). However, smallholders face many uncertainties linked to weather events, crops diseases, and market fluctuations. In addition, on-farm conditions are often suboptimal because of low availability of inputs and lack of control/information to decide on their use. Although access to water is not the only factor influencing farming, improving water control for small-scale farming is a major option to secure smallholder production (Burney and Naylor, 2012). Pressure-based irrigation technologies, either introduced as a new choice or as the result of former gravity-based systems converted into (water-saving) drip and sprinkler irrigation, are one option. Another option is to use the pumping technologies to allow water delivery to fields that used to be otherwise unirrigated.

Pumped irrigation is ruled worldwide by electricity- and diesel-based systems. They bear high operation and maintenance costs because of continuous use of electricity from the grid and expensive fuels, respectively. As a consequence, these technologies might be eventually (too) cost-intensive for most smallholders—which makes them less accessible and/or suitable for small farmers. Furthermore, they are strongly linked to air pollution due to their gaseous emissions and noise (Aliyu et al., 2018; Chandel and Naik, 2015). More environmentally sound and, at times, less-expensive alternatives would be pumping systems based on renewable energy (RE) sources, i.e., solar power, wind power, biomass/biogas, and hydropower (Gopal et al., 2013).

HPP technologies, namely those driven by the energy contained in the water they lift, correspond to a concept as ancient as effective (Rossi et al., 2009; Yannopoulos et al., 2015). Non-direct lifting (i.e., pressure-based) HPP devices started being envisaged by Al-Jazari in the early 13th century (Al-Jazari et al., 1974), and later on by Taqi Al-Din, Agricola, Ramelli, and other authors during the 16th century. These pumping systems pose further advantages over their other RE-based counterparts: (1) Their energy source is generally available 24 hours a day, 7 days a week, relatively concentrated and more predictable; (2) they have a higher power-to-size ratio, and thus are more cost-effective; (3) they are mechanically simpler and more robust, hence less maintenance-demanding and long-lasting; and (4) they are typically more efficient (up to 85%) (Fraenkel, 1986).

Nevertheless, and despite their advantages and long history in water lifting, HPP systems seem to be largely disregarded nowadays. On the one hand, there are some contemporary studies (Meah et al., 2008; Purohit and Kandpal, 2005; Purohit, 2007; Kumar and Kandpal, 2007; Ali, 2018; Argaw et al., 2003; Zhang et al., 2019) and literature reviews (Aliyu et al., 2018; Chandel and Naik, 2015; Gopal et al., 2013; Becenen and Eker, 2005; Mohammed Wazed et al., 2018) on RE-based water pumping systems. However, none of them address hydropower as a sound source of energy. On the other hand, there are several old publications (Fraenkel, 1986; Tiemersma and Heeren, 1984; Kristoferson and Bokalders, 1986; Hofkes and Visscher, 1986; Collett, 1981; Johansson and Nilsson, 1985; Fraenkel, 1979; Wood, 1976; Wood et al., 1977) that considered it to a bigger or lesser extent, though completely overlooking many other then-contemporary HPP technologies that were relevant—and, in some cases, even predominant—for other (non-Western) contexts. Therefore, and considering such knowledge fragmentation and consequent gap, this review constitutes the first worldwide-scale depiction of the past and present trends on the documented research, development, application, and commercialization of the HPP technologies. In turn, such information provides a general yet solid basis for scholars, (industry) researchers, managers, manufacturers, and users, with respect to the future uses these technologies (as well as new ones derived from them) might have under different sets of physical and social conditions.

It is so that two universities, namely Delft University of Technology and Comillas Pontifical University, from The Netherlands and Spain, respectively, are currently carrying out the DARE-TU (Developing Agriculture and Renewable Energy with the turbine pump) project (Intriago et al., 2018). It aims to research the cocreation and implementation of affordable clean irrigation systems, based upon novel HPP technologies (Intriago et al., 2018) developed in collaboration with the Dutch start-up company aQysta. Within this context, the objectives of the present chapter are:

- To summarize and classify the HPP technologies researched, applied, and eventually commercialized globally over time;
- To define their state-of-the-art by synthesizing their respective storylines and highlighting the highest level of their developments;
- To identify global spatial and temporal patterns on the (re)invention, application, and spread of the HPP technologies.

16.2 Types of Irrigation Systems

Successful agriculture is dependent upon farmers having sufficient access to water. Looking back to the middle of the last century, the common perception was that water was an infinite resource. Today, there is an awareness that water is a resource that needs to be managed. This is not only a question of more mouths to feed, people today consume more calories and eat more meat, and this requires more water to produce food. To meet future demands, world food production must double by 2045 (Grundfos Holding, 2020).

Irrigation is the artificial application of water to the soil through various systems of tubes, pumps, and sprays. Irrigation is usually used in areas where rainfall is irregular or dry times or drought is expected. There are many types of irrigation systems, in which water is supplied to the entire field uniformly. Irrigation water can come from groundwater, through springs or wells, surface water, through rivers, lakes, or reservoirs, or even other sources, such as treated wastewater or desalinated water. As a result, it is critical that farmers protect their agricultural water source to minimize the potential for contamination. As with any groundwater removal, users of irrigation water need to be careful in not pumping groundwater out of an aquifer faster than it is being recharged.

Sources of irrigation water can be groundwater extracted from springs or by using wells, surface water drawn from rivers, canals, lakes or reservoirs, or non-conventional sources like treated wastewater, desalinated water, drainage water, or reclaimed water generally (Grundfos Holding, 2020).

The location of the irrigation water makes a difference to the pump that should be selected. Deep-well submersible pumps and turbine pumps are specially designed to lift water from several 100 ft underground, and a variety of pumps can be used when drawing surface water (Grundfos Holding, 2020).

If submersible pumps are used when drawing water from a reservoir or lake, advantages are improved theft protection, because the pumps are submerged, and reduced noise, because noise is limited to that from the pipes and the valves. The majority of North American farmers use oil-lubricated vertical turbine pumps in locations where noise is not a factor. These pumps also provide easy access to the motor for any service or replacement that may be required (Grundfos Holding, 2020).

There are many different types of irrigation systems, depending on how the water is distributed throughout the field. Some common types of irrigation systems are as follows.

16.2.1 Surface Irrigation

Water is distributed over and across land by gravity, no mechanical pump involved. There are four general phases in surface irrigation, including advance phase, storage phase, depletion phase, and recession phase. The advance phase progresses when water enters the earth and continues until the water reaches the bottom of the earth, and this time is the time of progress. From the end of the advance phase until the input current is cut off, it is called the storage phase. In the storage phase, the required water becomes the farthest points of the field that have less penetration time in the advance phase. After the inflow, the water is drained from the beginning of the field or percolates the soil and the soil surface is exposed, which is called the depletion phase. After the soil surface appears at the beginning of the field, the regression phase begins, which continues until the water completely disappears from the soil surface. In low-slope farms, phases and regrowth occur simultaneously throughout the field (Valipour et al., 2015).

16.2.2 Localized Irrigation

Localized irrigation is the slow distribution of water on the surface or under the soil in the form of separate, continuous, narrow streams or fine sprays through the final dropper located along the water transmission line. Local irrigation system is usually used to irrigate trees, shrubs, vines, orchards, etc. Usually all systems have the following main components:

- Pumping station and central control
- Main water supply lines
- Sub-water supply lines
- Lateral or drainage tubes
- Dripper.

16.2.3 Drip Irrigation

The drip method is a pressurized irrigation system, so in the first place, a water pump must be received from the source and inserted into the drip irrigation pipe system with the necessary pressure. In some cases, the water may be piped from the main source to the farm, and the height difference may provide the energy needed to pump water. In this case, there will be no need for a pump. But in most conventional drip irrigation systems, a pump is required to supply pressure.

The water enters the main pipe with the pressure supplied by the pump and this pipe continues to the beginning of each irrigation piece. Each piece of irrigation is divided into a number of sub-pieces that supply the water they need from a semi-main pipe, also called a manifold. The semi-main pipe, the length of which is equal to the width of the irrigation sub-unit, branches from the main pipe. In this way, the main pipe feeds a number of semi-main pipes from which they branch. Depending on the distance between the plant rows, a number of sub-pipes or lateral pipes branch from the semi-main pipe, the length of each of which is proportional to the dimensions of the irrigation sub-unit or the length of the rows. The main and semi-main pipes can be made of steel, asbestos cement, PVC, or polyethylene, but the sub-pipes are usually made of polyethylene plastic. Sub-pipes that pass through the plant rows are equipped with drops. They are emitters or outlets by which water flows out at very low pressure and spreads on the ground. One of the main functions of the dripper is to reduce the water pressure when leaving it. The discharge of the drippers used in this system varies from 2 L/h to about 24 L/h, and in normal conditions, 4-L drippers are used per hour.

16.2.4 Sprinkler Irrigation

Sprinkler irrigation is a method in which water flows by pressure into a piped network and then outlets the sprinklers installed on this network. The sprinklers are manufactured in such a way that when the water is forced out of it, it becomes small and large droplets and pours on the surface of the field like rain. For this reason, this irrigation system is called sprinkler method.

In this method, water is spread as rain on the ground at a speed equal to or less than the soil permeability so that the soil has the percolation opportunity. If the intensity of rainfall is more than the rate of infiltration into the soil, it will cause runoff on the surface of agricultural land and reduce irrigation efficiency.

16.2.5 Center Pivot Irrigation

This system, as its name implies, is a method of irrigating farms in which equipment rotates around an axis and is irrigated by sprinklers. In this way, a circular area is irrigated and usually when you look at it from above, it creates a circular pattern in the crops. The stimulants of this system are able to use fertilizers, chemicals, poisons, and herbicides, and this multi-application capability increases the productivity and quality of the product. The radius of the central axes of this device is usually less than 1,600 ft (500 m), the most common of which is 400 m, and each device with a radius of 400 m can cover 125 ha of land.

16.2.6 Lateral Move Irrigation

Water is distributed through a series of pipes, each with a wheel and a set of sprinklers, which are rotated either by hand or with a purpose-built mechanism. The sprinklers move a certain distance across the field and then need to have the water hose reconnected for the next distance. This system tends to be less expensive but requires more labor than others.

16.2.7 Subsurface Irrigation

One of the effective irrigation systems is the subsurface drip irrigation system. In this system, irrigation is done by means of drip pipes or strips that are formed inside the soil and the task of supplying water to the plant roots is directly treated.

In subsurface irrigation, the plant has more air and water and as a result will grow better. This method can be used to irrigate orchards such as pistachio, olive, date, citrus, black root trees, a variety of crops such as potatoes, corn, cotton and alfalfa, urban forest green space and boulevards, areas with limited water, roof garden, and used areas with high evaporation.

16.3 Types of Water Pumps for Irrigation Systems

Pumps commonly used for irrigation systems fall into two broad categories: displacement pumps and centrifugal pumps. Within those categories there are sub-categories that further define the type of pump. This chapter will focus on those types of pumps most often used for irrigation systems (Nourbakhsh et al., 2010) (Figure 16.1).

16.3.1 Displacement Pump

Displacement pumps force the water to move by displacement. This means pumps such as piston pumps, diaphragm pumps, roller-tubes, and rotary pumps. The old fashioned hand pumps, the ones you operate by moving a long-lever handle up and down, are piston displacement pumps. So are those grasshopper-like oil well pumps. Displacement pumps are used for moving very thick liquids, creating very precise low volumes, or creating very high pressures. In addition to oil wells they are also used for fertilizer injectors, spray pumps, air compressors, and hydraulic systems for machinery. With the exception of fertilizer injectors (used for mixing fertilizer into irrigation water) you will not see them typically used for irrigation systems, so we'll move along to centrifugal pumps (Nourbakhsh et al., 2010).

Positive displacement water pumps create a pressure difference by changing the available space (volume) within the pump system. Using components such as pistons, closed chambers, and valves, the pressure is decreased on the inlet side which draws water into the pump. Then the pressure is increased forcing (displacing) the water through the outlet side of the pump system.

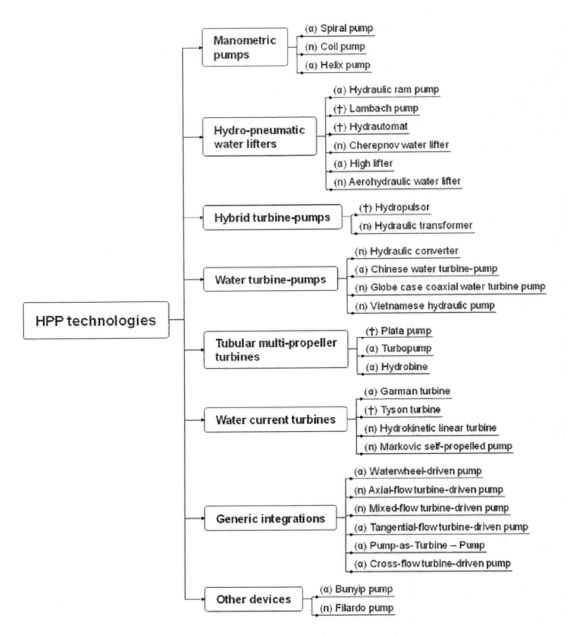

FIGURE 16.1 Classification of hydro-powered pumping (HPP) technologies and their latest development/production stage. The symbols (α), (†), and (n) stand for commercially available, commercially extinct, and noncommercial technologies, respectively (Intriago Zambrano et al., 2019).

16.3.2 Centrifugal Pumps

Pumps used for irrigation include centrifugal, deep-well turbine, submersible, and propeller pumps. Actually, turbine, submersible, and propeller pumps are special forms of a centrifugal pump. However, their names are common in the industry (Scherer, 1993). In this chapter, the term centrifugal pump refers to any pump that is above the water surface and uses a suction pipe.

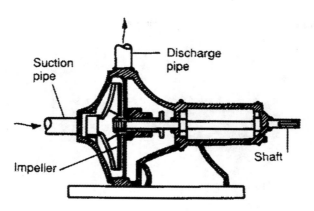

FIGURE 16.2 A horizontal centrifugal pump (Scherer, 1993).

Centrifugal pumps are used to pump from reservoirs, lakes, streams, and shallow wells. They are also used as booster pumps in irrigation pipelines. All centrifugal pumps must be completely filled with water or "primed" before they can operate. The suction line and the pump have to be filled with water and free of air. Air tight joints and connections are extremely important on the suction pipe. Priming a pump can be done by hand-operated vacuum pumps, internal combustion engine vacuum, motor-powered vacuum pumps, or small water pumps that fill the pump and suction pipe with water (Scherer, 1993).

Centrifugal pumps are designed for either horizontal or vertical operation. The horizontal centrifugal pump has a vertical impeller connected to a horizontal drive shaft as shown in Figure 16.2. Horizontal centrifugal pumps are the most common in irrigation systems. They are generally less costly, require less maintenance, easier to install, and more accessible for inspection and maintenance than a vertical centrifugal. There are self-priming horizontal centrifugal pumps, but they are special purpose pumps and not normally used with irrigation systems (Scherer, 1993).

On irrigated areas where raising the water level or the pressure is necessary to make modern irrigation systems work, centrifugal pumps are almost exclusively used. In these cases, a large part of the cost of water is due to the energy bill and, sometimes, is the only charge paid directly by farmers. The pump efficiency is dependent on the operating conditions and the wear and tear on the machine. Any deviation from the optimum operating conditions, at maximum efficiency, leads to considerable energy expenditure which is rarely estimated due to lack of measurement or use of reliable indicators (Luc et al., 2006).

Experience shows that very few managers have continuous control of the actual efficiency of the pumping station. But using a theoretical reminder of the efficiencies of centrifugal motor pumps, it is possible to define easily accessible indicators as well as reference values (adapted to each type of pump) (Luc et al., 2006).

Centrifugal pumps enter the dynamic pump group used for irrigation (Figure 16.3). They continuously give energy to the fluid. Fluid movement is continuous. It is not discontinuous. On the other hand, the fluid is absorbed while on the other hand is pressed. As the flow rate of centrifugal pumps increases, the pressure remains to a certain extent. Speeds are higher. The speed of these pumps can be 6,000 rpm and larger. In centrifugal pumps, the relationship between the flow rate and the speed is not linear. There is an inverse relationship between flow rate and head. Centrifugal pumps are simple machines. The efficiency in centrifugal pumps is highly dependent on the ratio between flow and pressure. With centrifugal pumps, the flow rate can be easily adjusted with a regulating valve. Centrifugal pumps have virtually no air absorption capabilities. The suction pipe must be filled with water for the pump to operate. Centrifugal pumps take up less space, are light and cheap. Their efficiency is low at low flow rates and high pressures (Keskin and Güner, 2012).

FIGURE 16.3 Centrifugal pump.

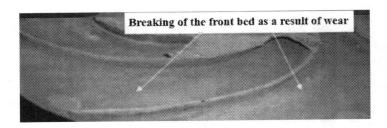

FIGURE 16.4 Wear pump impeller as a result of wear and bearing problems (Tural, 2011).

In centrifugal pumps, there are impeller, pump shaft, casing, chasis, suction and discharge port, diffuser, sealing elements (stuffing box), impeller, and wearing rings connected to the body. The impeller provides transmission of the energy from the engine to the water by means of the shaft and sends the water in the suction mouth to the discharge line. The impeller increases the kinetic energy of water. Water from the impeller loses its kinetic energy in the expanding body and converts its kinetic energy into pressure energy. The diffuser allows the kinetic energy of the water from the impeller to be changed more effectively to the pressure energy between the impeller and the volute. The wearing rings are located on the impeller, on the body, or on both the impeller and the body. These rings prevent fluid from passing through the discharge line into the suction line. Stuffing box prevents leakage between air and liquid leaks between shaft and body (Keskin and Güner, 2012).

The centrifugal pump elements are subject to wear due to friction, chemical and electrochemical effects, and high pressure formation (Figure 16.4). Pump elements wear, break, break down and as a result yield decreases, pump life is shortened, manufacturing and operating costs increase. It puts endangering human health and life in addition to the significant harm to the country's economy. It causes waste of metal resources. Labor's knowledge leads to loss of experience. A centrifugal pump is subjected to numerous harmful physical and chemical processes including service degradation during service. Each factor that damages the pump is wear.

The requirements for a successful pumping plant are performance and life. The performance is related to the pressure, flow rate, and efficiency of the pump. Life is the sum of hours of operation before the replacement of one or more parts for the pump to operate with acceptable performance. The pump manufacturer is responsible for the initial performance of the pump. It is connected to the design of the pump. The life of the pump affects the operating conditions of the pump. It depends on the resistance of the pump elements to wear under different operating conditions. Factors affecting the long life of pumps in centrifugal pumps are as follows (Tural, 2011):

- Neutral liquids at low temperatures;
- The absence of abrasive particles;
- The pump works at the maximum efficiency point, i.e. the pump operating point; and
- NPSHA (the absolute pressure at the suction port of the pump) must be greater than NPSHR (the minimum pressure required at the suction port of the pump) for the pump system to operate without cavitating.

A pumping facility providing all these conditions will be long-lasting. Pumps for water distribution networks are a typical example. Some water mains pump with bronze wheels and cast iron bodies last 50 years or longer. On the other hand, transmission of fluids with abrasive particles can shorten pump life considerably. The biggest factor that reduces wear, corrosion, and cavitation as well as pump life due to abrasive particles in the liquid is the operation of the pump outside the operating point. Damage of the impeller blades resulting from cavitation is particularly evident at the entrance of the impeller. The reason is that the impeller input is the lowest pressure point in the pump and the vapor bubbles formed in the suction line explode and generate high pressure (Karassik et al., 1985). The types of wear in centrifugal pumps can be classified under four headings (Tural, 2011):

a. Mechanical wear
b. Corrosion
c. Cavitation
d. Fatigue

16.3.3 Deep-Well Turbine Pump

Vertical turbine pumps were originally developed for pumping water from wells and have been called "deep-well pumps," "turbine-well pumps," and "borehole pumps." As their application to other fields has increased, the name "vertical turbine pumps" has been generally adopted by the manufacturers. (This is not too specific a designation because the term "turbine pump" has been applied in the past to any pump employing a diffuser. There is now a tendency to designate pumps using diffusion vanes as "diffuser pumps" to distinguish them from "volute pumps." As that designation becomes more universal, applying the term "vertical turbine pumps" to the construction formerly called "turbine-well pumps" will become more specific.) (Karassik and McGuire, 2012).

The largest fields of application for the vertical turbine pump are pumping from wells for irrigation and other agricultural purposes, for municipal water supply, and for industrial water supplies, processing, circulating, refrigerating, and air conditioning. This type of pump has also been used for brine pumping, mine dewatering, oil field repressurizing, and other purposes (Karassik and McGuire, 2012).

These pumps have been made for capacities as low as 2.5 or 3.5 m/h (10 or 15 gpm) and as high as 6,000 m/h (25,000 gpm) or more, and for heads up to 300 m (1,000 ft). Most applications naturally involve the smaller capacities. The capacity of the pumps used for bored wells is naturally limited by the physical size of the well, as well as by the rate at which water can be drawn without lowering its level to a point of insufficient pump submergence (Karassik and McGuire, 2012).

Vertical turbine pumps should be designed with a shaft that can be readily raised or lowered from the top to permit proper adjustment of the position of the impeller in the bowl. An adequate thrust bearing is also necessary to support the vertical shafting, the impeller, and the hydraulic thrust developed when the pump is in service. As the driving mechanism must also have a thrust bearing to support its vertical shaft, it is usually provided with one of adequate size to carry the pump parts as well. For these two reasons, the hollow-shaft motor or gear is more commonly used for vertical turbine pump drive. In addition, these pumps are sometimes made with their own thrust bearings to allow for belt drive or to use drive through a flexible coupling by a solid-shaft motor, gear, or turbine. Dual-driven pumps usually employ an angle gear with a vertical motor mounted on its top (Karassik and McGuire, 2012).

16.3.4 Submersible Pumps

A submersible pump is a turbine pump close-coupled to a submersible electric motor. Both pump and motor are suspended in the water, thereby eliminating the long drive shaft and bearing retainers required for a deep-well turbine pump. Because the pump is located above the motor, water enters the pump through a screen located between the pump and motor (Scherer, 1993).

The submersible pump uses enclosed impellers because the shaft from the electric motor expands when it becomes hot and pushes up on the impellers. If semi-open impellers were used, the pump would lose efficiency. The pump curve for a submersible pump is very similar to a deep-well turbine pump. Submersible motors are smaller in diameter and much longer than ordinary motors. Because of their smaller diameter, they are lower efficiency motors than those used for centrifugal or deep-well turbine pumps. Submersible motors are generally referred to as dry or wet motors. Dry motors are hermetically sealed with a high dielectric oil to exclude water from the motor. Wet motors are open to the well water with the rotor and bearings actually operating in the water (Scherer, 1993).

Submersible pumps are used for a wide range of applications. If you have a small pump around your property or a well pump, it is probably a submersible pump. Submersible pumps are also common in sewage applications, raw water pumping applications, and to remove storm water from structures like parking garages. A submersible pump is a type of centrifugal pump that operates while completely submerged in the liquid it is pumping (Scherer, 1993).

A centrifugal pump combines the hydraulic design of an end-suction pump with a submersible motor. Submersible pumps are always of the close-coupled type—meaning that the impeller mounts directly on the end of the motor shaft, and the pump casing attaches directly to the motor frame (Scherer, 1993).

Submersible pump motors are designed with a water-cooled jacket. In most cases, the liquid being pumped is circulated around the motor to provide cooling for the motor. The internal parts of the motor are protected by a water-tight enclosure, which prevents the entry of any liquid. In addition, most larger submersible pumps include a sensor placed between the pump and motor which senses when the liquid has gotten past the motor seals and prevents the unit from further operation until repairs have been made (Scherer, 1993).

16.3.5 Submersible Turbines

Another type of submersible pump, submersible turbines, are also used for irrigation and well pump applications. Submersible turbines incorporate a standard turbine bowl assembly with a submersible motor—a motor designed to operate submerged in water (Figure 16.5). The entire assembly is then bolted to the end of a length of pipe and lowered down into the well, lake, or river out of which water is to be drawn.

If you have ever seen a length of pipe rising out of a source of water, followed by an elbow, and additional piping, and you see a cable running down into the water, you are most likely looking at a submersible pump installation, and quite possible a submersible turbine installation.

Submersible turbines are common in deep well applications and are also used raw water pumps for irrigation, industrial plants, and municipal water supply.

16.3.6 Propeller Pumps

Propeller pumps are used for low-lift, high-flow-rate conditions (Figure 16.6). They come in two types, axial flow and mixed flow. The difference between the two is the type of impeller. The axial flow pump

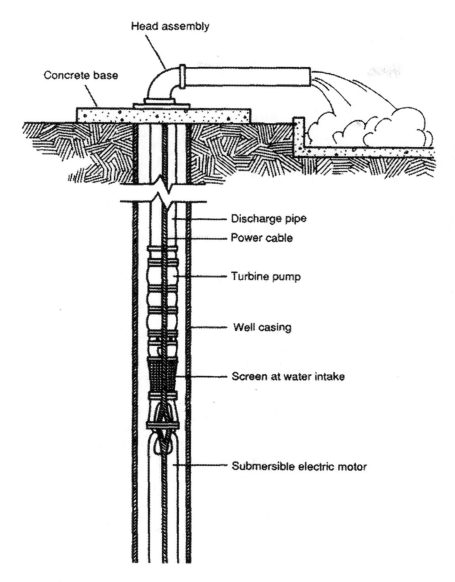

FIGURE 16.5 A submersible pump installed in a well (Scherer, 1993).

uses an impeller that looks like a common boat motor screw and is essentially a very low head pump. A single-stage propeller pump typically will lift water no more than 20 ft. By adding another stage, heads from 30 to 40 ft are obtainable. The mixed-flow pump uses either semi-open or closed impellers similar to turbine pumps (Scherer, 1993).

In permanent installations, propeller pumps are mounted vertically as shown in Figure 16.6. For portable pumping platforms, they are mounted on trailers or they are mounted on pontoons for use as floating intakes. Portable propeller pumps are commonly mounted in almost horizontal positions (low angles) to allow them to pump into pipelines easily as well as to be backed into a water source (Scherer, 1993).

Portable propeller pumps are commonly powered by the power-take-off on tractors. On many farms, propeller pumps are used to pump out waste storage lagoons. Power requirements of the propeller pump

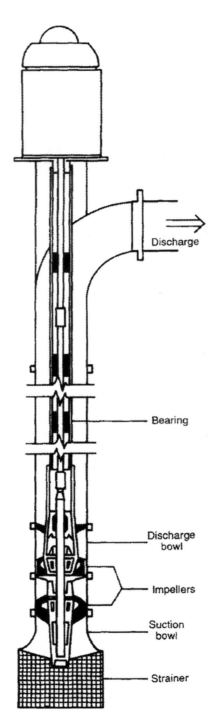

FIGURE 16.6 A propeller pump (Scherer, 1993).

TABLE 16.1 Selection of a Pump for Irrigation

Pump Type	Advantages	Disadvantages
	Factors to Consider in Selecting an Irrigation Pump	
Centrifugal	1. High efficiency 2. Easy to install. 3. Simple, economical, and adaptable to many situations. 4. Electric, internal combustion engines, or a tractor power can be used. 5. Does not overload with increased TDH. 6. Vertical centrifugal may be submerged and not need priming.	1. Suction lift is limited. It needs to be within 20 vertical feet of the water surface. 2. Priming required. 3. Loss of prime can damage pump. 4. If the TDH is much lower than design value, the motor may overload.
Vertical turbine	1. Adapted for use in wells. 2. Provides high TDH and flow rates with high efficiency. 3. Electric or internal combustion power can be used. 4. Priming not needed. 5. Can be used where water surface fluctuates.	1. Difficult to install, inspect, and repair. 2. Higher initial cost than a centrifugal pump. 3. To maintain high efficiency, the impellers must be adjusted periodically. 4. Repair and maintenance are more expensive than centrifugals.
Submersible	1. Can be used in deep wells. 2. Priming not needed. 3. Can be used in crooked wells. 4. Easy to install. 5. Smaller diameters are less expensive than comparable sized vertical turbines.	1. More expensive in larger sizes than deep-well vertical turbines. 2. Only electric power can be used. 3. More susceptible to lightning. 4. Water movement past motor is required.
Propeller	1. Not suitable for suction lift. 2. Cannot be valved back to reduce flow rate. 3. Intake submergence depth is very critical. 4. Limited to low (less than 75 ft) TDH.	1. Simple construction. 2. Can pump some sand. 3. Priming not needed. 4. Efficient at pumping very large flow rates at low TDH. 5. Electric, internal combustion engine, and tractor power can be used. 6. Suitable for portable operation.

increase directly with the total dynamic head (TDH), so adequate power must be provided to drive the pump at maximum lift. Propeller pumps are not suitable under conditions where it is necessary to throttle the discharge to reduce the flow rate. It is important to accurately determine the maximum TDH against which this type of pump will operate (Scherer, 1993).

Propeller pumps are not suitable for suction lift. The impeller must be submerged and the pump operated at the proper submergence depth. The depth of submergence will vary according to various manufacturer's recommendations, but generally, the greater the diameter of pump, the deeper the submergence. Following recommended submergence depths will ensure that the flow rate is not reduced due to vortices. Also, failure to observe required submergence depth may cause severe mechanical vibrations and rapid deterioration of the propeller blades (Scherer, 1993).

16.4 Conclusions

The selection of an irrigation water pump is based almost entirely on the relationship between pump efficiency and the TDH the pump will provide at a specific flow rate. As shown before, these parameters are also the basis of the pump characteristic curve. Table 17.1 can be used to narrow the selection of a pump type for a broad range of flow rates and total dynamic heads in irrigation purpose.

References

Al-Jazari, I.A.-R. 1974. Pump driven by a water-wheel. In *The Book of Knowledge of Ingenious Mechanical Devices*. Springer, Dordrecht, The Netherlands, pp. 186–189.

Ali, B. 2018. Comparative assessment of the feasibility for solar irrigation pumps in Sudan. *Renew. Sustain. Energy Rev.* 81, 413–420.

Aliyu, M.; Hassan, G.; Said, S.A.; Siddiqui, M.U.; Alawami, A.T.; Elamin, I.M. 2018. A review of solar-powered water pumping systems. *Renew. Sustain. Energy Rev.* 87, 61–76.

Argaw, N.; Foster, R.; Ellis, A. 2003. Renewable Energy for Water Pumping Applications in Rural Villages; Period of Performance: April 1, 2001–September 1, 2001; National Renewable Energy Laboratory, Golden, CO.

Becenen, I.; Eker, B. 2005. Powering of water pumps by alternative energy sources in Thrace Region. *Trakia J. Sci.*, 3, 28–31.

Burney, J.A.; Naylor, R.L. 2012. Smallholder irrigation as a poverty alleviation tool in Sub-Saharan Africa. *World Dev.*, 40, 110–123.

Chandel, S.; Naik, M.N. 2015. Chandel, R. Review of solar photovoltaic water pumping system technology for irrigation and community drinking water supplies. *Renew. Sustain. Energy Rev.*, 49, 1084–1099.

Collett, J. 1981. Hydro powered water lifting devices for irrigation. In *FAO/DANIDA Work. Water Lifting Devices Asia near East*. Food and Agriculture Organization of the United Nations, Bangkok, Thailand.

Fraenkel, P. 1979. *The Power Guide. A Catalogue of Small Power Equipment*. Intermediate Technology Publications, London, UK.

Fraenkel, P. 1986. *Water Pumping Devices: A Handbook for Users and Choosers*. Intermediate Technology Publications, London, UK.

Gopal, C.; Mohanraj, M.; Chandramohan, P.; Chandrasekar, P. 2013. Renewable energy source water pumping systems—A literature review. *Renew. Sustain. Energy Rev.*, 25, 351–370.

Grundfos Holding A/S. 2020. *Irrigation Pump Handbook*. www.grundfos.com.

Hofkes, E.H.; Visscher, J.T. 1986. *Renewable Energy Sources for Rural Water Supply*. IRC, The Hague, The Netherlands, Volume 23.

Intriago, J.C.; Ertsen, M.; Diehl, J.-C.; Michavila, J.; Arenas, E. 2018. Co-creation of affordable irrigation technology: The DARE-TU project. In Proceedings of the International Conference Water Science for Impact, Wageningen, The Netherlands, 16–18 October; p. 1.

Intriago Zambrano, J.C., Michavila, J., Arenas Pinilla, E., Diehl, J.C.; Ertsen, M.W. 2019. Water lifting water: A comprehensive spatiotemporal review on the hydro-powered water pumping technologies. *Water*, 11(8), 1677.

Johansson, S.; Nilsson, R. 1985. *Renewable Energy Sources in Small-Scale Water Pumping Systems*. Allmänna Ingenjörsbyrån AB, Stockholm, Sweden.

Karassik, I.; McGuire, J.T. eds. 2012. *Centrifugal Pumps*. Springer Science & Business Media, Berlin.

Karassik, I.; Krutzsch, W.C.; Fraser, W.H.; Messina, J.P. 1985. *Pump Handbook*. McGraw-Hill Book Company, New York.

Keskin, R.; Güner, M. 2012. Sulama Makinaları. Ankara Üniversitesi Ziraat Fakültesi Yayını, Ders Kitabı: 339, No: 1587, 292, Ankara, Turkey.

Kristoferson, L.A.; Bokalders, V. 1986. *Renewable Energy Technologies: Their Applications in Developing Countries*, 1st ed. Pergamon Press Ltd., Oxford, UK.

Kumar, A.; Kandpal, T. 2007. Renewable energy technologies for irrigation water pumping in India: A preliminary attempt towards potential estimation. *Energy*, 32, 861–870.

Lowder, S.K.; Skoet, J.; Raney, T. 2016. The number, size, and distribution of farms, smallholder farms, and family farms worldwide. *World Dev.*, 87, 16–29.

Luc, J.P.; Tarhouni, J.; Calvez, R.; Messaoud, L.; Sablayrolles, C. 2006. Performance indicators of irrigation pumping stations: application to drill holes of minor irrigated areas in the Kairouan plains (Tunisia) and impact of malfunction on the price of water. *Irrig. Drain.*, 55(1), 85–98.

Meah, K.; Ula, S.; Barrett, S. 2008. Solar photovoltaic water pumping—opportunities and challenges. *Renew. Sustain. Energy Rev.*, 12, 1162–1175.

Mohammed Wazed, S.; Hughes, B.R.; O'Connor, D.; Kaiser Calautit, J. 2018. A review of sustainable solar irrigation systems for Sub-Saharan Africa. *Renew. Sustain. Energy Rev.*, 81, 1206–1225.

Purohit, P. 2007. Financial evaluation of renewable energy technologies for irrigation water pumping in India. *Energy Policy*, 35, 3134–3144.

Purohit, P.; Kandpal, T.C. 2005. Renewable energy technologies for irrigation water pumping in India: Projected levels of dissemination, energy delivery and investment requirements using available diffusion models. *Renew. Sustain. Energy Rev.*, 9, 592–607.

Rossi, C.; Russo, F.; Russo, F. 2009. *Ancient Engineers' Inventions. Precursors of the Present*. Springer, Berlin/Heidelberg, Germany, Volume 8.

Scherer, T.F. 1993. Irrigation water pumps, NDSU Extension Service, https://www.ag.ndsu.edu/publications/crops/irrigation-water-pumps.

Tiemersma, J.J.; Heeren, N.A. 1984. *Small Scale Hydropower Technologies: An Overall View of Hydropower Technologies for Small Scale Appliances*. Stichting TOOL, Amsterdam, The Netherlands.

Tscharntke, T.; Clough, Y.; Wanger, T.C.; Jackson, L.; Motzke, I.; Perfecto, I.; VanderMeer, J.; Whitbread, A. 2012. Global food security, biodiversity conservation and the future of agricultural intensification. *Boil. Conserv.*, 151, 53–59.

Tural, H.N. (2011). Pompalarda Malzeme Kaynaklı Hasarlar ve Malzeme Seçimi. 7. Pompa ve Vana Kongresi–Mayıs http://www.standartpompa.com/sites/default/files/techdocs/pompalarda_malzeme_kaynakli_hasarlar_ve_malzeme_secimi-pomsad_7._kongre-mayis_2011_0.pdf, Erişim tarihi: 14.01.2018.

Valipour, M., Gholami Sefidkouhi, M.A., Eslamian, S. 2015. Surface irrigation simulation models: a review, *Int. J. Hydrol. Sci. Technol.*, 5(1), 51–70.

Wood, A.D. 1976. *Water Lifters and Pumps for the Developing World*. Colorado State University, Fort Collins, CO, USA.

Wood, A.D.; Ruff, J.F.; Richardson, E.V. 1977. *Pumps and Water Lifters for Rural Development*. Colorado State University, Fort Collins, CO.

Yannopoulos, S.I.; Lyberatos, G.; Theodossiou, N.; Li, W.; Valipour, M.; Tamburrino, A.; Angelakis, A.N. 2015. Evolution of water lifting devices (pumps) over the centuries worldwide. *Water*, 7, 5031–5060.

Zhang, Y.; Gao, Z.; Jia, Y.L. 2019. A bibliometric analysis of publications on solar pumping irrigation. *Sustainable Development of Water Resources and Hydraulic Engineering in China*, 303–315.

17
Pumps for Irrigation Systems

17.1	Introduction 356	
	Local Water Lifts	
17.2	Pump Characteristics 361	
	Terminology • Relations of Speed and Impeller Diameter on Pump Performance	
17.3	Centrifugal Pumps 366	
	Principle of Operation of Centrifugal Pumps • Priming of Centrifugal Pumps • Classification of Centrifugal Pumps • Centrifugal Pumps Used in Canals and Rivers • Portable Centrifugal Pumps	
17.4	Operation, Maintenance, and Troubleshooting of Centrifugal Pumps 368	
	Maintenance of Centrifugal Pumps • Troubleshooting of Pump • Cavitation	
17.5	Vertical Turbine Pumps 371	
	Pump Construction • Pump Lubrication • Pump Characteristics • Operation, Maintenance, and Troubleshooting of Vertical Turbine Pumps • Pump Maintenance • Pump Troubleshooting and Their Remedies	
17.6	Submersible Pumps 377	
	Pump Construction and Operation • Installation, Operation, and Maintenance of Submersible Pumps • Common Troubles and Their Remedies of Submersible Pumps	
17.7	Propeller and Mixed-Flow Pumps 379	
	Construction and Operation of Propeller Pump • Operating Characteristic of Propeller Pump • Propeller Pump Installation • Mixed-Flow Pump	
17.8	Jet Pumps 381	
	Uses and Adaptability of Jet Pump	
17.9	Air-Lift Pumps 382	
17.10	Pump Selection 382	
	Criteria and Procedure for Selecting a Pump for Irrigation • Determination of Discharge Capacity of the Pump	
17.11	Capacity of Pump Based on Crop Water Requirement 383	
	Selection of Pump by Well and Pump Characteristics Curves	
	References 385	

Mahbub Hasan and
Aschalew Kassu
Alabama A&M University

Saeid Eslamian
Isfahan University of Technology

DOI: 10.1201/9780429290152-23

17.1 Introduction

Application of irrigation water either from a surface or sub-surface sources needs lifting of water from the source to deliver into the fields (Albaji et al., 2020). In any water-lifting project, the overall efficiency of the pumping system depends on the following points:

- Application of sound principles in the design and construction of the irrigation well, and
- The characteristics of the lifting device (pumps) in relation to the source of water.

There are a variety of pumps ranging from indigenous types to modern and highly efficient types of pumps. Electric motors or engine-driven pumps have been prominent in all ranges of lift irrigation systems and the selection of the kind of pump to be used is dependent on specific requirements. It is evident that a higher output and level of efficiency can easily be controllable and attainable while using the mechanically powered water-lifting pumps.

The basic parameters that involved in pump operation are:

1. Atmospheric pressure,
2. Positive displacement,
3. Centrifugal force, and
4. Water column movement due to the difference in specific gravity.

Different types and classifications are shown in Figure 17.1. Pump selection for a specific requirement is a critical issue and needs the following considerations:

1. Characteristics of the source of water and the lifting device,
2. Flow size or amount of water to be lifted,
3. Pumping water level depth, and
4. Energy or power availability and the economic status of the users.

17.1.1 Local Water Lifts

In different developing and underdeveloped countries, farmers use some indigenous lifting devices. The sources of energy for lifting water are either manually operated or animal-operated. Based on the height of the lift, they are grouped into low, medium, or high lift.

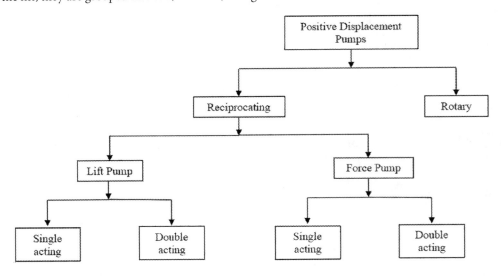

FIGURE 17.1 Classification of positive displacement-type pumps (Church and Jagdish, 1973).

17.1.1.1 Low Head Water Lifts

As explained before, a grouping of indigenous water-lifting method tools are in use in places where the mechanization of irrigation or agricultural sector is not yet established or in the transition stage. When the water lift doesn't exceed more than 1.2 m, the indigenous tools are usually in operation for irrigating the croplands. In India and some other Asian countries, a shovel-like basket is tied with a pair of ropes in each side and two persons facing each other swing the basket to collect water and throw to the field in a rhythmic manner. This is called *swing basket*. This is one of the old traditional methods of irrigating the croplands manually. The swing starts to collect water by the basket from a lower head (water source) and discharge into the channel or crop field at the upper head.

Irrigation water lift can be broadly divided into two major methods, which are as follows:

1. Indigenous water lift method, and
2. Pumps.

The indigenous water lift devices again consist of:

1. Low head,
2. Medium head, and
3. High head.

The indigenous water-lifting devices are still in use in different underdeveloped or developing countries where the agriculture sector is not yet mechanized. In the case of pumps, the basic types are as follows:

1. Positive displacement pumps, and
2. Variable displacement pumps.

17.1.1.2 Medium Head Water Lifts

When the height of the water lift is more than 1.2 m and less than 10 m, the water lifts can be classified as *medium headwater lifts*. Examples of medium headwater lifts are:

a. Leather bucket lift with self-emptying bucket,
b. Chain pump,
c. Circular two bucket lift,
d. Counterpoise-bucket lift, and
e. Persian wheel.

Above are the old and ancient types of water-lifting devices that were used when irrigating crop fields. The above-mentioned devices were used at the places where the crop fields are a little elevated compared to the source of water. As these were all operated manually or by animals, the discharges were lower compared to modern types of mechanical pumps (Figure 17.2).

17.1.1.3 High Head Water Lifts

There is only a single indigenous type of high head water-lifting device available, which is known as *rope-and-bucket* lift. It is known as a Charasa locally in India and other Asian countries. This is operated by animals, especially bullock. Depending on the irrigation water requirement and water yield, the device may be consisting of single or multiples *Charasa*. This device consists of a bag of capacity ranges from 150 to 200 L and made of leather or a bucket made of galvanized iron sheet. The bucket or bag tightly fixed on a secured iron ring, which is with an iron framework at the top and is knotted to a long rope. The rope passes over a pulley set on a wooden frame and fixed on top of the well. A pair of animal (usually bullock) is hitched at the other end of the rope delivering power to draw water from the well and deliver to the trough or irrigation canal through an inclined ramp sloped at an angle of 5°–10° to allow the gravity flow into the discharge canal.

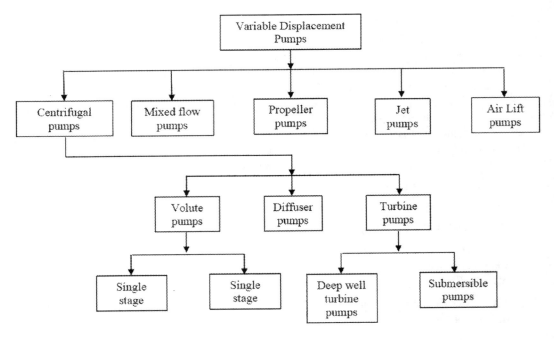

FIGURE 17.2 Classification of variable displacement-type pumps (Church and Jagdish, 1973).

17.1.1.4 Wind Mills

The utilization of natural power for lifting water from the well is taken place by a wind mill. It consists of a large-diameter vane wheel (or rotor) fixed on the top of a high steel or wooden tower, and there is a reciprocating pump at the bottom of the frame. The upper portion of the upper piston rod is connected with the vane wheel. The piston ring is connected eccentrically. The lower part of the suction pipe and the pump cylinder are always to be under water in the well. If a strong wind starts blowing, the vane wheels start rotating, and ultimately the water is pumped up. The delivery pipe is connected to a reservoir, and the discharge of water is controlled or regulated to distribute whenever water is required for uses. The operation, storage, and distribution of water are intermittent and seasonal. As the basic energy requirement comes from wind, the capacity is dependent on wind velocity and frequency. The windmill is appropriate for lifting water from shallow to moderately deeper wells. Also, the lifting capacity is associated to wind velocity (Michael, 1997).

Wind is a free resource of energy, and it can be well suited in areas with the following characteristics:

1. Wind has sufficient speed,
2. The catchment of wind has no obstruction, and
3. The availability of sufficient wind can be relied on during water necessity for water storage in the reservoir.

Wind velocity of 4 mph or more is essential for the suitable running of a windmill. As wind is a naturally available resource of energy, the operating cost of a windmill is almost zero once the initial development or construction cost is covered.

17.1.1.5 Positive Displacement Pumps

The pump which supplies the same volume of water irrespective of the head against which the pump is functioning is called the *positive displacement pump*. Maximum power to meet the highest load yielded

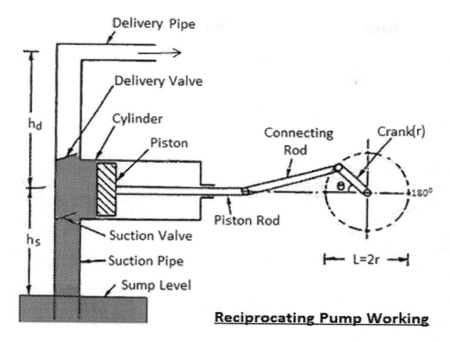

FIGURE 17.3 Diagram of a reciprocating pump showing different parts (Reciprocating Pump Working Principle, 2020).

from the capacity of the pump and the utmost head under which the pumps are anticipated to work. Deepwater irrigation needs a huge volume of water; these types of pumps are not used for irrigation nor for drainage because the discharge capacity of *Positive Displacement Pumps* is low. However, they can be used for home water supply systems (storage in the overhead tanks or reservoirs) and in vegetable irrigation systems.

17.1.1.6 Reciprocating Pumps

Different names of *reciprocating pumps* are *piston pump* or *displacement pumps*. In the reciprocating pump, a piston is under movement that displaces water in a cylinder and pumping in water is being regulated by valves. This piston (or sometimes it is called a plunger) moves forward and backward inside a hollow-shaped cylinder. The capacity of reciprocating pumps depends on:

1. Chamber size,
2. Difference between the extreme points of forward and backward movement, and
3. Speed of stroke (speed of the movement the extreme two points from backward to forward).

Figure 17.3 describes the function of a reciprocating pump.

Reciprocating pumps in the figure above shows two heads. They are h_s and h_d, which are representing suction and discharge head, respectively. The piston moves forward and backward by the action of a crank of radius r and is operated by a connecting rod. The movement of the piston creates a distance of 2r, which can be represented as *L*. When the piston moves backward, water from the sump starts moving toward upward and stored in the chamber and it continues to be stored in the chamber till the crank completes the half cycle of 180°. The crank starts moving forward, the suction valve gets closed and the discharge valve gets opened with the discharging water, and finally gets the water its way to the delivery pipe.

17.1.1.7 Variable Displacement Pumps

The typical characteristic of the discharge of variable displacement pumps is inversely proportional to the pressure head. If the pumping head increases, the rate of flow size decreases. Variable displacement pumps need higher power input at the lower head, but a positive displacement pump is opposite to this. Variable displacement pumps are usually used in pumping water for irrigation purposes. They are of different impeller types, including centrifugal, mixed-flow, and propeller pumps. Variable displacement pumps use a rotating impeller, and they produce small discharges with higher pressure heads and larger discharges with smaller pressure heads.

17.1.1.8 Specific Speed

Specific speed is an index that represents the operating characteristics of pumps. It expresses the relationship between speed, discharge, and pressure head of pumps. The speed is expressed as revolutions per minute, or simply it is called *rpm* while using the British units (Foot-Pound-Second (FPS) units). The rpm is the speed at which a geometrically and theoretically similar pump would run if proportioned to deliver one gallon per minute against 1 ft total pressure head at its best efficiency. Initially, the equation for the specific speed of pumps was developed using the following relationship:

$$n_s = \frac{nQ^{\frac{1}{2}}}{H^{\frac{3}{4}}},$$

where:

n_s = specific speed, rpm
n = pump speed, rpm
Q = pump discharge, U.S. gallon per minute
H = pressure head, ft.

In the case of the metric unit system, specific speed is defined as a geometrically and theoretically similar pump when discharging 1 m³ of water per second against a pressure head of 1 m (Church and Jagdish, 1973). Mathematically, it can be shown as follows:

$$n_s = \frac{nQ^{\frac{1}{2}}}{H^{\frac{3}{4}}},$$

where:

n_s = specific speed, rpm
n = pump speed, rpm
Q = pump discharge, m³/s
H = pressure head, m.

Specific speed in dimensionless form can be shown as follows:

$$n_s = \frac{nQ^{\frac{1}{2}}}{(gH)^{\frac{3}{4}}}.$$

Here, g is the acceleration due to gravity. However, the above two equations are used in practical purposes.

Example

A centrifugal pump with its highest efficiency is delivering 0.04 m³ of water per second. The total pressure head is 42 m, and the speed is 1,500 rpm. Calculate the specific speed of the pump.

Solution: $n_s = \dfrac{nQ^{\frac{1}{2}}}{H^{\frac{3}{4}}}$

$$= \dfrac{1{,}500 \times (0.04)^{\frac{1}{2}}}{(42)^{\frac{3}{4}}} = 16.498 \approx 16.5 \text{ rpm}$$

The normal specific speed of a single suction impeller ranges from 500 to 15,000 in FPS unit system, and this is 10–300 when measured in the metric unit system. Specific speed for a multi-stage pumping system is calculated based on the head in each stage. If the specific speed of a double suction impeller pump is compared with that of a single suction impeller pump, the capacity of the double suction impeller pump should be divided by $\sqrt{2}$ or its specific speed should be divided by. Typically, the lower the specific speed, the higher the head developed in each stage of the pump.

17.2 Pump Characteristics

A pump operates most satisfactorily under a head and a speed for which it is designed. Conditions for operating the pumps should be determined very accurately to select the pump for adapting to the specific conditions of operation.

17.2.1 Terminology

Capacity: volume of water pumped per unit time. This is generally expressed as liter per second. If the value of capacity is very small, it is expressed as liter per minute or per hour. For convenience of measurement, the larger value of capacity is expressed as cubic meter per second (m³/s).

Suction lift: source of water is below the centerline of the pump.

Static suction lift: vertical depth of the surface of the water to be lifted to the centerline of the pump.

Total suction lift: this is the sum of static water lift, friction, and entrance in the suction pipe system.

Suction head: when the source of the water supply is above the centerline of the pump as seen in turbine pumps.

Static suction head: vertical distance between the centerline of the pump and the free level of water to be pumped.

Total suction head: vertical distance between the centerline of the pump and the free level of the liquid to be pumped minus all friction losses in suction pipe and fittings, plus any pressure head existing on the suction supply.

Total discharge head: sum of the static discharge head, friction and exit losses in the discharge piping plus the velocity head and pressure head at the point of discharge.

Total static head: vertical distance between suction water level and discharge water level or the static suction lift and static discharge head.

Friction head: equivalent head expressed in meters of water required to overcome the friction, caused by the flow through the pipe and pipe fittings.

Pressure head: the pressure of water in a closed vessel from which the pump takes its suction or against which the pump discharges. This is expressed in meters. Pressure head is mathematically shown as follows:

$$H_p = \frac{P}{\omega},$$

where:

H_p = pressure head, m
P = pressure inside the vessel, kg/m^2
ω = specific weight of water, kg/m^3.

Total head: energy imparted to the water by pump. It is defined as the sum of the total discharge head and total suction lift when suction lift exists. This is the total discharge head minus the suction head where suction head exists.

Velocity head: pressure that creates velocity to move water. This is also expressed in meter. Mathematically it is shown as follows:

$$H_v = \frac{v^2}{2g},$$

where:

H_v = velocity head, m
v = velocity of water, m/s
g = acceleration due to gravity, m/s^2 (9.81 m/s^2).

Net positive suction head (NPSH): the total suction head determined at the suction nozzle (measure with respect to the pump centerline) minus the vapor pressure of water corresponding to the water temperature. Both are expressed in meters. The pressure at any point of the pumping liquid in the suction line must not be reduced to the vapor pressure of the liquid. Vapor pressure of a liquid at any temperature is the pressure at which it will vaporize if heat is added to the liquid, or it is the pressure at which the vapor at the given temperature will condense into liquid if the head is removed. NPSH is a characteristic of the pump and is typically furnished with a pump *characteristic curve*.

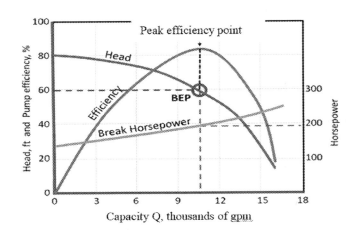

FIGURE 17.4 The characteristic curve of a centrifugal pump (Church and Jagdish, 1973).

Pumps for Irrigation Systems 363

The pump *characteristic curves* can be defined as "the graphical representation of a particular pump's behavior and performance under different operating conditions" (Characteristic Curves of Centrifugal Pumps, 2020). The operating properties of a pump are established by the geometry and dimensions of the pump's impeller and casing. The system characteristic curve is the response in head H (m) of the installation to a given liquid flow rate (Q m³/h, L/s, or gal/min).

The characteristic curve is also named as performance curve representing the interrelationship between capacity, head, power, and efficiency of the pump. The use of a characteristic curve may offer the choice of the best pump under particular conditions of operation and requirement with a lower operating cost.

Maximum practical suction lift of pumps: suction lift and all other losses must have to be less than theoretical atmospheric pressure while a centrifugal pump is in operation without any cavitation. The maximum practical suction lift can be shown and computed by the following equation:

$$H_s = H_a - H_f - e_s - NPSH - F_s,$$

where:

H_s = maximum practical suction lift (elevation of the pump centerline minus the elevation of the water surface), m
H_a = atmospheric pressure at water surface, m (10.33 m at the sea level)
H_f = friction losses in the strainer, pipe, fittings, and valves on the suction pipe, m
e_s = saturated vapor pressure of water, m
$NPSH$ = net positive suction head of the pump including losses at the impeller and velocity head, m
F_s = factor of safety (usually 0.6 m).

The correction for H_a for altitude is a reduction considered to be 0.36 m for every 300 m of altitude. Suction lift and friction losses are to be maintained as minimum as possible, and this is the reason why the suction pipe is longer than the discharge pipe. Also, the pump itself has to be positioned as near as possible to the water surface. Consideration for preventing cavitation should always be in mind, which is possible by taking the vapor pressure of water into consideration. The head loss due to the vapor pressure of water is not added to the total suction head when the pump is in operation.

Water horse power (WHP): It is the theoretical horsepower required for pumping operation. This is expressed as the multiplication of discharge and the head divided by a constant based on the unit of discharge. The factor is 75 if the discharge is in liters per second and it is 273 if the discharge is in cubic meter per second.

Hence, the equations can be shown as follows:

$$WHP = \frac{Q \times h}{75},$$

where:

Q = discharge in liter per second
h = pressure head, m,

and

$$WHP = \frac{Q \times h}{273},$$

where:

Q = discharge in cubic meter per second
h = pressure head, m.

Shaft horsepower (SHP): This is the power needed to rotate the shaft of the pump. This is defined by the following equation:

$$SHP = \frac{\text{Water horsepower}}{\text{Pump efficiency}}.$$

SHP is needed by the pump in WHP, disc friction, circular losses, shifting box, and bearing friction and hydraulic losses. It is always greater than the value of WHP. The ratio between WHP and SHP is called the pump efficiency.

$$\text{Pump efficiency} = \frac{\text{Water horsepower}}{\text{Shaft horsepower}}$$

Brake horsepower (BHP): actual horsepower needed to be supplied by the motor or electric source for driving the pump. When the pump is driven by a mechanical motor, it is called 100% drive efficiency means BHP is equal to SHP. But if the pump is driven by an indirect mover like driven by a belt, then the equation for BHP can be written as follows:

$$\text{Brake horsepower (BHP)} = \frac{\text{Water horsepower}}{\text{Pump efficiency} \times \text{drive efficiency}}.$$

Horsepower needs to be available to the electric motor can be shown as follows:

$$\text{Horsepower input to the motor} = \frac{\text{Water horsepower}}{\text{Pump efficiency} \times \text{drive efficiency} \times \text{motor efficiency}}.$$

The electrical energy in kilowatt input to the motor can be shown as follows:

$$\text{Energy input to the electric motor} = \frac{\text{Brake horsepower} \times 0.746}{\text{Motor efficiency}}.$$

17.2.2 Relations of Speed and Impeller Diameter on Pump Performance

Changing the speed or impeller diameter causes a change in the performance of the pump. The relationships in the performance of the pump due to changes in speed are shown below:

1. The capacity of the pump is directly related to the pump speed, and it is shown as

$$Q = Q_1 \left(\frac{n}{n_1}\right),$$

where:
Q = capacity at the desired speed n, L/s
Q_1 = capacity at speed n_1, L/s
n = new desired speed, rpm
n_1 = speed at which the characteristics are known, rpm.

2. The head varies as the square of the speed. The relationship can be shown as follows:

$$H = H_1 \left(\frac{n}{n_1}\right)^2,$$

where:
H = head at the desired speed n for capacity Q, m
H_1 = head at speed n_1 for capacity Q, m.

3. The BHP varies with the cubic value of the speed. The relationship is shown below:

$$P = P_1 \left(\frac{n}{n_1}\right)^3,$$

where:
P = BHP at the desired speed n, head H, and capacity Q
P_1 = BHP at speed n_1 at H_1, and capacity Q_1.

Changing the speed or impeller diameter causes a change in the performance of the pump. The relationships on the performance of the pump due to change in impeller diameter is shown below:

1. The capacity of the pump is directly related to the impeller diameter and it is shown as

$$Q = Q_1 \left(\frac{D}{D_1}\right),$$

where:
D = changed diameter of the impeller, mm
D_1 = original diameter of the impeller, mm.

2. The capacity of the pump varies as the square of the impeller diameter. The relationship can be shown as follows:

$$Q = Q_1 \left(\frac{D}{D_1}\right)^2.$$

3. The BHP varies with cubic value of the impeller diameter. The relationship is shown below:

$$P = P_1 \left(\frac{D}{D_1}\right)^3.$$

Now the equivalency in the ratios of the relations can be shown as follows:

$$\left(\frac{D}{D_1}\right) = \left(\frac{Q}{Q_1}\right) = \left(\frac{H}{H_1}\right)^{\frac{1}{2}} = \left(\frac{P}{P_1}\right)^{\frac{1}{3}}.$$

17.3 Centrifugal Pumps

To have an efficient irrigation system that maintains the principles of irrigation scheduling, a pump must be (What Is a Centrifugal Pump, 2020):

1. simple to construct,
2. easy to operate,
3. lower initial cost, and
4. producing a constant steady discharge.

The above-stated criteria are existing in centrifugal pumps. They can be very easily coupled with the motor or engine drives without the use of expensive gears. This pump is well suited for irrigation, water supply, and sewage purposes. This pump doesn't have a valve, and can handle water with suspended solid particles, provided it is constructed to suit such conditions.

17.3.1 Principle of Operation of Centrifugal Pumps

A centrifugal pump consists of two major parts. One is the rotary element and it is also called an impeller, and the other is the static part and it is a stationary part, casing of the pump. Impeller is a wheel or disc mounted on a shaft and made of a number of vanes having a curvature form. Figure 17.5 shows a *volute*-type centrifugal pump, and Figure 17.6 shows a *diffuser* type of centrifugal pump. The vanes in the pump are organized in a spherical array around the opening at the inlet at the center. In other types of pumps, a diffuser has a series of guide vanes surrounding the impeller, as shown in Figure 17.5. In this case, the impeller is protected on a shaft mounted on appropriate bearings. The shaft remains inside a staffing box and passes through the casing wall. The staffing box and its packing are usually made with asbestos or organic fiber. The casing surrounds the impeller and usually in the form of a spiral or volute curve with a cross-sectional area increasing toward the discharge opening.

The working principle of a centrifugal pump is that an impeller rotating inside a close-fitting case draws in water at the center and, due to the action of centrifugal force, throws out the water through an opening at the side of the casing. The basic hydraulic principle in the design of an impeller is the production of high velocity and a part of what is transformed into pressure head. Operation takes place starting with water intake into the pump casing, water rotation with the impeller, producing a higher velocity,

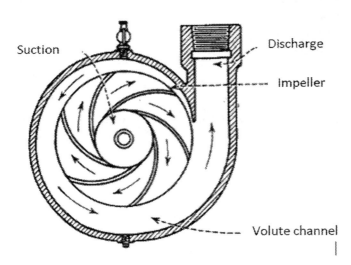

FIGURE 17.5 Volute-type centrifugal pump (Finkel, 2018).

Pumps for Irrigation Systems

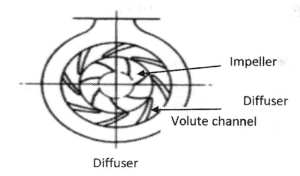

FIGURE 17.6 Diffuser-type centrifugal pump (Merkle, 2014).

and the part of this higher velocity is transformed into pressure head that helps the water causes throwing of water outward from the impeller into the casing. The outward flow of water through the impeller reduces pressure at the inlet, allowing more water to be drawn in through the suction pipe by the pressure due to atmosphere or an external pressure. Water passes into the casing where the high velocity is reduced and converted into pressure, and water is pumped out through the discharge pipe. Conversion of velocity into pressure head takes place either in volute (Figure 17.5) or in a diffuser chamber casing (Figure 17.6).

There is another kind of centrifugal pump, called a *diffuser* centrifugal pump. Figure 17.5 shows a picture of the cross section of the diffuser type of pump.

17.3.2 Priming of Centrifugal Pumps

Piston pump or positive displacement pumps can move and compress all types of fluids, including air. Centrifugal pumps are not capable of doing this operation except in very little quantity. They need to be primed, and the centrifugal pumps are filled with water up to the top of the pump casing prior to starting the pumping operation. Water filling up to the top of the casing of the pump initiates the pumping operation. Priming consists of either or the combination of the following (Centrifugal Pumps Working Principles and Characteristics, 2020):

1. A foot valve to grip the water in the pump,
2. A provision of filling with water the casing and the suction pipe, usually done by an auxiliary piston,
3. A connection to the outside source of water under pressure for filling the pump, and
4. Use of a self-priming construction.

Self-priming construction helps in retaining water in the auxiliary chamber, which is an additional part of the pump. The water retained in an auxiliary chamber or by the valve installed at the bottom of the suction pipe can eliminate the need of priming the pump for its initial start and then smooth operation. An extra strainer is attached at the bottom before the valve to prevent the intake of water along with any foreign material.

17.3.3 Classification of Centrifugal Pumps

There are a variety of centrifugal pumps. They can be classified according to their type as follows:

1. type based on the energy conversion
   a) volute type, and b) turbine or diffuser type

2. type based on staging
    a) single-stage, and b) multi-stage
3. type based on impeller
    a) open, b) semi-open, c) closed, and d) non-clog
4. type based on suction inlet
    a) single suction, and b) double suction
5. type of construction of the casing
    a) spit vertically, and b) spit horizontally
6. type of axis of rotation
    a) rotate vertically, and b) rotate horizontally
7. type based on method of driving
    a) connected directly, like
        (i) coupled, (ii) closed-coupled or uni-built, and
    b) driven by belt.

If a centrifugal pump driven by an engine is at a longer distance away and is coupled with a belt, transmitting power by the mechanical engine to the centrifugal pump, the excessive energy loss due to the longer vertical belt is not efficient. A better arrangement of replacing the long belt by a vertical shaft to transmit power from the engine placed at the ground to the pump placed as close as possible to the static water level in the well. Usually, a short cross belt is connected with the engine to the vertical shaft horizontally and the vertical shaft is connected with the pump. Construction of a deep watertight pit is difficult and expensive and if the drawdown is beyond 6 m suction lift, it is better to avoid the use of the centrifugal pump in this type of circumstance.

17.3.4 Centrifugal Pumps Used in Canals and Rivers

In most cases, centrifugal pumps are used for pumping water from a pond, lake, or canal. The pump is coupled to an electric motor or an engine to operate it. The electric motor or engine is fixed on a permanent foundation. The pumping unit is installed on a float and used as a portable unit with flexible suction and delivery pipes so that the pump can be easily movable. The flexible pipes are supported at suitable points. The foot valve is located at the minimum level of at least 60 cm of water during the pumping period.

17.3.5 Portable Centrifugal Pumps

The basic requirement of pumping water level (PWL) for a centrifugal pump must have to be maintained within 6 m depth. To achieve that condition, centrifugal pumps may be used as a portable device for pumping water from different wells, one after another, or from stream, lake, or open channel. Portable centrifugal pumps are generally engine-driven. The pumps and the engines are mounted on a common platform which is secured on a trolley and made with steel girders or wooden beams. The trolley is mounted on the steel wheels or rubber-tired pneumatic wheels.

17.4 Operation, Maintenance, and Troubleshooting of Centrifugal Pumps

Efficient and prolonged services are possible to be obtained by appropriate operation, care, and maintenance of centrifugal pumps (Michael, 1997). The pump should be started for its operation after the following investigations:

1. Alignment between the pump and the driver must be checked and misalignment (if any) must be corrected before starting the pump.

2. Check the direction of rotation of the pump and the motor or engine as it is indicated in the pump and the machine or motor's body.
3. Make sure that the gland is evenly and lightly adjusted and the pump shaft revolves turned freely when rotated by hand.
4. Check if there is any leakage in the suction pipe or in the foot valve before starting the pump.
5. Fill the suction pipe with water and remove air from the pump casing.
6. Check if the bearings are filled with recommended engine oil.

The centrifugal pumps need to be stopped immediately if it doesn't deliver water, which may cause excessive wearing in the moving parts of the pumps.

17.4.1 Maintenance of Centrifugal Pumps

Frequent and scheduled maintenances of the pump are required periodically to ensure the best possible life of the pump. The operating conditions of the pump vary broadly and so does the maintenance. Usually, the scheduled maintenances are as follows:

1. *Every monthly maintenance*: Check the bearing temperature. Excess or less lubricant may cause a temperature rise in the bearings.
2. *Every three monthly maintenance*: Using kerosene, drain out the lubricants from the ring oil-bearing and wash out oil wells and bearings. The oil rings in the sleeve bearings are to be checked if they are free to turn with the shaft. Lubricant should be refilled as recommended by the manufacturer. Check the wear in the bearings and to be replaced if it is excessive.
3. *Every six monthly maintenance*: Gland packings are to be replaced. The alignment of the pump with the driver should be checked. Add shims if essential. If there is frequently misalignment occurring, the entire piping system may have to be carefully checked and correction must be done.
4. *Every yearly maintenance*: A detailed inspection of the total unit is needed. Bearings should be removed, cleaned, and examined for any faults. The bearing chamber should be cleaned, packing removed, and examine the wear in the shaft sleeve or shaft itself. Coupling valves need to be disconnected and check the alignment. Foot valve and the check valves are required to be checked. They should be replaced if there is any wear found in any of the valves.

17.4.2 Troubleshooting of Pump

There may be drop of discharge or pressure in the pump when it is very urgently needed to be investigated. The necessary steps should be taken to get rid of the situation. A list of most commonly occurring troubles is mentioned below:

1. There is no water discharge.

    a. lack of priming the pump
    b. speed is very low
    c. head of discharge is very high
    d. suction head is very high
    e. suction pipe of the impeller is completely plugged
    f. direction of rotation is wrong
    g. suction pipe is having an air pocket
    h. leakage of air in the suction pipe of stuffing box
    i. NPSH is insufficient

2. Insufficient discharge.
    a. air leakage in the suction pipe or in the stuffing box
    b. speed is very low
    c. predicted discharge head is higher than actual discharge head
    d. suction lift is very high
    e. impeller or suction pipe is partially plugged
    f. direction of rotation is wrong
    g. NPSH is insufficient
    h. foot valve is very small than it is required
    i. submergence of suction inlet is insufficient
    j. wearing rings are worn

3. Pressure development is not enough.
    a. speed is very low
    b. amount of air or gas is excess in the liquid
    c. direction of rotation is wrong
    d. viscosity of liquid is higher than the predicted value
    e. wearing rings are worn
    f. the diameter of the impeller is very low

4. Priming is lost after a little operation.
    a. leakage or air in the suction pipe or in the stuffing box
    b. amount of air or gas is excess in the liquid
    c. the suction pipe is having air pocket
    d. water seal tube is clogged
    e. the water seal ring is improperly located
    f. suction lift is very high
    g. submergence of the suction inlet is insufficient

5. Pump needs excessive power.
    a. speed is very high
    b. the suction pipe is having air pocket
    c. viscosity or specific gravity of the fluid is very high
    d. the direction of rotation is wrong
    e. not correct alignment
    f. the stuffing box is very tight
    g. rubbing of rotating elements or binding
    h. shaft is bent
    i. wearing rings are worn

6. Stuffing box leaks excessively.
    a. improperly lubricated or the packing is worn
    b. packing is improperly installed
    c. type of packing is incorrect
    d. shaft sleeve scored
    e. shaft is bent

7. Pump vibrates or noisy.
    a. suction lift is very high
    b. NPSH is insufficient
    c. impeller or suction pipe is partially plugged
    d. alignment is not correct
    e. foundation is not rigid
    f. lubrication is insufficient
    g. bearings are worn
    h. rotating elements are in imbalance
    i. shaft is bent

17.4.3 Cavitation

Formation of cavities that are filled with liquid vapor due to the local pressure drop and their collapse as soon as the vapor bubbles, reach the zone of higher pressure. Typically, in a centrifugal pump, water enters into the suction pipe and then into the eye of impellers where the water increases its velocity. Simultaneously, a drop of pressure occurs due to the increment of the velocity of the water. If the pressure falls below the vapor pressure corresponding to the temperature of the water, the water will vaporize, and the flowing stream of water will accompany vapor pockets. Flowing further through the impeller, water reaches into a higher pressure region, and the cavities of vapor collapse. The collapse of vapor pockets creates the noise and occurrences of cavitation.

Cavitation is caused due to improper design, installation, and operation of pumps. Cavitation varies from mild to severe. A pump may be operating with mild cavitation but with higher noise. The ultimate effect is the lower in efficiency only. The other case a pump may be operating with severe cavitation and causing vibration and noise, ultimately destroying the impeller and other parts in contact with the pump. Therefore, the prevention of cavitation is a very important issue that needs to be taken care of. The following precautionary measures can be taken to prevent cavitation of a centrifugal pump while in operation.

 1. Avoid operating the pump at heads and discharge capacities far below the designed values.
 2. A higher value of water velocity and a higher suction lift should be avoided.
 3. The recommended speed of the pump by the manufacturer should be maintained, and higher speed shouldn't be allowed.

17.5 Vertical Turbine Pumps

Vertical turbine pump, which is also called a deep-well turbine pump, is a vertical-axis centrifugal or mixed-flow type of pump consisting of stages. The name signifies the meaning of a deep-well turbine pump; the suction lift is higher and it needs to be multi-staged. This deep-well turbine pump consists of rotating impellers and fixed bowls casing the vanes. The bowl with the assemblies is always situated beneath the water surface. A deep-well turbine pump is designed to adapt to the seasonal fluctuation of water level in the well. They are specially adapted to the PWL well below the actual limit of a volute-type centrifugal pump. Comparatively, a smaller diameter of the turbine pump matches well with their installation in tube wells.

Vertical turbine pumps ensure higher lifts of water with higher efficiencies under optimum operations. But vertical turbine pumps have a higher initial cost. These pumps are more difficult to install and cost a higher amount of money for any repair. Figure 17.7 shows the cross-sectional view of a vertical turbine pump. This figure clearly shows the different parts of a vertical turbine pump starting from vertical hollow-shaft motor down to the bottom with a secured strainer.

Pressure heads developed in centrifugal pumps depend on the diameter of the impeller and the speed of impeller rotation. In deep-well turbine pump, the diameter of the bowl and its impeller is restricted

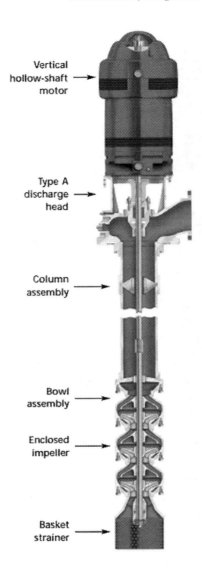

FIGURE 17.7 Cross-sectional view of a vertical turbine pump (Vertical Turbine Pumps, 2020).

by the relatively smaller diameter of the tube well. Therefore, the pressure head developed by a single unit of the impeller is not enough until and unless the additional pressure head is developed by multi-staging of the pumps.

17.5.1 Pump Construction

Deep-well turbine pump or vertical turbine pump consists of three main parts as follows:

1. *Pump element*

    The pump element is made up of one bowl (single-stage) or more bowls (multi-stages). Each bowl assembly has an impeller, a diffuser, and a bearing. The pump element has a secured screen at its bottom to keep uneven sized sand and gravel inflowing to the pump. The impeller is closed or may be semi-open type.

2. *Discharge column*

The discharge column is connected with the bowl assembly and the pump head assembly. It conveys water from the impeller bowl to the pump head assembly. It consists of a discharge or column pipe and the line shaft with coupling and bearings.

3. *Discharge head*

The head assembly comprises a base from which discharge column, bowl assembly, and shaft assembly are suspended. The discharged head elbow directs the water to be conveyed into the delivery pipe. Electric motors, right-angle gear drives, flat or V-belt pulleys, and matching combinations of these drives may be used on the same discharge head. The discharge elbow is equipped with flanges to the discharge pipe.

17.5.2 Pump Lubrication

Vertical turbine pumps are classified into two categories based on methods of lubrication. They are as follows:

1. Water lubricated, and
2. Oil lubricated.

The pumps with column pipes are made of steel and are joined by thread sleeve-type couplings. In this case, the drive shaft is positioned at the center of the discharge pipe. This is made of cold-finished, carbon steel of correct size to function the pump without vibration or distortion.

In the oil-lubricated pumps, the drive shaft is bounded in a concealing pipe, which directs the lubricants to the appropriate parts. Typically, bearings made with bronze are provided at every 1.5 m. The bearings themselves aid as couplings for the cover pipe. The shaft and its cover pipe assembly are braced and aligned by the reinforced rubber spider usually spaced in the column pipe at about every 15 m. The oil-lubricated column pipe has a plus point that the shafting is guarded against any abrasive or corrosive parts which may enter the pump. Oil lubricants system offers a more advantageous system of shaft and tube alignment.

Bearings are made of natural or synthetic rubber, set in bronze retainers in a water-lubricated pump. Water is pumped and lubricates the bearings in water-lubricated pumps. A renewable stainless-steel sleeve eliminates shaft wearing, and the shaft is surrounded by this steel sleeve. The shaft is held in alignment by the rubber bearings, which are placed usually at 3-m intervals. Care must be taken, and the rubber bearings should be lubricated when they are wet. A pre-lubrication system is provided in the water-lubricated column. The pre-lubricated system consists of a water storage tank which is placed near the discharge head. Water is conveyed to the inlet provided in the discharge head down to the discharge column before starting the pump each time. It will allow the rubber bearings to stay wet avoiding the damage due to overheating.

The condition of operation is the main point of consideration for selecting the type of lubricating pump (Singh, 2011; USDA-NRCS, 2010). If the pump is used and contains fine soil particles, oil-lubricated pumps are chosen. Logically, oil-lubricated pumps are not preferred for drinking water since there is a possibility of getting the oil mixed with discharged water. If the pump has to be kept idle frequently for a considerable length of time, it is advisable to use an oil-lubricated pump, as the water-lubricated pump having rubber bearings may get damaged due to a considerable length of time as inactive. Generally, the oil-lubricated pumps are slightly costlier than water-lubricated pumps.

17.5.3 Pump Characteristics

The design of the impeller, the bowl, and the speed of rotation (rpm) are the main related factors to determine the operating characteristic of a turbine pump. If the pump is operated with the designed speed, the characteristics are very similar to the volute-type centrifugal pumps. Unlike the volute-type centrifugal pump, the turbine pump can't function with higher efficiency in a wider range of speed, because the volute-type centrifugal pumps have their vanes in the same line of water flow when the water leaves the tip of impeller. Changing the speed of the impeller, direction of water flow leaving the impeller is also changed, and this may cause turbulence against the vane and will yield a reduced efficiency of the pump.

The operating characteristics of a vertical turbine pump may be determined by an inspection of the characteristic curves that are supplied by the manufacturer. The action of the single-stage pump is determined by the characteristic curve under various operating conditions. If a requirement of head is to be considered more to be a single stage, additional stages may be added as a series connection.

17.5.4 Operation, Maintenance, and Troubleshooting of Vertical Turbine Pumps

Before starting the pump, the sluice valve is to be kept at a closed position. Tools or any rugs must be removed from the pump surroundings. The mains are to be switched on to start the pump. The starting handle of the starter should be at the "start" position till the pump picks up till the full speed and then in a single motion thrown over to the "run" position. The points mentioned below are to be checked while the sluice valve is totally closed.

1. Is the pump properly getting oil? If not check if there is any dirt in the sight feed valve or anything wrong with the solenoid coil.
2. Variation in electric line voltage should be within the range of ±5%.
3. Amperage should be less than the rating stamped on the motor body.
4. Nuts and plates (staffing box in case of water-lubricated pumps) should be checked to see if there is any leakage of water at the tube section. Also, check the pressure gauge line connection, connection to discharge head and sluice valve, etc. For any leakage (if existing), it must be sealed off by packing, lead compound, and jute.

The flow of water discharge is allowed to be delivered by opening the sluice valve of the pump slowly. The content of any dirt or sand particles are present, the sluice valve needs to be opened partially and keep the pump operating for 10–15 minutes at the throttled flow condition. The flow of water should be clear by this range of time. Once the flow is properly developed by opening the valve fully and the flow becomes stable, the following related points are to be checked:

1. the line voltage
2. the amperage
3. any leakage in the tube tension plates
4. any leakage in the staffing box (for water-lubricated pumps)
5. any flow of oil to the pump
6. excessive vibration or noise from the pump.

The sluice valve should be totally closed before stopping the pump.

17.5.5 Pump Maintenance

To ensure a non-hazardous, safe, and economic operation of pump, flow and availability of lubricants at different parts are to be maintained strictly. Most of the turbine pumps need the following checks and maintenances:

a. *Drive head with a vertical belt:* after 2,000 hours or each year running
   1. Oil must be drained out by removing the drain plug and filter plug from the oil filter cap.
   2. Recommended lubricating oil should be filled up to the top of the oil filter cup after replacing the drain plug and drive head.

3. After a few minutes when the oil becomes stable, add (if necessary) some more oil to the top of the oil filter cup.
4. Replace the filter plug.
  b. *Right-angle gear head:* after 2,000 hours or each year running
1. Oil filter cup should be totally emptied by removing the drain plug and filter plug.
2. Recommended lubricating oil should be filled up to the top of the oil filter cup after replacing the drain plug and drive head.
3. After a few minutes when the oil becomes stable, add (if necessary) some more oil to the top of the oil filter cup.
  c. *Drive head of electric motor:* after 500 hours running
1. Thrust bearing cap should be unscrewed and fill the cap with grease.
2. Cap is replaced and it is down one turn.
  d. *Shaft packing gland*: if required
1. Unscrew the cap from the shaft packing gland lubricator and fill the cap with grease.
2. Lubricator cap is replaced and screw it for down two turns. Each day of operation, screw the cap down one turn.

17.5.6 Pump Troubleshooting and Their Remedies

Mostly, turbine pumps offer trouble-free service, if they are properly installed and routine maintenances are done. Common troubles and their remedies are explained below:

17.5.6.1 Pump Doesn't Start

1. Fault in electric connection.
2. Misalignment of the pump and excessive friction in the bearing, inappropriate grade of oil or insufficient lubricant, or dry rubber bearing (in water-lubricated pump).
3. Improper impeller adjustment or impellers are locked.
4. Sand-locked in the pump.
5. Shaft is broken or disengaged.
6. Packing in the discharge side is very tight.

The remedies are evident from the cause of trouble. Adjustment of impeller height (either lowering or raising) may release them. Cleaning the bowls may be done by backwashing with clear water.

17.5.6.2 Pump Doesn't Deliver Water

1. Suction line of the pump is broken.
2. Pumping speed is very slow due to poor electric voltage.
3. Lower the water table such that the pump bowls are not submerged.
4. Impellers are plugged (happens when the tube well is not developed fully, and the pump discharges clay or sand at the starting time). The bowl assembly is to be flashed with a pressure water jet. Sometimes, for severe clogged condition, the bowl may need to be opened and clean thoroughly.
5. Impellers jammed due to foreign matter in the bowls. Open wells, sometimes, may cause a jam due to smaller sized pieces of wood or other materials getting through the strainer that are caught in the impellers.
6. Strainer is clogged.
7. Direction of rotation is wrong.
8. Pumping head is too high.
9. Mechanical failures like a broken shaft or broken bowl assembly or loose impellers or leakage in the column pipe or its joints.

17.5.6.3 Insufficient Discharge of Water

1. Speed is too slow.
2. Lower water table such that the pump bowls are not submerged.
3. Impellers and pump bowls are worn out.
4. Impellers loose or damaged.
5. Pipe column joint is defective or leaking.
6. Strainer of impellers are partially clogged.
7. Improper impeller adjustment.

17.5.6.4 Insufficient Pressure

1. Speed is too low.
2. Air is mixed in water.
3. Pump bowls are insufficiently submerged.
4. Bearing is worn out.
5. Impellers are damaged.
6. Pipe column joint is leaking.
7. Impeller adjustment is inappropriate.

17.5.6.5 Excessive Power Requirement

1. Discrepancy between pump design condition and the actual operating condition.
2. Insufficient lubrication.
3. Impellers are improperly adjusted. They may be very high and binding against the top of the bowls or too low and getting rubbed against the bowl seats.
4. Sand partially clogged the impellers.
5. Excessive oil in the oil tube or oil used too heavy.
6. Dis-alignment of pump.
7. Bending in shaft.
8. Bearings are worn out due to tightness.
9. Dirt or impurities bind the shaft against the bearings.
10. Improper greasing or oiling the pump drive.

17.5.6.6 Vibration of Pump

1. Speed is very high.
2. Pump is not leveled or not aligned properly.
3. Crooked well.
4. Bend in shaft.
5. Rubbing against bowl seats due to poor impeller adjustment.
6. Pump is not properly bolted with the foundation.
7. Pump is not on a stable or solid foundation.
8. Improperly lubricated bearings in drive.
9. Impellers are loose.
10. Impellers or bowl passage is clogged due to foreign materials or excessive wear in the rotating parts.
11. Suction condition is poor.

Pumps for Irrigation Systems

17.5.6.7 Lower Discharge

1. Head is too high.
2. Speed is too low.
3. Poor suction conditions due to turbulence, eddy or vortex formation at the suction, or air leakage in the suction pipe.
4. Leakage due to loose threads or if the flange or packing or gasket is worn out.
5. Impeller or bearings or wearing rings are worn out.
6. Impeller is loose on the shaft.

17.5.6.8 Discharge Contains Oil or Shaft Is Wet

1. Tube joints are not tightened.
2. Line shaft bearings broken.
3. Discharge head gland is improperly tightened.
4. Cuts or dents exist on the tube faces.
5. Tube enclosing the shaft is leaking, or broken.

17.5.6.9 Components Wear Out Readily

1. Insufficient lubrication or bearings are dry.
2. Pump bearings are not aligned.
3. Tube well crooked.
4. Bending in shaft.
5. Water is mixed with sand.
6. Vibration due to improper alignment or bad leveling of discharge head.

17.6 Submersible Pumps

A small-diameter submersible electrical motor close-coupled with a vertical turbine pump is called a *submersible pump*. The motor is fitted below the intake of the pump. The motor and the pump element entirely under the water level means submerged condition. This type of installation doesn't require any vertical shaft in the column pipe. The performance characteristics of a submersible pump are very similar to the vertical turbine pump. Efficiency is higher because this type of pump is directly coupled with the motor, doesn't have any loss due to the non-existence of the vertical shaft, and has effective cooling rate due to submergence in water.

The main advantages of this pump are that the long shaft is not used and this pump can be used in a very deep tube well. Therefore, a dis-alignment of shaft and pump is not a practical issue. A flooded area can also be suitable to using this pump because almost all parts of this pump are submersible. It can also be used in the open or any public space if it is used as over-the-ground. A suitable place to construct a pump house can be a good choice to accommodate the pumping unit.

17.6.1 Pump Construction and Operation

A submersible pump is comprised of a motor and a pump, a discharge column, a head assembly, and a waterproof cable to connect the motor, which is submerged under water, e.g., aquarium with gold or other decorative fishes having a water recirculation system by a submersible pump. The parts of the submersible pumps are as follows:

1. *Pump element*

    Pump element construction is similar to the vertical turbine pump. The short propeller shaft is made with stainless steel and is mounted to the bronze impellers in the shaft. Impellers may be closed or semi-open. The closed type is for developing a higher pressure. Water gets into the pump through the screen, which is located between the motor and pump.

2. *Electric motor*

    The submersible electric motor has the same diameter as the pump bowl. It is longer than an ordinary motor. This is an induction motor with a squirrel cage and it may be of wet or dry type. In the case of a dry motor, it is contained inside the enclosed steel case filled with a light oil of higher dielectric strength. The oil leakage through the armature is prevented by a mercury sealing placed directly above the armature. It also prevents the leakage of water entrance at the point when the drive shaft passes through the case to the impellers.

    The motor has the access of well water. Bearings and rotor are actually operating in water. In the wet-type motor, the winding of the stator is firmly sealed off from the rotor by a thin stainless-steel inner liner. There is a filter surrounding the shaft to prevent the entrance of the foreign materials into the motor. The wet-type motor must be filled with water during installation so that the bearings are not having a lack of lubricants when it is first started.

    The stator windings are continuous, and it covers the total length of the motor. The rotors are made in sections on a continuous shaft with bearings between them to guide the shaft and assure correct alignment. The waterproof electric cable runs from the motor to the starting switch box situated at the ground surface are placed outside the discharge pipe.

17.6.2 Installation, Operation, and Maintenance of Submersible Pumps

The outstanding feature of the submersible pump is the ease in installation. Firstly, the pump motor assembly is lowered into the water well and pipe length is decided by its requirement. Extra care must be taken for the outside sheathing of waterproof cable against damage, including the top edge of the well casing while submerging the pump. The cable is tied with tapes with the column pipe at every 2 m. A pipe clamp is fastened at the top of the well to support the pump and the delivery pipe. A sluice valve is fitted at the discharge end of the pump. Pump house for a submersible pump is not required as the pump and the motor assembly are coupled and submerged under water together. A control board consisting of the switch, starter, and meter need to be enclosed in a waterproof box.

The pump is started at the initial starting time with the sluice valve closed or slightly open. At the initial run, it is necessary to observe whether the water is clear or muddy and if there are any impurities or foreign materials are being pumped. If the water contains sandy or gritty particles or other impurities, care should be taken not to stop pumping so the unwanted particles will be settled down inside the pump and on top of the non-return valve (if existing) and may choke or seize the pump. The technique of initial starting of the pump is to keep the sluice valve a little closed at the beginning and open slowly after the water shows no more sand or any foreign particle. If the sluice valve is fully opened and any impurities are found, the valve should be adjusted so that the impurities are minimum. Once the water becomes clear, the pump may be stopped and then restarted, if required (USDA-NRCS, 2016).

When a pump is first started, most of the new tube wells, naturally, deliver water mixed with some mud, sand, and debris from drilling operations. The quality of the water gets better slowly. In some cases, if the well is commissioned in a place of sandy areas, a certain amount of sand is continuously delivered along with the water. This issue could be troublesome as it may cause undue wear on the pump parts, but could be controlled with an operation of the pump at a rate so that the sand content is kept to a minimum.

Once the submersible pump is installed and operated as per the manufacturer's instruction and under suitable working conditions, it needs very little maintenance. After about 2 years or 6,000 hours of operation, the pump would be necessary to be withdrawn from the borehole and do the overhauling following the instructions of the manufacturer.

17.6.3 Common Troubles and Their Remedies of Submersible Pumps

The usual troubles in the operations of submersible pumps may be due to the following causes:

1. *Pump fails to start*
   a. Fuses blow.
   b. Overload trip causing short circuit.
   c. Heavy load on the pump.

2. *Pump starts and discharge is steady but the amount of water is less or not at all*
   a. Direction of motor running may be reverse.
   b. Operating head of the pump may be greater than the designed head.
   c. Suction may be obstructed by salt deposit or block by foreign matter.
   d. Pump is air locked (when there is no discharge at all).
   e. Reflux valve above the pump is jammed, or riser pipe is closed by obstruction or by valve.
   f. Voltage is considerably lower.

3. *Pump runs and discharge are steady, but the amount of water is below normal*
   a. May be fault in power supply.
   b. Friction in pump of motor.
   c. Riser pipe has developed a hole or a leakage has developed below ground level.
   d. Pump is worn out by sand and the mechanical friction increases as well.

4. *Pump runs and discharge intermittently and existence of air bubble in the discharge*
   Pumping rate is greater than the rate of flow into the bore. If it is continued to operate with this condition, it will cause damage to the pump, shaft keys, couplings, and ultimately the motor as well.

17.7 Propeller and Mixed-Flow Pumps

In some cases, selection of a pump for irrigating the land needs to consider a lower head but a higher discharge. At that time, propeller and mixed flow are the best choices. If a head is within or even less than a meter, the propeller pump offers a higher efficiency compared to a centrifugal pump. Centrifugal and mixed-flow pumps are better for a slightly higher head and give a higher efficiency. Both these pumps are extensively used for drainage purposes. They are also suitable for irrigating the crop lands adjacent to the canals, rivers, or streams.

Pressure head in centrifugal pumps is developed mainly by the action of the centrifugal force, whereas a propeller pump develops most of the head by propelling or lifting action of the vanes of impellers on the liquid. The mixed-flow pump combines some of the features of both the turbine and propeller pump. A part of the total head is developed by centrifugal action and part by lifting action of vanes of impeller on the liquid.

The main parts of mixed-flow pump and propeller pumps are similar to the deep-well turbine pump. They have head, impeller, and a discharge column. A shaft runs from the head to the center of the column and finally drives the impeller. Propeller and mixed-flow pumps are of oil- and water-lubricated types like vertical turbine pump.

17.7.1 Construction and Operation of Propeller Pump

Flow of propeller pump takes place through the impeller parallel to the drive axis of the drive shaft. This doesn't take place radially like in a centrifugal pump. This is also, therefore, called an *axial flow pump*. The principle of operation is that the propeller pump is enclosed inside a housing. Propeller pumps are not easily clogged by foreign materials or by any suspended sediments with water. The available propeller pump has a variety of heads of size ranging from 1 to 2.5 m. The impeller, in the case of a propeller pump, is called a *propeller* that operates in a cylindrical casing which is an extension of the pump discharge column. A flared entrance below the propeller is used to cut down the entrance losses, and to guide vanes above smooth out the disturbances caused by the propeller. Water is moved up by the lift up of the propeller blades. The individual blade of the propeller imparts a velocity in the direction of the shaft. Propeller pumps usually consists of three to five blades. They are attached with the shaft at an angle as designed for the specific speed and head.

17.7.2 Operating Characteristic of Propeller Pump

It is observed that a small increase in head causes a higher decrease in discharge by the propeller pump. To ensure the best efficiency, a propeller pump has to be operated with rated head as far as possible. The power requirement is increased in a propeller pump with the increase of head and the capacity gets decreased, whereas, in centrifugal pumps, the power requirement is decreased with increasing head and decreasing discharge. Overload occurs when the valve is nearly closed instead of open in contrast to the centrifugal pump. Propeller pump is not suitable under the circumstances when the pump is required to throttle to ensure reduced delivery of water. The head against which water is pumped using a mixed-flow or propeller pump affects its efficiency greatly.

17.7.3 Propeller Pump Installation

A strong and firm foundation is necessary for propeller pump installation. The foundation should be able to carry the load of the pump uniformly throughout the foundation surface. In the case of drainage installation, the pump is installed in a sump with an automatic float control. Start and stop collars on a float rod drive the auto-operation of the electric motor. Automatic switches are there for a start and stop of the pump during essential flood stages. Two pumps of different capacities may be necessary during drainage installation, one for managing the surface runoff during rainy seasons and the other one is for seepage or flow from tile drains.

To control the intake of any floating debris (if any) into the propeller, a strain at the entrance of the intake pipe may be attached to avoid the damage of the pump. A small strain is provided with the suction bowl and it works satisfactorily when water pumped is comparatively free of floating small debris and any vegetation. If the source of water contains these types of material, the small strainer is apt to become clogged. An increase in the surface area in the screen may increase the catching of the foreign materials. Trash racks may be provided at some distance away in the drainage channel for an improved and efficient system.

If a suction lift is necessary, a propeller pump is not a good choice. The impeller bowl has to be submerged with the pump operating at the proper submergence depth, and the submergence is the distance from the pumping water level to the lowest part of the suction pipe. Failure to maintain the required submergence of 65–90 cm for pump of sizes 20–35 cm may result in cavitation. A clear space of 30–50 cm between the end of the suction pipe and the side walls and the bottom of the pit or pump intake bay, a space between 20 and 30 cm, must be maintained. If the pump is run without maintaining the specifications of clear spaces requirement, it may result in poor pump efficiency.

17.7.4 Mixed-Flow Pump

Mixed-flow pump consists of some of the common features of a vertical turbine pump and the propeller pump. The head is made partially by the centrifugal force as a traditional centrifugal pump and partially by the lift of the impeller vanes on the liquid as observed in a propeller pump. This pump is suitable for *higher discharge and medium head* conditions. The operating head varies from 5 to 10 m.

The basic difference between propeller and mixed-flow pump is in their impeller design and construction. Mixed-flow pump impeller is designed to direct the flow of water outward in addition to imparting the flow an upward velocity. The top of the impeller, the vanes with curve shape, leads the water to attain the nature of flow of water to be straight in the water column. Mixed-flow pump has been experimented and found to be satisfactory two or three stages below the bottom of the shaft. Column assembly and the pump drive of mixed-flow pump are similar to the propeller pump. A water-lubricated column assembly is suitable when pumping clear water, whereas an oil-lubricated pump is suitable if the flow of water contains suspended sediments and impurities.

17.8 Jet Pumps

From the name, it is evident that a centrifugal pump in combination with a jet mechanism or ejector is called the *jet pump*. An electric motor or engine coupled with the centrifugal pump is placed at the ground surface and constitutes the driving head and capacity for the jet placed at the well below the water surface. A nozzle and a venturi are the main parts of a jet pump. A portion of high pressure water at the delivery pipe returns through the pressure pipe into and activate the nozzle in the ejector. The nozzle has a shape that is slow and smooth but abruptly reduces its diameter where the water passes through. Therefore, an increase in velocity and thereby a decrease in pressure take place in that narrow space. This shape of narrow space is termed as venturi. Later, the narrow space gets into an enlarged area with the full diameter like in suction pipe. This section of the enlarged diametric area helps in creating a higher velocity and a lower pressure with minimum turbulence. The vacuum created by the impeller rotation of the pump placed at the ground level draws the flow through the suction pipe. The delivery of water takes place with a desired pressure. Additional supply of water which is obtained from the well is discharged past the control valve while the volume required for delivering the flow is recirculated through the pressure pipe. The pressure is controlled by the valve to produce flow at the existing pumping head. The main governing factor for the capacity and the efficiency of a jet pump is the selection of the jet that corresponds to both the type of pump and depth of low water level of the source. Typically, jet pumps are less efficient and the highest possible efficiency is about 35%. The efficiency is directly related to nozzle throat ratio.

17.8.1 Uses and Adaptability of Jet Pump

Jet pumps are used where a low-capacity deep-well pumping is existing and it is necessary to locate moving parts of the pump and the prime mover at the ground surface. The advantages of the jet pump are as follows:

1. It is adaptable to be installed in a well as small as 5 cm inside diameter.
2. It can be used in a place where a high-suction lift is necessary, which cannot be obtained by a traditional centrifugal pump.
3. All moving parts are accessible which are located at the ground level.
4. This is simple and low cost, and maintenance is easy.
5. Installation is possible with the moving parts offset from the well. Jet pumps are not suitable for the conditions as follows:
    a. if there is a large seasonal variation in water levels, and
    b. if the nozzle is plugged or enlarged due to corrosion incrustation.

TABLE 17.1 Pipe Sizes for Air-Lift Pumps (Michael, 1997)

Pumping Rate (L/s)	Size of Wells Casing (cm)	Size of Eductor Pipe (cm)	Size of Air Line (cm)
2–4	10.0 or larger	5.0	1.25
4–5	12.5 or larger	7.5	2.50
5–6	15.0 or larger	8.7	2.50
6–9	15.0 or larger	10.0	3.13
9–16	20.0 or larger	12.5	3.75
16–25	20.0 or larger	15.0	5.00
25–44	25.0 or larger	20.0	6.25

17.9 Air-Lift Pumps

An air-lift pump is operated by the compressed air injected directly into inside a discharge pipe at a point below the water level in the well. The injected compressed air leads to a mixture of air bubbles and water. This combination of compressed air and water is lighter in weight than water so that the heavier column of water around the pipe displaces the lighter mixture forcing at upward and out of the discharge pipe.

The piping assembly used for air-lift pumping from a well consists of a vertical discharge pipe, which is called the eductor pipe, and a smaller air pipe. An air-lift pump in a well is arranged commonly with the air pipe inside the eductor pipe. It is possible to locate the air pipe outside the eductor pipe if there is a space hole, as otherwise, there is considerable friction loss when the air pipe is located inside a small-diameter eductor pipe. Both the eductor pipe and the air pipe must be submerged in water in the well with 40% or more of their lengths extending below the pumping level. The energy needed to operate the air-lift pump is available from the compressed air.

The well casing itself can be used for the eductor pipe if the diameter of the casing is not very large compared to air line. It provides an extra advantage of pumping sand or mud from the bottom of the well during development and cleaning operations. The most important factors in the air-lift pump are as follows:

1. percent submergence of air line
2. relative size of the air and eductor pipes.

Table 17.1 shows the sizes of air lines for various sizes of eductor that are used under most of the conditions:

Air-lift pumps are extensively used in developing and preliminary testing and cleaning of the tube wells. Places where vertical turbine pumps or submersible pumps cannot be installed, air-lift pump may be a good option in those areas and used in crooked tube wells. The advantages of air-lift pumps are as follows:

1. It is very simple to install, to operate, and easy to maintain.
2. Tube well doesn't need to be perfectly straight or vertical.
3. Impure water is not a problem and doesn't damage the pump.

The only main disadvantage is its lower efficiency. This initial cost of this pump, including the air compressor, is a little high, and it requires an extra depth of water for appropriate submergence.

17.10 Pump Selection

Pumps and related tools are an expensive asset in an irrigation project. An efficient use of this equipment is always desired with respect to the timeliness and amount of water requirement in that specific time for the highest possible crop yield. Energy requirement for pump operation occupies a significant amount of money. Efficient utilization of the limited amount of energy resources has to be compromised by selecting the most efficient and suitable pump. With that point in mind, the irrigation water requirement has to be economized with consideration of the characteristic of the well or other sources of water, kind of power available, farmers' economic status, and other related factors (Singh, 2011).

17.10.1 Criteria and Procedure for Selecting a Pump for Irrigation

The factors influencing the irrigation pump selection are:

1. irrigation water requirement means the amount of crop water requirement,
2. yield of the source of water (surface water or groundwater, lake, ponds, river, etc.),
3. type of pump that can be considered for serving the specific area and crops,
4. cost and availability of the pump, and
5. kind and cost of energy,

17.10.2 Determination of Discharge Capacity of the Pump

The data on the safe discharge rate of the well (or other source of water) and the discharge rate required for the crops to be irrigated with a particular cropping pattern are estimated. If needed, the cropping pattern can be adjusted based on the safe yield of the well during various cropping seasons. Again, in some cases, cropping pattern adjustment is not possible due to the fragmentation of the land. In these cases, any of the following options could be implemented:

1. Pumping plant should be designed based on the safe yield of the well and use the pump to provide irrigation water to the farm of an individual farmer, keeping the spare capacity idle. But a pump leads to high investment and under-utilization of installed capacity and invested funds.
2. Pumping plant should be designed based on the maximum crop water requirement for the area of an individual farmer and relinquish the extra capacity of the well. This option will help a lower investment in pumping system compared to the one stated in option number (1).
3. Pumping plant should be designed based on the safe yield of the well and using the installed capacity of water for irrigating the crop land of the pump owner of the well and selling the remaining water to irrigate the crop lands of other farmers, if they have the demand, this option is better than options (1) and (2), that allows full utilization of the lift irrigation potential, which requires a higher amount of monetary investments.

Therefore, a pumping system needs to be designed based on two factors. They are as follows:

1. the safe yield of the well
2. the maximum crop water requirement by the crop, usually crop water requirement in each day during peak water requirement periods. After determining the dominating factor(s) from the above, the pump size is to be calculated, and consequently, efforts are to be made to ensure the other related factors are within the allowable limits so that the efficient utilization can be achieved. Computing the size of the required pumping unit needs the consideration on the conveyance efficiency of water into the fields.

17.11 Capacity of Pump Based on Crop Water Requirement

A pump must satisfy the water demand or requirement of crop and the size of the discharge of the pump must be based on the peak water requirement of the crops and selected cropping pattern. The pumping rate on the area is dependent on the following factors:

1. area under different crops
2. water requirement by the crops
3. crop rotation period (interval between two successive irrigation of a crop)
4. duration of pumping operation every day.

Discharge of a pump may be calculated by the following equation:

$$q = 27.78 \frac{Ay}{RT},$$

where:

q = discharge of the pump in L/s
A = area needs to be irrigated in hectares
y = depth of irrigation in cm
R = rotation period in days
T = duration of pumping in hours/day.

Example

A farmer wants to irrigate his lands using his own pumping unit for the following cropping pattern. Table 17.2 shows the required information for determining the correct size of the centrifugal pump he should select.

Solution:

Discharge of the pump:

$$q = 27.78 \frac{Ay}{RT}$$

$$= 27.78 \left[\frac{2 \times 7.5}{12 \times 10} + \frac{0.4 \times 7.5}{20 \times 10} + \frac{0.4 \times 7.5}{10 \times 10} + \frac{2.2 \times 5.0}{40 \times 10} \right]$$

$$= 5.49 \, \text{L/s}$$

Therefore, a pump should have the capacity of 5.50 L/s is required.

17.11.1 Selection of Pump by Well and Pump Characteristics Curves

Drawdown–discharge characteristic established curve is the main resource for selection of a pump for a specific well. The pump that matches best a well can be chosen by a good agreement between the characteristics of the pump and the well.

TABLE 17.2 Selection of pump size based on the area and crop to be irrigated, depth of water for irrigation, rotation in days, and period of work in hours per day

Crop	Area of Irrigation, A (ha)	Intensity of Irrigation, y (cm)	Rotation Period, R (days)	Period of Irrigation, T (hours/day)
Wheat	2.0	7.5	12	10
Cotton	0.4	7.5	20	10
Vegetables	0.4	7.5	10	10
Mustard	2.2	2.2	40	10

Source: Michael (1997)

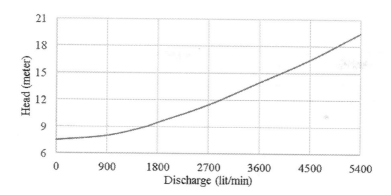

FIGURE 17.8 A typical head–discharge curve of a well.

17.11.1.1 Well Characteristics

Yield of a well can be defined as the volume of water discharged from a well in unit time. The yield of a well is dependent on the drawdown due to pumping. The well characteristics are designated by the discharge and drawdown relationship. The well test data for discharge are drawn at the x-axis, whereas the depth of the corresponding water level is drawn in y-axis. Level of water during different discharges is measured from the ground surface.

17.11.1.2 Selection of Pump Based on Well Characteristics

The pump is selected based on the characteristics curve of the well and the pump enable the selection of the pump which suits best the well. A characteristic curve is drawn on a tracing paper with the same scale of the pump curve. The tracing paper is held on the characteristic curve of the well. These two characteristic curves one (traced curve) of the pump hold above the other (well characteristics curve) (Figure 17.8).

References

Albaji, M., Eslamian, S., Eslamian, F. (2020). *Handbook of Irrigation System Selection for Semi-Arid Regions*, Taylor and Francis, CRC Group, Boca Raton, FL, 317 Pages.
Centrifugal Pumps Working Principles and Characteristics. (2020). http://generalcargoship.com/centrifugal-pumps.html on 02/21/2020.
Characteristic Curves of Centrifugal Pumps. (2020). http://ecoursesonline.iasri.res.in/mod/page/view.php?id=1864 on 02/17/2020
Church, A.H. and Jagdish L. (1973). *Centrifugal Pumps and Blowers*. Metropolitan Book Co., Pvt. Ltd., Delhi, India.
Finkel, H.J. (2018). *CRC Handbook of Irrigation Technology*. Taylor & Francis Group, CRS Press, Boca Raton, FL.
Merkle, T. (2014). *Damages on Pumps and Systems: The Handbook for the Operation of Centrifugal Pumps*, 1st Edition. Elsevier, Oxford.
Michael, A.M. (1997). *Irrigation Theory and Practices*. Vikas Publishing House Pvt. Ltd, Noida, India.
Reciprocating Pump Working Principle. (2020). https://mechanicallyinfo.com/reciprocating-pump-working/ on 01/26/2020.
Singh, P. (2011). *Handbook of Irrigation*. Oxford Book Company, Oxford, UK.
USDA-NRCS. (2010). Chapter 7: Farm Distribution Component. National Engineering Handbook - Irrigation Guide, Part 652, 210-vi-NEH, Part 652, Amend. AL1, USA.

USDA-NRCS. (2016). Chapter 8: Irrigation Pumping Plants. National Engineering Handbook - Irrigation Guide, Part 653, 210-VI-NEH, Amend. 78, USA.

Vertical Turbine Pumps. (2020). http://unitedstates.xylemappliedwater.com/brands/ac-fire-pump/vertical-turbine-pumps/ Accessed on 03/30/2020.

What Is a Centrifugal Pump. (2020). https://www.introtopumps.com/pumps-101/what-is-a-centrifugal-pump/ Accessed on 02/18/2020.

18

Inverted Siphon Implementation Method in Karun River for Farm Water

18.1	Introduction 387
18.2	Different Options for Transferring Water to Khorramshahr Lands...388
18.3	Siphon Characteristics 389
18.4	Siphon Implementation Processes 389
	Establishing Dry Ponds • Assembling and Welding Pipes • Making 45 Degree Elbow Joints
18.5	Connecting Elbows and Lower-Hand Dollies to Structure 390
18.6	Preparing the Pipe Establishment Bed.. 391
18.7	Establishing Four Winch Units 391
18.8	Establishing Cathodic Protective System, Making and Mounting Siphon Indices and Lower-Hand Dolly Caps........... 391
	Establishing Cathodic Protective System • Making and Mounting Siphon Indices • Manufacturing and Establishing Lower-Hand Dollies
18.9	Floating..392
18.10	Transferring Structure inside the River 392
18.11	Waterlogging in the Pipes 393
18.12	Assembling, Welding and Painting Upper-Hand Dollies..........393
18.13	Establishing Upper-Hand Dollies 395
18.14	Conclusions...395
References...396

Faramarz
Ghalambaz
Khuzestan Water and Power Authority

Saeid Eslamian
Isfahan University of Technology

18.1 Introduction

Alhusseini (2008) found that optimum hydraulic design for siphon had been studied depending on the method of optimization (Modified Hooke and Jeeves) with some modifications. These modifications are, firstly, modification in the assumed initial base points; secondly, modification in the value of step length; and thirdly, modification in the value of the reduced step length at each trial. The siphon shapes used in this study are circular, square and rectangular. The materials used are concrete and steel for designing an inverted siphon. A computer program depending on the method of Modified Hooke and

Jeeves has been written for optimum hydraulic design of the inverted siphon structure with Quick-Basic language (Sadeghi et al., 2012).

Boudreaux (2014) has reported the improvements in the $6 million Oklahoma River Inverted Siphon System. The Phase II project included the potential rehabilitation of an existing three-barrel inverted siphon, and the construction of a new inverted siphon, a sewer flow distribution vault and a sewer line connecting the new inverted siphon to the collection system. This project increased the city's capacity to transport sewage from south Oklahoma City, USA to the North Canadian Wastewater Treatment Plant. The existing siphon did not have the adequate capacity to handle the flows during wet weather events. The project was developed to help reduce the sanitary sewer overflows in the Brock Creek and Lightning Creek sewer sheds. The new inverted siphon was installed under the Oklahoma River with 587 ft of 24-in fiberglass pipe and 587 ft of 48-in fiberglass pipe. Inverted siphons allow stormwater or wastewater sewers to pass under obstructions such as rivers.

This chapter details a project in the Khuzestan Province, located in the southwest of Iran during 2011–2013. The project execution place is located in furthermost southwestern Iran, Khuzestan Province, northeastern Khorramshahr, over Karun river, 2.5 km from Haffar village, 234084.1–233909.7 N and 3371982.2–3372201.8 E. This siphon would be constructed as a part of the Valiasr canal route (Mahab Qods Consulting Engineering Company, 2012)

The project is going to transfer the water in 13.5 m^3/s from the left Karun bank to the right one. This is to be constructed as a part of the Valiasr canal route, the function of which is to transfer water from Marred canal to Shalamcheh area, west Khorramshahr (Naseri et al., 2009; Behzad et al., 2009; Albaji et al., 2010; Boroomand Nasab et al., 2010; Albaji and Hemadi, 2011; Jovzi et al., 2012, Albaji et al., 2015; Hooshmand et al., 2019; Neissi et al., 2020).

18.2 Different Options for Transferring Water to Khorramshahr Lands

According to the studies conducted, the primary siphon execution option was to make use of polyethylene pipes as inverted siphon, and a value engineering workshop was established to finalize this option. Taking into account this and other options, the following methods were technically, executively and economically considered:

- Base option of using single-walled polyethylene pipes of 1,200 mm in 10 strings as inverted siphon and establishing concrete blocks at intervals of 3 m.
- Utilizing steel pipes.
- Utilizing double-walled polyethylene pipes.
- Utilizing Glass Reinforced Plastic (GRP) pipes.
- Increasing the diameter of pipes and reducing the number of inverted siphons.
- Prefabricated concrete siphon.
- Constructing deviating dam and pumping station at the right bank in the siphon location.
- Concreted siphon with river diversion.
- Mechanized piping – pipe.
- Making use of Tunnel Boring Machine (TBM) and/or direct tunnel.
- Direct siphon.

After discussing the options (steel and polyethylene pipes), steel pipes were prioritized for the following reasons:

- Resistance against hits and fractures.
- Pipe implementation and drown easiness without using the concreted blocks.
- Possibility for implementing in different diameters due to the existence of pipes in the state production market.

Inverted Siphon Implementation Method in Karun River

TABLE 18.1 Siphon Characteristics

Siphon Type	Pipe
Pipe material	Steel
Pipe diameter	1,600 mm
Pipe line numbers	5 series
Pipe thickness	15 mm
Siphon length	160 m
Total passed discharge	13.5 m³/s
Each pipe discharge	2.7 m³/s
Speed within pipes	1.34 m/s
Siphon inlet and outlet height difference	0.3 m

18.3 Siphon Characteristics

In the Karun siphon project, spiral steel-type pipes (1,600 mm diameter and of 15 mm thickness) have been used, with epoxy polyamide internal cover of 500 μm thickness and polyethylene external cover of 3 mm thickness.

Siphon characteristics like siphon length, diameters of pipes, etc. are mentioned in Table 18.1.

18.4 Siphon Implementation Processes

18.4.1 Establishing Dry Ponds

A drilled space along the river was used to assemble and weld the pipes. Karun river was closed with embankment. Its helpful dimensions were 18 × 170 m, and its floor was −1.8 m. For the ease of assembling the pipes, transverse concreted pillows of 40 cm height were used at intervals of 6 m.

Water-level elevation in Karun river in the pond location is approximately +0.1 achieving +0.8 during high tide and −0.3 during low tide.

The pipe settlement amount after floating was estimated about 50 cm therefore the pipes are of approximately 60 cm with top of pillows when there is completely high tide.

18.4.2 Assembling and Welding Pipes

Operations of assembling and welding pipes in dry pond are performed in five parallel lines and at a distance of about 80 cm from each other. Processes of these operations are as follows:

- Transferring pipes within pond utilizing crane.
- Establishing pipes on concreted pillows.
- Welding and assembling pipes.
- Externally welding pipes.
- Grinding and completing weld inside pipes.
- Performing the weld joint radio graphical tests.
- Repairing the pipe color and protective cover.

Integrated connecting structures were conducted between five pipes strung by ten metallic connector series of IPE 24 and IPE16 materials including approximately 2,500 hardening metallic slab pieces, following completion of the above processes and ensuring correction of welding pipes to integrate the pipe collection (Figure 18.1).

FIGURE 18.1 Implementing connecting structures.

18.4.3 Making 45 Degree Elbow Joints

Pipes should be directed within a gredient 1:1 (45 degree) to ground surface in order to connect them to canal inlet and outlet converter with respect to establishment of pipes in river floors. Joints were manufactured in workshops because of the lack of proper joints in the market. To do so, joints of interest were manufactured with characteristics of interest, in ten numbers performing proper cutting and welding them. Welding quality was controlled using ultrasonic test method following their manufacturing termination.

18.5 Connecting Elbows and Lower-Hand Dollies to Structure

Although 15 m pipes are needed to connect siphon to bank with regard to depth of establishing pipe sets, in this step, a 6 m pipe branch (lower-hand dollies) was connected to each of the elbows and another end of a 6 m pipe was equipped to flange till rest to be assembled as bolt and nut for preventing overheaviness of set bolt ends and possibility for its deflection in the middle.

Elbows and lower-hand dollies were finally connected to the horizontal part of the siphon structure following painting (Figure 18.2).

FIGURE 18.2 Connecting elbow to lower-hand dolly.

FIGURE 18.3 Dredging operations.

18.6 Preparing the Pipe Establishment Bed

It was necessary to dredge the river bed to approximately −13 m depth of water level as the pipe establishment location was at that depth. Dredging depth ranged from about −12 m at the deepest river points to about −3 m at its banks. Trench width should be at least 5 m more than the pipe set's width. In this step, preparing establishment location of pipe sets in trench of interest, making use of dredging device in order to discharge the sediments resulting from dredging is done (Figure 18.3). In this step, other operations such as river bed dredging as well as structure harness primary operations were also performed as follows along with completing the siphon set.

18.7 Establishing Four Winch Units

Four winch units and two crane units (two winch units and a crane unit in each bank) were provided in order to harness structure collection as well as to control it when drowning. The force resulting from fluid on pipes was about 23 tons with respect to pipe length of about 160 m. Therefore cables, winches and cranes should have this force capacity with a reliability coefficient of about 2.

18.8 Establishing Cathodic Protective System, Making and Mounting Siphon Indices and Lower-Hand Dolly Caps

18.8.1 Establishing Cathodic Protective System

Total collection was equipped for cathodic protection to prevent form rottenness with respect to the pipe (steel) materials and their continuous adjacency to water.

18.8.2 Making and Mounting Siphon Indices

Six indices were established on four sides of the structure while providing and painting them so that they would be controllable using map-making cameras (Figure 18.4).

FIGURE 18.4 Mounting indices.

18.8.3 Manufacturing and Establishing Lower-Hand Dollies

The caps were made and mounted on lower-hand dollies to prevent uncontrollable water entrance within the pipe sets. These caps were so made that each has two holes—one for water injection and another for air discharge.

18.9 Floating

With the termination of whole processes such as assembling and welding of pipes in dry pond, dredging siphon establishment at trench location, mounting winches as well as establishing cathodic system and floating pipe collections, the weir between river and pond was ripped slowly.

Two points were important in this regard; firstly, pipes should be harnessed by cables to ensure the locations in order to control the pipe situations when floating in pond and secondly, it was necessary to rip the weir slowly to prevent sudden surge of water attack and the possible damages.

18.10 Transferring Structure inside the River

The pipes were separated from their establishment location on the concreted D-shapes and the whole collection was floated the following whole weir between the pond and the river that was removed, and finally the water inside the pond made iso-level with the river water. In this step, the total structure collection was directed inside the river and established in location of interest (in river width) helping two motorized boat units.

After structure establishment in the location of interest, the whole structure was connected to winches and cranes by cables. Water inlet and outlet hoses were connected to lower-hand dolly caps and the system was prepared to fill the pipes by water and as a result it's floating. But it was found that the collection middle part to have about 10–12 cm deflection, because of the structure both end heaviness and this concern was emerged that the water that would first concentrated in both pipe ends and increasing in heaviness of these both areas, reduced the structure middle deflection over the considerations and monitoring conducted. Therefore, the whole collection was transferred to the bank, separating whole harnessing connections (cables and hoses) in order to reduce risks (Gian Constructing Company, 2010a, 2010b).

FIGURE 18.5 Bags filled with sand.

Pipes were loaded using plastic bags including 230 bags filled with sand in order to remove this problem, thereby increasing the safety aspect (Figure 18.5). These bags exercise a weight approximately 460 tones on the structure middle part and as a result deflection creation within was prevented. Structure was re-established in river width, making use of a motorized boat unit following these operation terminations.

18.11 Waterlogging in the Pipes

In this step, the following processes were performed to control and keep the structure afloat while carrying out these operations after ensuring lack of deflection in the structure's middle part during waterlogging and transferring it to a place of interest in the river.

- Harnessing whole structure using winches established on both banks.
- Making use of two air bags on both sides of the structure (Figure 18.6).
- Avoid to synchronous water logging for five pipe strings and regarding such order in this operation that three pipe strings (middle pipe and two side pipes) were first waterlogged and pipe numbers 2 and 4 were then waterlogged. Waterlogging inside these two pipe strings continued till the structure was submerged and established at the location of interest in the trench designed to do so, while input water to each pipe was measured and controlled through meters mounted on input hoses (Figures 18.7 and 18.8).

18.12 Assembling, Welding and Painting Upper-Hand Dollies

Assembling and providing the parts of proper length and angle (upper-hand dollies) were performed to connect the structure to the banks and transferring the canals of which their manufacturing processes are as follows:

Cutting pipes to proper pieces, welding, assembling and testing welds, equipping both ends of the hand dollies to flanges, sandblasting, repairing internal cover and painting the pipe hand dollies were done.

FIGURE 18.6 Establishing air bags.

FIGURE 18.7 Starting water logging pipes.

FIGURE 18.8 Floating structure.

Inverted Siphon Implementation Method in Karun River

FIGURE 18.9 Termination of establishing upper-hand dollies' operations.

18.13 Establishing Upper-Hand Dollies

Each hand dolly (9 m length) was connected to the structure using bolts and nuts following the termination of upper-hand dollies and transferring them inside the river helping skillful divers (Figure 18.9).

Emerging upper elbows on water level implementing operations were continued in performing the following processes;

- Soil faced were embanked in order to access and to elbow the head flange which is necessary for the rest of horizontal hand dollies, which unfortunately outlet pipe descriptions differed to 1.92 m and exited water over implementing soil faced embankment. Returning pipes of outlet horizontal hand dollies to the previous descriptions by dredging underneath and around pipes through long boom backhoe established on the floated tower and 100 tones crane from bank and inside the water were successfully done.
- After that, the rest of the horizontal hand dollies, each of 24 m, were connected to structures directly and through armed concrete.
- Concreted supports were manufactured in horseshoe form and placed at 4 m distance from the structure inlet and outlet, and pipes were placed over them.
- The siphon enters the water and the pipes are separated using prefabricated concrete walls and the caps are installed on them.

18.14 Conclusions

The structure type and the associated implementation method may change across regions. In this chapter, an inverted siphon implementation method used to supply Karun river water for irrigation to the right hand side of the Karun river and also for the Khorramshahr urban area has been discussed. The project was intended to transfer water at a rate of 13.5 m^3/s and was constructed in a part of the Valiasr canal route during 2011–2013. Its intended purpose is to transfer water from the Marred canal to Shalamcheh area, west Khorramshahr, Khuzestan Province, southwestern Iran. The sustainability evaluation of the project is now being recommended after about one decade of operation.

References

Albaji, M., Hemadi, J. 2011. Investigation of different irrigation systems based on the parametric evaluation approach on the Dasht Bozorg Plain. *Transactions of the Royal Society of South Africa* 66 (3), 163–169.

Albaji, M., Nasab, S.B., Kashkoli, H.A., Naseri, A. 2010. Comparison of different irrigation methods based on the parametric evaluation approach in the plain west of Shush, Iran. *Irrigation and Drainage* 59 (5), 547–558.

Albaji, M., Nasab, S.B., Golabi, M., Nezhad, M.S., Ahmadee, M. 2015. Application possibilities of different irrigation methods in Hofel Plain. *Yüzüncü Yıl Üniversitesi Tarım Bilimleri Dergisi* 25 (1), 13–23.

Alhusseini, T.R. 2008. Optimum hydraulic design for inverted siphon. *Al-Qadisiya Journal for Engineering Sciences* 1 (1), 45–59.

Behzad, M., Albaji, M., Papan, P., Boroomand Nasab, S. 2009. Evan region qualitative soil evaluation for wheat, barley, alfalfa and maize. *Journal of Food, Agriculture and Environment* 7 (2), 843–851.

Boroomand Nasab, S., Albaji, M., Naseri, A.A. 2010. Investigation of different irrigation systems based on the parametric evaluation approach in Boneh Basht plain, Iran. *African Journal of Agricultural Research* 5 (5), 372–379.

Boudreaux, E. 2014. Going under the oklahoma river. http://editiondigital.net/article/Going+Under+the+Oklahoma+River/1609950/0/article.html

Gian Constructing Company, 2010a. Karun inverted siphon establishment instruction, GCC, Iran. (In Persian)

Gian Constructing Company, 2010b. Report of stress analysis when establishment, GCC, Iran. (In Persian)

Hooshmand, M., Albaji, M., Zadeh Ansari, N.A. 2019. The effect of deficit irrigation on yield and yield components of greenhouse tomato (Solanum lycopersicum) in hydroponic culture in Ahvaz region, Iran. *Scientia Horticulturae* 254, 84–90.

Jovzi, M., Albaji, M., Gharibzadeh, A. 2012. Investigating the suitability of lands for surface and under-pressure (drip and sprinkler) irrigation in Miheh Plain. *Research Journal of Environmental Sciences* 6 (2), 51–61.

Mahab Qods Consulting Engineering Company, 2012. Report of the second steps of Khoramshahr inverted siphon project (following drowning), Tehran, Iran. (In Persian)

Naseri, A.A., Albaji, M., Boroomand Nasab, S., Landi, A., Papan, P., Bavi, A. 2009. Land suitability evaluation for principal crops in the Abbas Plain, Southwest Iran. *Journal of Food, Agriculture & Environment* 7 (1), 208–213.

Neissi, L., Albaji, M., Boroomand Nasab, S. 2020. Combination of GIS and AHP for site selection of pressurized irrigation systems in the Izeh plain, Iran. *Agricultural Water Management* 231, 106004.

Sadeghi, S.H., Mousavi, S.F., Eslamian, S.S., Ansari, S., Alemi, F. 2012. A unified approach for computing pressure distribution in multi-outlet irrigation pipelines. *Iranian Journal of Science and Technology* 36 (C2), 209–223.

Index

Note: **Bold** page numbers refer to tables and *italic* page numbers refer to figures.

abscisic acid (ABA) 5
accessible water (AW) 305, *306*
acoustic Doppler devices 40
adaptive weight paths 280–281
add-on ET controllers 308
adequacy of application
    conventional irrigation scheduling 24
    crop water requirements 14
    cumulative frequency curve 24
    vs. distribution efficiency 27, *27*
    function **25**
    normal distribution 25
    water distribution *26*
adjustable dripper 151, *152*
Agbots 277
agricultural marketing 276
agricultural policies 215
agricultural production management 286
agricultural production systems 24, 113
Agricultural Revolution 276
agricultural society 215
agricultural technology 275
agriculture
    automation 316
    big data 277–279, *279*
    blockchain 279, *280*
    climatic and population demands 275
    computer vision 286–287
    developments 276–279
    irrigation 295
    rain-fed/dryland farming 142
    stabilization 214
    sustainable development 8
    variable soil landscapes 120
    water-consuming sector 3
    water resources 210
Agriculture 4.0 275
AgriTech *see* agricultural technology
agronomy water-saving techniques 4
agrotech/agritech *see* agricultural technology
air-lift pumps 382, **382**

alternative irrigation systems 130
application efficiency (AE) 18–20, 85, 91, *93*, 126, 296
artificial intelligence (AI)
    ANNs 280–281
    defined 279–280
    expert systems 281
    fuzzy logic 281
    IISs 302–303
    techniques 280
artificial neural networks (ANNs) 280–281, 303, *303*, 312–313
artificial neurons 303
ASCE On-Farm Irrigation Committee 18
automatic control 43–44, 47
automatic control optimization 44–46
automation kit for farm pond irrigation
    connect irrigation application to outlet pipe *178*
    connect water supply to inlet pipe *177*
    contents 174
    double pump, inlet filter *175*
    float switch *175*, *176*
    gravity-feed drip irrigation *174*
    installation 176–178
    pump controller, ON Auto position *177*
    smallholders 173
    unpowered irrigation controller, solenoid valves *178*
    unpowered terracotta valve *175*
    waterproof pump controller *174*
automation of irrigation systems 297–298
axial flow pump 380

backpropagation algorithm 281
base unit (BU) 313
basin irrigation
    advance and recession curves *96*
    agricultural irrigation survey data 67
    animation results window *90*
    definition 65
    examined characteristics **96**
    experimented characteristics **85**
    factors 70

397

basin irrigation (*cont.*)
   infiltration uniformity and water management 65–66
   intake opportunity time 70
   limitations 69
   methods 69–70
   modeling tools 73–106
   optimization algorithms 106–107
   performance 66
   principles 69–73
   recession and depletion 70
   SCS method 70–73
   simulation results *89*
   SIRMOD model 68
   spatial variability 67
   systems 65–66, *66*
   time–space hybrid numerical method 67
   2D computer simulation model (COBASIM) 67
   two-dimensional hydrodynamic simulation model 66
   upstream and downstream 65
Bayesian models 284
best management practices (BMPs) 36, 46
big data 277–279, *279*, 290, 302
biological water-saving techniques 4
Birch effect 5, 6, 8
blockchain 279, *280*
Blue River technology 276
borehole pumps 347
brake horsepower (BHP) 364
broadacre irrigation system design
   categories, measurable performance 133
   factors, irrigation interval 133–135
   guidelines 133
   methods and soil condition parameters **135**
   monitoring irrigation performance 135–136, *136*, **136**
   parameters and metrics **134**
   soil and crop parameters *134*
   soil moisture content *135*
broadacre sprinkler delivery systems
   advantages and disadvantages 120–122, **121**
   agricultural production 115
   centre pivot and automatic linear move irrigation 117–118, *119*, *120*
   components *115*
   cooperative/self-suppliers 115
   factors 115
   floppy (flexible) 118–120
   inefficiencies and losses 116
   lateral/periodically moved irrigation 116–117, *118*
   solid-set/fixed 116, *117*
   system components and controllers 122
   water scarcity 114
   water sources and on-farm water supplies 115
   wind speed and direction *121*
Buneh system

characteristics 212–213
classification in terms of assistance 210–212
classification in terms of irrigation 212
definitions 210, **211**, 215
varieties 210–212
Bureau of Meteorology (BOM) 151, 186, 198
bypass irrigation system 309

canal cleaning 207
canal control
   algorithm 41–43
   automatic control 43–44
   methods 41
   water supply 40
canal irrigation
   automatic control techniques 34
   control (*see* canal control)
   design 34–36, 46
   distributing water 33
   India *34*
   model 36–40
   objectives 33, 46
   optimal management 44–46
CanalMod 38
canal structure design 34–35, **35**
cathodic protective system 391
cause-and-effect analyses 14, 283
center pivot systems 22, 117–118, *119, 120*, 343
Central California Irrigation District (CCID) 45
centrifugal pumps
   canals and rivers 368
   cavitation 371
   characteristic curve *362*, 363
   classification 367–368
   defined 344, *346*
   diffuser-type 366–367, *367*
   efficiency 345
   elements 346
   factors 347
   fluid movement 345
   horizontal 345, *345*
   maintenance 369
   operation 368–369
   performance 346
   portable 368
   priming 345, 367
   principles 366–367
   troubleshooting 369–371
   vertical 345
   volute-type 366, *366*
   wear pump impeller 346, *346*
   wear types 347
Charasa 357
Christiansen's uniformity coefficient (CUC) 20, 22, 25
climate-based intelligent irrigation system
   add-on ET controllers 308
   ET controllers *307*, 307–308

Index

evaporation and transpiration 304
evaporation–transpiration 306–307
irrigation planning 304–305
parameters 304
signal-based controllers 308
standalone controllers 308
water-based elements *304*
water content of soil 305–306, *306*
climate measurement on site 308
clock/timer 143
cloud-based thermal imaging 313
cloud technology 298
clustering 283
collapsed (shattered) Buneh 210, 212
communication networks 301–302
complete Buneh 210
computational methods 284
computational rationality 279
computer vision
    agriculture 286–287
    autopilot features 286
    CNN 286
    facial recognition software 286
    irrigation 287–288
    ML 276
    terminology and definitions 286
constant-head method 263, 264
consumed fraction 14
contour furrows 53
control dripper 152
control methods
    feedback and feedforward control 41
    local, distant and remote control 41
    manual and automatic control 41
    upstream and downstream control 41
control volume 166, **167**, 184, **185**
convolutional neural networks (CNNs) 286
corrugations 53
COTFLEX 312
cotton management expert (COMAX) 312
crop production 315, *316*
crop water requirement
    factors 383–384
    head–discharge curve *385*
    irrigation requirement 56–57
    well and pump characteristics curves 384–385
cumulative probability density/frequency distribution 18, *19*

Darcy–Weisbach equation 146
DARE-TU project 340
data transmission unit (DTU) 316
decision support and automation system (DSAS) 313
decision support system 68
decision tress 283
decomposition index (DI) 226, *227,* 228
deep ANNs 281

deep learning (DL) 281
deep neural networks (DNNs) 281
deep penetration percentage (DP) 85
deep percolation losses (DPLs) 59–60
deepwater irrigation 359
deep-well turbine pump; *see also* vertical turbine pumps
    applications 347
    diffuser pumps 347
    driving mechanism 347
    dual-driven pumps 348
    volute pumps 347
deficit irrigation (DI)
    agricultural water productivity 8
    crop yield 4, 7
    disadvantage 7
    soil amendments 7
    soil salinization 4
    techniques 4
    tomato quality 6
    valuable and sustainable strategy 4
deficit irrigation factor method 331
demand-oriented systems 41
depleted Buneh 210
depletion phase 341–342
digital agriculture 275
digital elevation model (DEM) 39
digital humidity and temperature (DHT) 315
dimensionality reduction 282–283
displacement pumps 343, *356, 358,* 359
distributed ledger technology (DLT) 279
distributed model predictive control (DMPC) 46
distribution efficiency (DE) 20–21, **25,** 126
distribution uniformity
    vs. deep percolation 26–27
    economic irrigation system 24
    efficiency 126
    micro-irrigation systems 23
    performance indicators **24**
ditch irrigation; *see also* furrow irrigation
    controlling water table, water logging and salinity 60–62
    design 53–55
    example *52*
    supply channels 52
    unsaturated zone *60*
    water management principles and crop water requirement 55–60
domestic sprinkler systems 131–133
double-ring infiltration test 40
drainage, soil profile 220
drainpipes
    ceramic *222*
    flow resistance *224,* 225
    measuring points *232*
    permeability 224–225
    vertical distribution *232*

drawdown–discharge characteristics 384
drip irrigation; *see also* micro-irrigation systems
    cucumber greenhouse cultivation 266
    field water efficiency 143
    pressurized/gravity feed 153
    scheduling paradigm 150
    unpowered terracotta valve 162
    water pumping systems 342
dripper control volume 190, *190*
drones 277
drought 3, 6, 61, 204, 208, 214, 247, 341
dry/wet motors 348
dumb irrigation clock 317
dust suppression/logging 143

eductor pipe 382
effective ratio of application (ERA) 21, 26
electrical conductivity 265
electric motor 348, 356, 368, 378, 381
electromagnetic linear transmission 331
embedded systems 316
energy saving 213
engineered surface-water distribution systems 136
engineering water-saving techniques 4
ensemble learning (EL) 284
environmental awareness 296
environmental demands 17
environmental sustainability 203, 269
ET-based technology 313
ethylene-propylene-diene monomer (EPDM) 36
EVALUE model 77–78
evaporator 151, *152, 153, 157,* 159, 160, 190
evapotranspiration (ET) 4, 15, 16, 18, 21, 57, 132, 151, 199, 233, 240, 285, 300, 304, *307,* 307–308, 332
event analysis tool
    advance/recession tab 81, *82*
    assumptions 84–85
    branch function 76
    characteristic infiltration time 75
    Elliot and Walker's two-point method 77
    flow depth hydrographs 78, *78*
    flow depth tab 81, *82*
    geometric specifications **78**
    infiltration equation 75
    infiltration tab 82–83, *83*
    inflow/runoff tab 81, *81*
    Kostiakov influence equation 76
    Merriam–Keller post-irrigation volume balance 77, **77**
    modified Kostiakov 76
    NRCS infiltration families 75
    probe penetration analysis 76–77
    roughness tab 83, *83*
    Saint Venant equation 84
    series of tabs 78
    slope options 79–80
    soil/crop properties tab 80, *80*
    soil water infiltration 75
    start window 78, *79*
    system geometry tab 79, *79*
    time-rated intake families 75–76
    volumetric equilibrium method, EVALUE model 77–78
expert systems 281
exploitable water (EA) 332, 333

falling-head method 263
farm management practices 66
farm pond technology 150
feedback-control algorithms 34
fertigation 74, 98, 143
Fertile Crescent 276
field capacity (FC) 305
field irrigation application efficiency 56
filtered soil content (FSC) 226, 227, *228*
financial resources 214, 312
finite element 37, 40, 265
finite volume 37
flood irrigation *see* surface irrigation methods
floppy sprinkler system 118–120
flow controllers 118, 122, **124**
flow control routing 43
flow/flow–pressure relationship in pottery 254–255, **255**
flow regime 257
flow uniformity of pottery 254
Food and Agriculture Organization (FAO) 68
food demand 3, 295
food security 113, 245, 288
foot-pound-second (FPS) 360
Fourth Agricultural Revolution 275, 288, 290
furrow irrigation; *see also* ditch irrigation
    advantages and disadvantages 52–53
    example *52*
    hydraulic phases 55, *55*
    irrigation requirements 56–59
    losses of water 59–60
    ridges 52
    slope 54
    soil types 51–52
    spacing and length 54–55
    surface irrigation methods 51
    types 53
    water application and management 55–56
    width, depth and shape 53–54
    WinSRFR model 68
fuzzy control 281
fuzzy K-means 283
fuzzy logic algorithm 259–260, 281

gapping void 227, *229*
genetic engineering 276
geographic information systems (GISs) 39, 298, 300–301, 317

Index

glucose 5
graded contour furrows 53
graded straight furrows 53
gravity-fed methods 51
greenhouse gas emission 7, 8, 213
Green Revolution 276
gross irrigation water requirement (GIR) 59

head works 115
Heisen–William's equation 258, **258**
hierarchical clustering 283
high head water lifts 357
high-transmission pipes 262
Hoogoudt's equation 222
horizontal water movement 222–223, *223*
human-machine interface (HMI) 43
humidity sensors 313, 314, 316
hybridisation 276
hydraulic conductivity 219, 223–224, 259–265
hydraulic model 36–38, 42, 47, 147
hydraulic simulation models (HSMs) 38
hydrodynamic model (HD) 84
hydrological model 38–39
hydrological response units (HRUs) 39
hydro-powered pumping (HPP) technologies 339, *344*
HYDRUS-2D model 264

IEEE 802.15.4 standard 302
indigenous/non-indigenous Bunehs 212
indigenous water-lifting method 357
Industrial Internet of Things 276
inference engine 281
infiltration opportunity time 55
infiltration parameters estimation 68
inflow method 81
inflow–outflow measurement method 40
information and communications technology (ICT) 313
instance-based models 284
integrated model 39–40
integrated reservoir-based canal irrigation model (IRCIM) 39
integrator-delay (ID) 42
integrator-delay zero (IDZ) 42
intelligent irrigation systems (IISs)
    applications 312–316
    artificial intelligence 302–303
    categories 303
    climate-based controllers 304–308
    implementation 295
    importance 297–298
    management equipment and techniques 298–303
    RS 299–301
    sensor and communication networks 301–302
    soil moisture sensors 308–312, *309*
    system diagram *314*
    watering program 303
    WUE 295–297
Internet Information Services (IIS) 313
Internet of Things (IoT) 276, 277, 299, 312, 316
Internet of Underground Things (IoUT) 276
Internet of value 279
interoperability 290
interviewees, benefits 290, *290*
inverted siphon implementation method
    assembling and welding pipes 389
    assembling, welding and painting upper-hand dollies 393–395
    bags filled with sand *393*
    characteristics 389, **389**
    dredging operations *391*
    elbows and lower-hand dollies 390, *390*
    establishing air bags *394*
    establishing cathodic protective system 391
    establishing dry ponds 389
    establishing four winch units 391
    establishing upper-hand dollies 395, *395*
    floating 392, *394*
    hydraulic design 387
    implementing connecting structures *390*
    making 45 degree elbow joints 390
    making and mounting siphon indices 391, *392*
    manufacturing and establishing lower-hand dollies 392
    pipe establishment bed 391
    shapes 387
    stormwater/wastewater sewers 388
    transferring structure inside river 392–393
    waterlogging pipes 393, *394*
irrigation adequacy 85
irrigation assessment
    adequacy of application 14, 24–26
    cause-and-effect analyses 14
    distribution uniformity *vs.* deep percolation 26–27
    efficiency indicators 16–21
    interfering factors 14
    system evaluation 13
    system performance 13, 14
    technical and financial assessment 13
    uniformity indicators 22–24
irrigation canals *see* canal irrigation
irrigation consumptive use coefficient (ICUC) 17
irrigation deficit *see* deficit irrigation (DI)
irrigation efficiency (IE) 16–17, 123, *125*, 126, 133
irrigation rehabilitation/automation 297
irrigation requirements
    calculation, water 58–59
    crop growth stages *57,* 57–58
    determination of K_c 58, *58*
    reference crop evapotranspiration 56–57
irrigation sagacity (IS) 17–18
irrigation scheduling strategies 13, 14
irrigation water *see* water

jet pumps
    defined 381
    nozzle and venturi 381
    uses and adaptability 381

Kaf shekani *see* tunnel deepening
kernel trick 284
Khuzestan Province 388
kinematic wave (KW) 68, 84
K-means technique 283
knowledge base 281
Kostiakov–Lewis equation 101

Lamal Kardan 209
land management 39, 215
Landsat's infrared thermal images (TIR) 300
landscape water requirements
    contaminating compounds 328
    corky roots, overwatered dead tree *328*
    deficit irrigation factor method 331
    density coefficient **330**
    garden coefficient *330*
    microclimate coefficient **330**
    naturalistic landscaping/xeriscape 328
    ornamental quality 328
    phytosanitary problems 328
    soil water sensors *331*
    water content in soil 331–332
    WUCOLS method **329**, 329–330
lateral move irrigation 343
lawn sprinkler system 143
leaching requirement 62
learning machine (LM) 44
learning methods 285
learning models
    Bayesian models 284
    clustering 283
    decision tress 283
    ensemble learning 284
    instance-based models 284
    ML methodologies 282
    regression 283
    support vector machines 283–284
level-basin irrigation *see* basin irrigation
level furrows 53
life cycle cost 213
linear move systems 117–118, *119*
linear-quadratic regulator control (LQR) 43
linear regression 283
lining canals 36
localized irrigation 143, 342
local water lifts
    high head 357
    indigenous lifting devices 356
    low head 357
    medium head 357
    positive displacement pump 358–359

    reciprocating pumps 359, *359*
    specific speed 360–361
    variable displacement pumps 360
    wind mills 358
logistic regression 283
low-cost innovative water-saving technology 150
low head water lifts 357
low-volume irrigation 143

machine learning (ML)
    algorithms *283*
    analysis of learning 282–284
    canal flow control scheme 44
    decision making 276
    defined 276
    implementation 302
    irrigation 284
    learning methods 285
    learning models 283–284
    tasks 282
    terminology and definitions 282, *282*
    water management 284–285
machine-to-machine (M2M) technology 314
Manning roughness coefficient 72, **73,** 76–78, 80, 81, 83, 87
Manning's equation 35, 37, 70
Manning's law 224
manual measured irrigation
    adjustable dripper 151, *152*
    evaporator 151, *152, 153*
    flowchart *154*
    installation 152–153
    root zone scheduling 190
measured irrigation (MI)
    angle grinder *188*
    automation kit for farm pond 173–178
    climate change 150
    conditions 150, 288
    conventional drip irrigation systems 150
    conventional volume control paradigm 150
    definition 151
    economic manual irrigation scheduling technique 288
    flowchart *289*
    gravity-feed 150
    hammer, steel pipe *189*
    low-cost approach 150
    manual 151–153
    PC drippers 199
    remove steel pipe *189*
    soil moisture and scheduling 188–190
    terracotta controller, latching solenoids 171–173
    unpowered evaporative valve 153–162
    unpowered irrigation controller for solenoid valves 178–188
    unpowered terracotta valve 162–170

Index

unpowered uniform drip irrigation on sloping land 191–198
water-efficient and energy-efficient solutions 199
medium head water lifts 357
memory-based models *see* instance-based models
Merriam–Keller method 77, **77**
microcontroller 313, 316
micro-irrigation systems 22, 23, 113, 141–143, 147, 150, 253, 254, **255**, 266
microtopography 66, 67
mineral nutrient accumulation 6
mini-bubbler irrigation
    advantages 142, 146
    design considerations 146–147
    design criteria and considerations 145–146
    diaphragm material 144
    disadvantages 146
    emitters 145, *145*
    feature 141, *142*
    friction head loss distribution 141
    high- and low-pressurized systems 141
    micro-irrigation 141, 142
    scheduling 145
    single/multiple port outlets 144
    steps 147
    system layout and components 144
    types 142–144
minor losses 259
mixed-flow pump 379, 381
model predictive control (MPC) 34, 43, 45–47
MODFLOW model 40
moisture distribution *254*, 264–266
moisture soil wetting pattern 255–257
mole drain 222–224
mono-proprietary/multi-proprietary Bunehs 212
moral commitment, water users 214–215
mulching 59
multilayer perceptron (MLP) 44
Multimedia Internet of Things (MIoT) 276
multi-sensor controllers 309
multi-tasking processes 287
Muskingum routing technique 40

Nash-Sutcliffe Efficiency (NSE) error index 83
National Agricultural Research Organization 224
Natural Resources and Conservation Service (NRCS) 75
net irrigation water requirement (NIR) 58–59
net positive suction head (NPSH) 362–363
neural networks 312
neuro drip 285, 312
neutral value 15
nodes 280, 303
non-consumed fraction 15
non-erosive stream size 54
nonpressure-compensating (NPC) drippers 168
Normalized Difference Vegetation Index (NDVI) 314
nutrient management policies 287

oil lubricants system 373
Omodei method
    defined 191
    dripper assemblies *193*, *196*
    example 194–197, *195*, **197**
    features 191
    installation 193–194
    schematic diagram 191–193, *192*
    sub-drippers 198
    unpowered terracotta valve *195*
    zone 1 with 2 drippers *196*
    zone 2 with 3 drippers *196*
on-demand SMS irrigation controller 309
on-site weather data 167
open channel flow 51
operating pressures of system 255–257
operation analysis tool 88–91, *90*, *92*
optimization
    algorithms, basin irrigation 106–107
    automatic control 44–46
    particle swarm 106–107
optimum scheduling 88

paddy fields; *see also* subsurface drainage
    ceramic drainpipe *222*
    core technologies 219
    performance diagnostics 231–233
    restoration/reconstruction project 236
    rice husk 219–220
    soil profile 220
    subirrigation 233–236
    supplementary drains 236
PalmHAND 287
partial root-zone drying (PRD) 4, 7, 8
particle swarm optimization (PSO) 106–107
PC dripline *vs.* NPC dripline 199
peers/non-peers Bunehs 212
Penman method 56
perforated pipe irrigation 249
periodically moved systems 116–117, *118*
permanent wilting point (PWP) 305
permeameter measurement method 40
pervasive computing 276
physical design tool 91–94, *94*, *95*
PIC16F877A processor 313
piston pump 359
pitcher irrigation
    adoption and sustainability 245
    agronomic practices 244
    data collection 242
    defined 239
    deserts shrubs 240
    employment opportunities 245
    food production 245
    fruit size 242, **243**
    high-tech methods 241
    justification 241

pitcher irrigation (cont.)
   materials and methods 242
   number of fruits per plant 242, **243**
   objectives 241–242
   observations 244
   performance 240
   porosity 240
   pressure gradient 240
   recommendations 245
   results 242–244
   seedling 239
   self-regulative system 240
   small-scale method 240, 241
   supply of moisture 244
   water conservation measures 239
   yield of vegetables grown 243, **244**
plant growth-promoting bacteria (PGPR) 7–8
plant root development 249, 259
plant typologies **333**
plunger 359
point measurement method 40
ponding tests 40
positive displacement pump 358–359
potential AE 21
pot irrigation
   agricultural activities 247
   benefits 252
   climate *248*
   components 253–254
   constant pipe pressure head *256*
   cost estimation **268**
   disadvantages 251–252
   economic, social and environmental issues 268
   flow/flow–pressure relationship 254–255, **255**
   flow regime 257
   flow uniformity 254
   fundamentals and criteria 254–265
   history, design criteria and hydraulics assessments 259–264
   leakage rate and reaction capability *261*, 261–262
   moisture and salinity distribution 264–265
   moisture soil wetting pattern 255–257
   performance *vs.* systems 265–267
   pitcher mounting *250*
   preparing and firing mechanism 267–268, *268*
   pressure loss 257–259
   specification, installation and performance 249–253, *251*
   subsurface irrigation systems 247–249, *249*
   world water resources *248*
precision farming *see* digital agriculture
pre-lubricated system 373
pressure-based irrigation technologies 340
pressure chamber method 262
pressure-compensating (PC) drippers 150
pressure fluctuations 122
pressure-independent dripper discharge 186–187, *187*

pressure loss
   pipe fittings 259
   pipes 257–258
   pivot system 118
pressurized drip irrigation *vs.* gravity-feed measured irrigation **151**
pressurized irrigation methods 51, 141
probability theory 281
product quality 5–6, 14, 24, 27
programmable logic controller (PLC) 41, 43
propeller, defined 380
propeller pumps
   axial flow and mixed flow 348–349
   construction and operation 380
   installation 380
   oil- and water-lubricated types 379
   operating characteristics 380
   permanent installations 349, *350*
   portable 349
   power requirements 349, 351
   submergence depth 351
proportional-integral (PI) controller 34
proportional-integral-filter (PIF) 46
puddling 220, 222
pump characteristics
   speed and impeller diameter 364–365
   terminology 361–364
   well curves 384–385
pump element construction 378
pumping water level (PWL) 368
pump operation parameters 356
pump selection
   criteria and procedure 383
   discharge capacity determination 383
   equipment 382–383
   size **384**
PVC plastic irrigation pipe (PIP) 36

Qanat-based cooperative Buneh system 210
Qanat irrigation systems
   benefits 204
   destructive effect 215
   extraction system 204
   goals 205
   history 204
   length 203
   maintenance 204
   maintenance, executive operations 206–209
   physical system 215
   political and structural changes 215
   reconstruction and restoration 216
   restoration 215
   social-cultural system 209–213
   structure 205–206, *206*
   sustainability of water resources 213–215
   techno-physical aspects 205–209
   water resources 205

water supply sources 204
waterway/canal dug 203

rain-fed Bunehs 212
rain-fed/dryland farming 142
rapidly available water (RAW) 305–306, *306*
reciprocating pumps 359, *359*
reference crop evapotranspiration 56–57
regression analysis 283
regulated deficit irrigation (RDI) 4
reinforced concrete pipe (RCP) 36
relevance vector machine (RVM) 44
remote sensing (RS) 299–301, 317
remote terminal unit (RTU) 43
removing sediments 209
renewable energy (RE) *315*, 315–316, 340
renovation process 297
requirement efficiency (RE) 296
reservoir-based irrigation system 46
Reynolds number 257, **257**
Richards equation 98
robots 277
root zone scheduling 190
rope-and-bucket lift 357

salinity
    distribution 264–265
    problems 61, 62
    reduction 214
satellite imagery 300
saturated–unsaturated flow model 40
seepage reduction design 36
self-motivated/intervention-based Bunehs 212
semi-main pipe/manifold 342
semi-supervised ML 282
sensor installation 310–312
sensor networks 301–302
sensor unit (SU) 313
shaft horsepower (SHP) 364
shallow-rooted crops 53
signal-based controllers 307–308
SIRMOD model
    design panel 104–105, **105**, *105*
    field topography/geometry 100–101, *101*
    hydrograph inputs 104, *104*
    infiltration characteristics 101–102, *103*
    inflow controls 101, *102*
    main window 99, *100*
    management practices 98
    methods 99
    outputs 106, *106*
    sub-windows 99
    surface irrigation simulation 102–104, *103*
smart farming 277, *278*, *279*
smart greenhouses 299
smart irrigation
    AI and robotics 277

    concept 327
    developments 288
    high-quality water 303
    IoT 285
    water management methods 284
smart manual irrigation 288
social-cultural system
    Buneh (*see* Buneh system)
    cooperative groups 210
    human activities and social entities 209
    resource management systems 210
    technological systems 209–210
social development 214
social–ethical–cultural system 210
social stratification 215
socio-technical system 210
soil and water assessment tool (SWAT) 39, 46
soil conservation service (SCS)
    calculation process 70–73
    curve number method 40
    infiltration equation 70
    irrigation efficiency in basins 72–73, **73**
    length of basin and inflow rate per unit 71
    Manning roughness coefficient 72, **73**
    maximum depth of flow 71–72
    metric system **71**
    net infiltration time 70–71
    net required irrigation depth 72
    WinSRFR model 68
Soil Conservation Service – United States Department of Agriculture (SCS-USDA) 23
soil humidity sensors function 309–310
soil moisture depletion (SMD) 18
soil moisture probe 188–189
soil moisture sensors (SMSs)
    automatic irrigation system *311*
    bypass irrigation system 309
    electronic timer *310*
    fuzzy logic 315
    multi-level 298
    on-demand irrigation controller 309
    sensor installation 310–312
    soil humidity sensors function 309–310
    switch controller *311*
soil moisture technologies 277
soil salinity 7, 214, 264
soil sensors 277
soil textures 255–257
soil water balance 60, *61*
soil water infiltration 75
soil water storage capacity 56
Solanum lycopersicum L. *see* tomato
solar energy 315
solid-set irrigation system 143
solid-set system 116, *117*
specific speed 360–361
spray irrigation 114

sprinkler irrigation systems
   advantages and disadvantages **124**
   agricultural/horticultural production 113
   automation and moving systems 114
   broadacre delivery 114–122
   broadacre design 133–136
   comparison 129–131
   design criteria 137
   domestic application 131–133
   efficiency 125–128, *127, 128*
   elements, site irrigation plan *131*
   garden use *124*
   global heating 113
   improvements 114
   openings/nozzles 114
   performance measures 122–129, **129**
   relative costs and water use efficiencies *130*
   research 136–137
   uniform and non-uniform water application *132*
   uniformity 128–129
   unregulated and regulated flows *123*
   water application rate and running time calculation *132*
   water pumping systems 342
sprinkler packages 118
sprinkler systems
   Christiansen's uniformity coefficient 20, 22
   components and controllers 122
   cumulative frequency distribution 18
   water droplets 16
stabilization of agriculture 214
standalone controllers 308
state-space model 45
statistical methods 284
statistical uniformity coefficient 22
steady-state evapotranspiration *234*
stomata 3
storage efficiency (SE) 21, **25**, 56
storage phase 341
stream gauging approach 40
submersible pumps
   advantages 377
   applications 348
   construction and operation 377–378
   defined 377
   dry/wet motors 348
   electric motor 348
   installation *349*
   installation, operation and maintenance 378–379
   semi-open impellers 348
   troubles and remedies 379
   water-cooled jacket 348
submersible turbines
   deep well applications 348
   irrigation and well pump applications 348
sub-Saharan Africa (SSA) 150
subsoiling 222–224

subsurface controlled drainage 234, *235*
subsurface drainage
   annual maximum drainage ability *230*
   concept 221–222
   construction 220
   degradation of filler materials 225–229
   drainage ability *230*
   drainpipes 224–225
   functional deterioration 224–231
   Japan *221*
   land/flood plain 219
   main drainpipe and supplementary 222–224
   needs 221
   rice husk filler *231*
   water flow resistance 229
subsurface drainpipe *235*
subsurface drip irrigation (SDI) 249, 269
subsurface irrigation systems 247–249, *249, 253,* 343
sucrose 5
summer cropping Bunehs 212
sunset scheduling 190
supervised learning 282
supervisory control and data acquisition (SCADA) 41
supply-oriented systems 41
support vector machine (SVM) 283–285
support vector regression (SVR) 285
surface irrigation methods
   agricultural lands 107
   border strip/basin irrigation 142–143
   design and management 68
   gravity flow 51
   hydraulic simulation software 74
   infiltration equation 75
   mathematical model 84
   single and multi-objective programs 106
   soil surface evaporation loss 59
   uniform flow 84
   versatility 133
   water distribution 66
   water pumping systems 341–342
surface irrigation modeling software (SIRMOD) 67
surface irrigation simulation model (SRFR) 67, 102
surface irrigation simulation model-Windows version (WinSRFR) 67
sustainable agriculture 150, 247
sustainable crop production 239, 241
sustainable management 40
sustained deficit irrigation (SDI) 4
swing basket 357
synthetic intelligence 279

Tahsoo Roosoo *see* tunnel doubling
tail water runoff (TWR) 59
tech farming *285*
terracotta irrigation controller, latching solenoids
   above-ground installation *172*
   colour-coded wires *173*

Index

control box 171, *173*
  in-ground installation *171*
  installation 171–172
thorn-based Buneh 212
time-domain reflectometry (TDR) 76
tomato 5–7
total dynamic head (TDH) 351
total soluble solids (TSS) 5
traveling sprinklers 143
trial-and-error method 77, 82, 105
trickle irrigation 143
t-test statistical method 242
tunnel branching 208
tunnel checking 207
tunnel deepening 208
tunnel doubling 208
tunnel extending 207–208
tunnel insulation 209
turbine-well pumps 347

uniformity 14, 22–24, 128–129
uniformity coefficient of Hart (UCH) 23
uniformity coefficient of Wilcox–Swailes (UCW) 22–23
uniformity distribution (DU) 85
universal asynchronous receiver-transmitter (UART) 302
Unmanned Ariel Vehicles 277
unpowered evaporative valve
  adjusting control dripper *157, 161*
  adjust irrigation frequency 159–160, **160**
  adjust water usage rate 160–161
  cylindrical float and float rings *159*
  drip irrigation system 153
  features 162
  fill evaporator with water, irrigation stops *157*
  float over float shaft *156*
  float ring over cylindrical float *159*
  float shaft, vertical *156*
  installation 155–158
  irrigation starts, float reaches low level *158*
  irrigation stops, float reaches high level *158*
  measuring container, irrigation dripper *161*
  push float down to start, irrigation manually *160*
  remove float to stop irrigating *161*
  root zone scheduling 190
  water supply connection 155
  water usage rate 153
unpowered irrigation controller for solenoid valves (UICSV)
  adjusting control dripper *184*
  calibrating 183–185
  check ring magnet activation *181*
  control dripper *186*
  evaporation *182*
  features 187–188
  high level, ring magnet and float *179*
  hunter solenoids, ring magnets *181*
  installation 180–183
  large area, polyester cloth exposed *185*
  low level, ring magnet and float *179*
  plastic saucer 180, *180*
  position, ring magnet *182*
  position the control dripper *183*
  pressure-independent dripper discharge 186–187, *187*
  rotate the magnet *182*
  rubber band *183*
  small area, polyester cloth exposed *185*
  sprinkler/dripper systems 178
  weather-based irrigation control 186
  wires removed *181*
unpowered terracotta valve
  above-ground installation 164
  adjustable control dripper *166, 168*
  adjust inlet and outlet pipes *165*
  adjust irrigation events *167, 167, 169*
  adjust water usage rate 168
  automatic sprinkler irrigation/drip irrigation 162
  connect irrigation application to outlet pipe *164*
  connect water supply to inlet pipe *164, 165*
  dig a hole midway, adjacent plants *165*
  features 170
  15 mm inlet and outlet *163*
  float and water level *163*
  flow rate 162
  in-ground installation 162, 164–165
  pressure-independent dripper discharge 168–169
  ring magnet, bottom of float *163*
  6 mm flexible tubing *170*
  usage 166–167
unpowered uniform drip irrigation on sloping land
  installation cost 198
  Omodei method 191–198
  pressure compensation 191
unsupervised learning 282
upwind conservation scheme 37
urban irrigation
  algae growth 326
  clogged emitter 327
  control and automatization systems 327–328
  doses and frequency determination 332–334
  drip irrigation laterals 325
  elements 323–327
  landscape water requirements 328–332
  macro-basin level 322
  perception of modernity 322
  quality of wastewater 323
  scheduling 328–334, *333, 334*
  treated wastewater 322–323
  waste/contaminated water 321
  water distribution network *324*
  water reservoir *325*
  zone valve cabinet *326*
urethane 36
USDA Natural Resource Conservation Service 33

USDA-NRCS surface irrigation system 80
US Environment Protection Agency (USEPA) 151

value chain 245
valve in head 144
valve unit (VU) 313
vapor pressure deficit (VPD) 3
variable displacement pumps 360
vertical turbine pumps; *see also* deep-well turbine pump
   application 347
   characteristics 373–374
   construction 372–373
   cross-sectional view *372*
   lubrication 373
   maintenance 374–375
   operation 374
   optimum operations 371
   pressure heads 371
   seasonal fluctuation 371
   troubleshooting and remedies 375–377
volume balance 68
volumetric efficiency 126
volumetric/hydrological technique 296
VS2D computer program 262

water; *see also* water management; water pumping systems; water use efficiency (WUE)
   balance computations 60
   Buneh 210
   chlorination 323
   conservation 213
   content diagram, root zone *306*
   conveyance efficiency 126
   demand 203
   distribution uniformity **25**
   efficiency 247
   environmental issues 3
   furrow 55
   humans' health and survival 3
   justice issues 214
   partitioning 15–16
   quantities 14
   requirement efficiency 56
   saving, mechanisms for 4–5
   scarcity problems 107
   shortage 302
   uniform distribution 52
   usage and consumption 14–15
   wastage 288, 302
water-borne pollutants 114
water horse power (WHP) 363
water-lifting project 356
water management
   application 3, 55–56
   automatic control techniques 34
   benefits 15
   infiltration uniformity 65–66
   issues and problems 203
   ML 284–285
   smart manual irrigation systems 276
   sustainable irrigation 215
   tools 210
water production section (WPS) 209
water productivity (WP) 295–296
water pumping systems
   electricity- and diesel-based systems 340
   HPP technologies 340, *344*
   irrigation types 341–343
   objectives 341
   on-farm conditions 340
   pressurized irrigation networks 339
   RE-based counterparts 340
   renewable energies 339
   types 343–351, **351**
water requirement
   crop (*see* crop water requirement)
   gross irrigation 59
   landscape (*see* landscape water requirements)
   net irrigation 58–59
water resources management 15, 203–205, 215, 276, 321, 322
water-saving potential 6–7
water scarcity 4, 91, 107, 114, 203, 212, 215, 269, 284
WaterSense 151
water transport section (WTS) 209
Water Use Classifications of Landscape Species (WUCOLS) 328
water use efficiency (WUE)
   agricultural 3
   DI effects 4
   estimation **265**
   factors 296–297, *297*
   field scope 7
   IISs 295–297
   instantaneous 6
   intrinsic trade-off 6
   irrigation strategies 6–7
   optimal distribution 113
   physiological/agronomic indices 6
   stomatal aperture 7
   subsurface irrigation 249
   vegetative growth 267
weather-based irrigation control 186
weather sensors 277
weighted connections 280–281
wind mills 358
WinSRFR model
   evaluation and modern design 85–94
   farm tools 74
   furrow irrigation system 68
   hydraulic simulation models 73
   infiltration parameters and roughness coefficient 75–85

irrigation efficiency 94–98, *97–99*
operation analysis tool 88–91, *90, 92*
physical design tool 91–94, *94, 95*
project management window *74*
simulation tool 85–88, *86, 87*
sub-programs 74
unsaturated flow hydraulic model 74
version options 98
wireless communication 299–301
wireless radio-frequency (RF) module 316
wireless sensor networks (WSN) 312
wireless support network 285
wireless technologies 277, 300
WUCOLS method **329,** 329–330

XBee communication protocol 302

zero-inertia (ZI) 68, 84
ZigBee communication protocol 301–302